Engineering Heat Transfer

Donatello Annaratone

Engineering Heat Transfer

Springer

Prof. Donatello Annaratone
Via Ceradini, 14
20129 Milano
Italy
donatello@annaratone.eu

ISBN 978-3-642-03931-7 e-ISBN 978-3-642-03932-4
DOI 10.1007/978-3-642-03932-4
Springer Heidelberg Dordrecht London New York

Library of Congress Control Number: 2009938931

Printed on acid-free paper

Springer is part of Springer Science+Business Media (www.springer.com)

Preface

After the publication of "Pressure Vessel Design" and "Steam Generators" by Springer Verlag which were written for field experts, I decided to write a generalist textbook.

It is designed for anybody interested in heat transmission, including scholars, designers and students, instead of experts of a specific field.

Two criteria constitute the foundation of all my publications including the present one.

The first one consists of indispensable scientific rigour without theoretical exasperation. The second criterion is to deliver practical solutions to operational problems.

Admirable theoretical studies stand out because of their scientific rigor and depth. Unfortunately, though, at times they have little impact on design requirements because they either refer to schemata too remote from actual phenomena, or because the obtained results differ from the outcome derived from experimental data.

In addition, heat transfer involves quite complex phenomena, and typically theory is either missing or too generic to be applied to operational scenarios. Therefore, it is sometimes necessary to adopt empirical computation criteria. These are less striking than a theoretical discourse, but certainly more useful to the designer.

Moreover, in my opinion the inclusion of these theoretical studies would have strengthened the scientific foundation of this publication, yet without providing the reader with further applicable know-how.

The second criterion is fulfilled through equations grounded on scientific rigor, as well as a series of approximated equations, leading to convenient and practically acceptable solutions, and through diagrams and tables. When a practical case is close to a well defined theoretical solution, we discuss corrective factors to offer simple and correct solutions to the problem.

After a brief introduction in Chap. 1, heat transfer by conduction in both steady and unsteady state is examined in Chap. 2 as well as Chap. 3.

Chapter 4 develops the dimensional analysis as an indispensable premise to Chap. 5 with its focus on heat transfer by convection.

Chapter 6 analyzes heat transfer by radiation including radiation by flame.

Chapter 7 illustrates the required behavior when examining heat transfer in heat exchangers, as well as tube banks.

Chapter 8 discusses pressure drops in detail.

Appendix A shows a series of Tables relative to thermal characteristics of the materials.

Appendix B includes a series of Tables about the corrective factors to be adopted to obtain the real value of the mean temperature difference in design computation.

Appendix C includes a series of Tables about the corrective factors to be adopted to obtain the real value of the exit temperature of the heating fluid in verification computation.

Hopefully the goal to offer the reader a relatively easy and useful reference and to provide the designer a valuable tool was partly accomplished.

Italy Donatello Annaratone

Contents

Notation

A = air per fuel mass unit (kg/kg, Nm^3/kg), cross sectional area (m^2)
a = thermal diffusivity (m^2/s)
B = black level
C = carbon mass percentage (%)
c = specific heat (J/kg K)
D, d = diameter (m)
E = efficiency factor, monochromatic radiation (W/m^3)
F = force (N)
f = arrangement factor, factor for coal
G = mass velocity (kg/m^2 s), flue gas per fuel mass unit (kg/kg, Nm^3/kg)
g = gravity acceleration (m/s^2)
H = hydrogen mass percentage (%), height (m), heat value (kJ/kg)
h = enthalpy (kJ/kg), distance (m)
I = electric current intensity (A)
k = thermal conductivity (W/m K)
L = length (m)
M = mass flow rate (kg/s)
m = mass moisture percentage (%)
N = number of rows, number of diameters
n = air index
P = wet perimeter (m)
p = pressure (Pa, bar)
Q = heat (J, kJ), volumetric flow rate (m^3/s, m^3/h)
q = heat per time unit (W, kW), thermal flux (W/m^2, kW/m^2)
R = thermal resistance (K/W), radius (m)
r = radius (m)
S = surface (m^2)
s = pitch (m)
T = absolute temperature (K)
t = temperature (°C)
U = overall heat transfer coefficient (W/m^2 K)
V = velocity (m/s), volume (m^3)
v = specific volume (m^3/kg)

x = thickness (m)
Fo = number of Fourier
Gr = number of Grashof
Nu = number of Nusselt
Pe = number of Peclet
Pr = number of Prandtl
Re = number of Reynolds
α = heat transfer coefficient (W/m^2 K)
β = cubic expansion coefficient (1/°C), characteristic factor
ε = emissivity, relative roughness
γ = characteristic factor
η = efficiency
λ = wave length (m), friction factor
μ = dynamic viscosity (kg/ms)
ν = kinematic viscosity (m^2/s)
ρ = density (kg/m^3)
ψ = characteristic factor
σ = constant of Stefan-Boltzmann ($W/m^2\ K^4$)
ζ = factor for concentrated pressure drops
θ = time (s)
Δt = temperature difference (°C)
Δp = pressure drop (Pa, bar)
ΔV = electric potential difference (V)

Superscripts

$'$ = heating fluid
$''$ = heated fluid

Subscripts

a = air, arrangement
ad = adiabatic
b = bulk
c = cavity, counter flow
d = depth
e = exit, heat exchange, equivalent
f = fuel, film, flame, fin
g = gas
hy = hydraulic
i = inside, inferior, inlet
l = logarithmic, longitudinal

m = mean, mass
o = outside
p = constant pressure (isobaric), parallel flow
r = radiation, row
s = steam, solar, surface
t = theoretic, tube, transverse
v = volume, constant volume
w = water, wall
0 = normal condition (for air and gas) or room condition
1 = inlet (for heating or heated fluid)
2 = outlet (for heating or heated fluid)

Chapter 1
Introduction to Heat Transfer

1.1 General Considerations

The laws governing heat transfer are of fundamental importance for the design of machines or parts of them used in numerous industrial sectors, as well as the civil sector.

We can think of steam generators in thermoelectric power plants, both in industries producing steam for technological purposes, as well as in industry and in living complexes for heating purposes.

We recall furnaces of different kinds in industry and heat exchangers, evaporators and condensers that are often part of it.

Heat recuperators became increasingly popular given the high cost of fuels that leads to solutions to recuperate the otherwise inevitable heat loss.

In the civil sector the correct study of heat transfer is crucial to resolve heating and air conditioning issues in the best possible way in order to maximize energy savings.

Finally, solar energy should not be forgotten since its diffusion will increasingly grow in the future to differentiate energy sources for both political and economic reasons.

The primary goal of the designer is to maximize the amount of transferred heat per surface unit at minimum cost.

As we already pointed out, another important goal is to recuperate heat from the exhaust gas produced by various machines by reducing heat loss in the environment. This is done through waste heat generators, exchangers, heat recuperators and heat regenerators.

Still with energy savings and cost reduction in mind, it is important to ensure proper insulation of all elements at high temperature as far as compatible with its cost.

D. Annaratone, *Engineering Heat Transfer*, DOI 10.1007/978-3-642-03932-4_1,

1.2 Modes of Heat Transfer

1.2.1 Conduction

Heat transfer through a homogeneous and opaque solid occurs by conduction and originates from the difference in temperature between the warm and the cold side of the solid.

The value for thermal conductivity which depends on the material used and varies greatly from material to material is crucial to determine the amount of transferred heat.

Thermal conduction consists of transfer of kinetic energy from one molecule to the adjacent molecule.

Heat transfer by conduction also occurs through liquid or gas but in this case under certain conditions heat can be transferred simultaneously by convection and radiation.

1.2.2 Convection

Heat transfer by convection in fluids usually occurs after mixing one part of fluid with another with a temperature difference between one area and the other, or between the temperature of the fluid and that of the wall licked by the fluid itself.

This particular movement of the fluid may be generated in turn by temperature difference only, or created through mechanical means.

The first situation occurs with natural convection. The difference in temperature produces a variation in terms of the fluid density from area to area that produces the movement itself.

In the case of forced convection velocity is imprinted onto the fluid mechanically, and this considerably increases the heat transferred from a warm wall to the fluid at lower temperature, or from the fluid to the wall at lower temperature, as we shall see later on.

If the fluid is transparent heat by radiation goes hand in hand with heat by convection.

1.2.3 Radiation

A warm body emits energy by radiation in all directions. If this energy hits another body part of it is reflected, whereas the remaining energy is absorbed by the body and transformed into heat.

The amount of emitted and reflected heat depends on the temperature of the body and on the characteristics of the material making it up.

We talk about a black body when the emitted heat reaches its peak at a certain temperature. Various materials get more or less close to this theoretical condition, as we shall see further on.

If two bodies face each other through insertion of an interspace there will be an ongoing exchange of energy between them. The warmer body radiates more energy than it absorbs, while the colder body absorbs more energy than it reflects. In other words, there is heat exchange between warm and cold body, i.e., heat transfer by radiation.

A few particular types of gas are able to radiate and absorb energy by radiation, too.

1.3 Laws of Heat Transfer

1.3.1 Conduction

Fourier's law with regard to heat transfer by conduction in one direction (x), is as follows:

$$\frac{dQ}{d\theta} = -kS\frac{dt}{dx} \tag{1.1}$$

The heat $dQ/d\theta$ is the heat instantly transferred through the wall, where θ represents time; it is proportional to surface S through which the heat transfer takes place, to the gradient of temperature $-dt/dx$ and to thermal conductivity k, a physical characteristic of the material being used.

Temperature varies with location but it can also vary over time.

In this case heat varies over time, as well, and this is called "unsteady state". This is, in fact, an unsteady phase that leads to a steady state over a variable period of time when temperatures stabilize and the transferred heat becomes constant.

If heat is constant and we are in "steady state" Eq. (1.1) may be written as follows:

$$q = -kS\frac{dt}{dx} \tag{1.2}$$

where q stand for the heat transferred within the time unit.

If the wall is flat, given that S is constant in that case, integrating (1.2) leads to

$$q = \frac{k}{x_w} S\Delta t \tag{1.3}$$

where Δt is the difference in temperature between the warm and the cold side and x_w is the thickness of the wall.

As we shall see, it is possible to adopt an equation similar to (1.3) even for curved walls; it is as follows:

$$q = \frac{k}{x_w}\frac{S_m}{\phi}\Delta t \tag{1.4}$$

where S_m is a particular average value of the surface where the heat transfer occurs, which depends on the shape of the wall, and ϕ is a corrective factor that may often be neglected, though, in many instances. The values of ϕ for the tubes (cylindrical wall) are listed on Table 2.1; those relative to the spherical wall are shown in Table 2.2.

If we introduce the thermal resistance of the wall given by

$$R = \frac{x_w \phi}{k S_m}; \tag{1.5}$$

(1.4) is simply reduced to

$$q = \frac{\Delta t}{R} \tag{1.6}$$

Introducing thermal resistance is useful in the case of a wall consisting of various layers of different materials. In that case they behave like electrical resistors in series. Total resistance is the sum of the resistances of the individual layers. In other words, we have

$$q = \frac{\Delta t_t}{\sum R} = \frac{\Delta t_t}{\sum \frac{x_w \phi}{k S_m}} \tag{1.7}$$

where Δt_t is the difference in temperature between the warm side of the first layer and the cold side of the last layer.

1.3.2 Convection

The law governing heat transfer from a wall to a fluid or viceversa goes back to Newton, and is as follows:

$$q = \pm \alpha S(t_w - t) \tag{1.8}$$

with a positive sign if the heat moves from the wall to the fluid, and a negative sign if it moves in the opposite direction. In (1.8) q is the transferred heat during the time unit, S is the surface of the wall, t_w and t are the temperatures of the wall and the fluid, respectively, and α is the heat transfer coefficient of the fluid.

The heat transfer coefficient is a characteristic quantity of convection, thus ruling out the presence of heat transfer by radiation. Nonetheless, in case of radiation concerning the fluid (special radiating gases), as we shall see, it is possible to adopt a particular value for α that takes it into account.

The value of α depends on various physical quantities of the fluid (density, isobaric specific heat, thermal conductivity and dynamic viscosity) beyond dimensional elements and velocity. In addition, it is influenced by temperature. A particular situation is created if a change of state takes place.

Even in this case, as with conduction, it is possible to introduce thermal resistance given by

$$R = \frac{1}{\alpha S}; \tag{1.9}$$

and this leads to

$$q = \frac{\Delta t}{R}. \tag{1.10}$$

If we consider heat transfer from one fluid to another through a wall, the phenomenon is characterized by global thermal resistance which is the sum of the resistances of both fluids and of the wall. This way it is easy to compute the overall heat transfer coefficient that characterizes the heat transfer from one fluid to another. In fact, the product of the overall heat transfer coefficient by the surface in question is equal to the reciprocal of the global thermal resistance.

The heat transfer coefficient varies considerably when switching from one fluid to another. For instance, it is extremely high for boiling water and high for warming up water, high on average for superheated steam and rather low for air and flue gas.

1.3.3 Radiation

Thanks to Stefan-Boltzmann the basic equation for radiation is as follows:

$$dq = \sigma dST^4 \tag{1.11}$$

where dq is the heat transfer by radiation from the "black body" via area dS ,T is the absolute temperature of the body and σ is the constant of Stefan-Boltzmann.

If the surfaces are grey, as is often the case, in other words if they emit less heat than the black body, their emissivity is not equal to 1, as is true for the black body, but lower. The heat exchange between two surfaces separated by an interspace consisting of a transparent medium is equal to

$$q = \sigma S_1 F_{12} \left(T_1^4 - T_2^4 \right) \tag{1.12}$$

where S_1 is the radiating surface, T_1 e T_2 the temperatures of the radiating surface and of the radiated one, respectively, and F_{12} is a function of the geometry and the emissivity of both surfaces.

Heat transfer by radiation is a very complex phenomenon. Using (1.11), (1.12) or similar equations it is possible to resolve only few elementary issues, but it becomes basically impossible to resolve complex cases, such as the heat transfer inside the furnace of a steam generator.

In these cases it is necessary to apply experimental data or empirical computation methods derived from experimental surveys.

1.4 Overall Heat Transfer Coefficient

If t' and t'' indicate the temperature of a heating fluid and that of a heated fluid separated by a wall, respectively, (Fig. 1.1), the heat transferred from the warmer to the colder fluid within the time unit (in W) is equal to

$$q = US\left(t' - t''\right) \tag{1.13}$$

where S stands for the surface of the reference section (in m^2), and U is the overall heat transfer coefficient (in W/m^2 K).

We see that U factors in globally the triple heat transfer from the heating fluid to the wall, through the wall, as well as from the wall to the heated fluid.

As mentioned earlier, the product US is the reciprocal of the global thermal resistance. If we consider the case where the heat crosses a plane wall with a surface S, and recalling what was said about conduction in the wall and about convection relative to fluids, we may write that

$$\frac{1}{US} = \frac{1}{\alpha' S} + \frac{x_w}{kS} + \frac{1}{\alpha'' S}; \tag{1.14}$$

then

$$U = \frac{1}{\dfrac{1}{\alpha'} + \dfrac{x_w}{k} + \dfrac{1}{\alpha''}} \tag{1.15}$$

where α' and α'' stand for the heat transfer coefficient of both heating and heated fluids.

If the wall consists of more layers (Fig. 1.2) we will write that

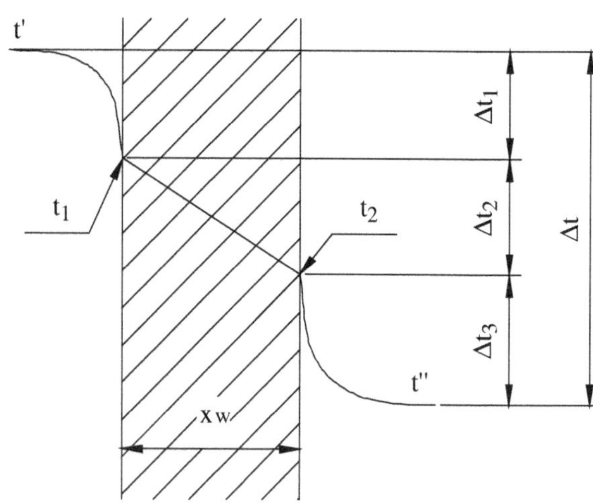

Fig. 1.1

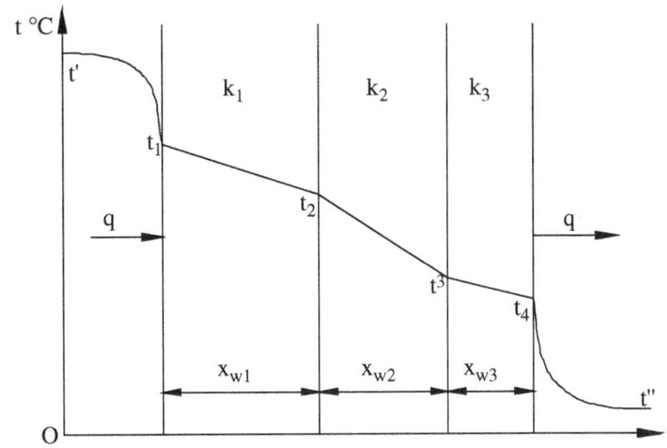

Fig. 1.2 Several walls in series

$$U = \frac{1}{\dfrac{1}{\alpha'} + \dfrac{x_{w1}}{k_1} + \dfrac{x_{w2}}{k_2} + \cdots \dfrac{x_{wn}}{k_n} + \dfrac{1}{\alpha''}} \tag{1.16}$$

where $x_{w1}, x_{w2} \ldots x_{wn}$ represent the thickness of the different layers (in m) and $k_1, k_2 \ldots k_n$ the thermal conductivity of the materials used for the layers (in W/mK).

Let us now consider a cylindrical instead of a plane wall (tube). In this case, in (1.13) the surface of reference may be the internal, as well as the external one, matched by two different values of U (of course, the product US is equal).

We consider the internal surface, thus indicating the relative overall heat transfer coefficient with U_i (Fig. 1.3). If we refer to a unitary length of the tube, and if we consider that the heating fluid is on the inside (it is advisable to adhere to the convention that the surface of reference is the one licked by the heating fluid), we may write that

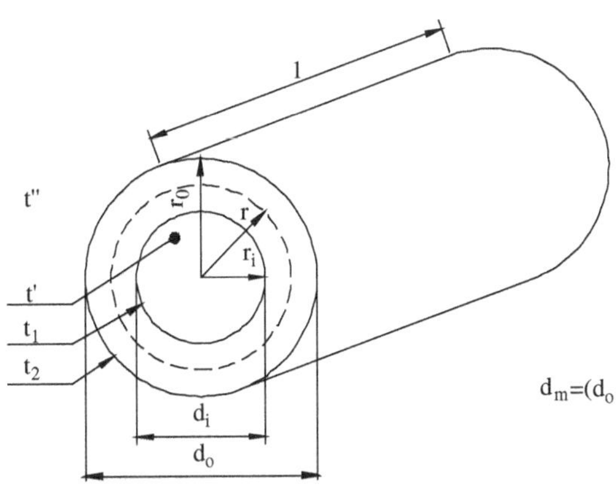

Fig. 1.3 Centrifugal heat flux

Fig. 1.4 Centripetal heat flux

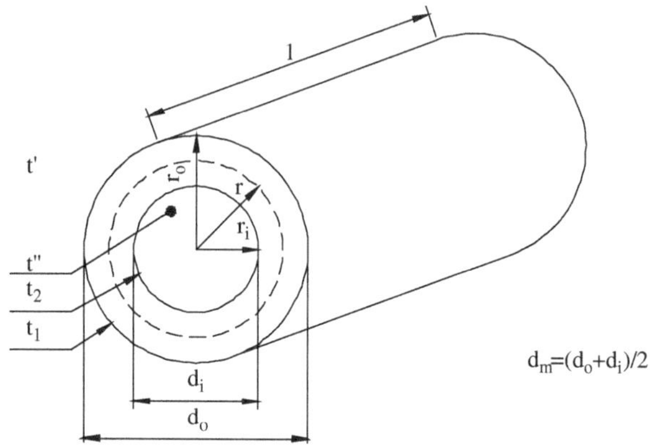

$$\frac{1}{\pi U_i d_i} = \frac{1}{\pi \alpha' d_i} + \frac{x_w \phi}{\pi k d_m} + \frac{1}{\pi \alpha'' d_o}; \tag{1.17}$$

then

$$U_i = \frac{1}{\dfrac{1}{\alpha'} + \dfrac{x_w \phi}{k} \dfrac{d_i}{d_m} + \dfrac{1}{\alpha''} \dfrac{d_i}{d_o}} \tag{1.18}$$

In (1.18) d_i is the inside diameter and d_o the outside one.

The values of ϕ are shown in Table 2.1

If the wall in question is that of a metal tube, the corrective factor ϕ can generally be neglected.

On the other hand, if we take the external surface licked by the heating fluid as reference (Fig. 1.4), we have

$$\frac{1}{\pi U_o d_o} = \frac{1}{\pi \alpha' d_o} + \frac{x_w \phi}{\pi k d_m} + \frac{1}{\pi \alpha'' d_i}; \tag{1.19}$$

and then

$$U_o = \frac{1}{\dfrac{1}{\alpha'} + \dfrac{x_w \phi}{k} \dfrac{d_o}{d_m} + \dfrac{1}{\alpha''} \dfrac{d_o}{d_i}}. \tag{1.20}$$

If the cylindrical wall consists of various coaxial layers, (1.18) will have to be written as follows:

$$U_i = \frac{1}{\dfrac{1}{\alpha'} + \dfrac{x_{w1} \phi_1}{k_1} \dfrac{d_i}{d_{m1}} + \dfrac{x_{w2} \phi_2}{k_2} \dfrac{d_i}{d_{m2}} + \cdots \dfrac{x_{wn} \phi_n}{k_n} \dfrac{d_i}{d_{mn}} + \dfrac{1}{\alpha''} \dfrac{d_i}{d_o}} \tag{1.21}$$

with an obvious meaning assigned to the symbols for $x_{w1}, x_{w2} \ldots x_{wn}$, for $\phi_1, \phi_2 \cdots \phi_n$ for $k_1, k_2 \ldots k_n$ and for $d_{m1}, d_{m2} \ldots d_{mn}$.

By analogy, Eq. (1.20) must be structured accordingly.

With reference to (1.18) note that in the case of a thin metal tube, as is usually the case, the term for the thermal resistance of the wall is irrelevant with respect to the resistance of both fluids.

With sufficient approximation for practical purposes it is possible to simply write that

$$U_i = \frac{1}{\dfrac{1}{\alpha'} + \dfrac{1}{\alpha''}\dfrac{d_i}{d_o}} = \frac{\alpha'\alpha''}{\alpha'' + \alpha'\dfrac{d_i}{d_o}}. \tag{1.22}$$

Similarly, and through the same simplifying process, based on (1.20) we obtain

$$U_o = \frac{1}{\dfrac{1}{\alpha'} + \dfrac{1}{\alpha''}\dfrac{d_o}{d_i}} = \frac{\alpha'\alpha''}{\alpha'' + \alpha'\dfrac{d_o}{d_i}} \tag{1.23}$$

In many instances the heat transfer coefficient of the heated fluid is much greater compared to the one of the heating fluid, as is the case, e.g., when the heating fluid is represented by flue gas while the heated fluid is boiling water, warming up water or superheated steam. Note, e.g., that in steam generators the value of α' usually varies from 40 W/m^2 K to 80 W/m^2 K, depending on the type of generator and on the geometry of the tube banks and other technical characteristics. The boiling water has a heat transfer coefficient of about 12000 W/m^2 K, the warming up water will have a heat transfer coefficient of about 4000/W/m^2 K, and finally, the superheated steam will generally have a heat transfer coefficient of about 1500–2000 W/m^2 K.

The great gap between the values of α'' for the heated fluid with respect to the value of α' of the flue gas is evident.

Under these conditions the impact of the corrective factor d_i / d_o or d_o / d_i present in (1.22) and in (1.23) is negligible; thus, with acceptable approximation for practical purposes, starting from (1.22) and from (1.23) it is possible to write that

$$U_i = \frac{\alpha'\alpha''}{\alpha' + \alpha''}; \tag{1.24}$$

$$U_o = \frac{\alpha'\alpha''}{\alpha' + \alpha''}. \tag{1.25}$$

As we see, we obtain the same equation for U_i and U_o which may seem absurd, given that the first value of the overall heat transfer coefficient is in reference to the internal surface of the tube, whereas the second one is in reference to the external surface.

In fact, this is not true since Eqs. (1.24) and (1.25) may be used under the conditions described above, only if the overall heat transfer coefficient is in reference to the surface licked by the heating fluid. Of course, the latter may be in reference to the internal or the external one, to which, in fact, the equations above are about.

Note that these simplifications must clearly be justified to avoid serious errors. In other words, there must be a thin wall thickness when confronted with not too small values for the heat transfer coefficient of the fluids to eliminate the term relative to thermal resistance of the wall. There must be a high ratio between the heat transfer coefficient of the heated fluid and the heating fluid in order to use either (1.24) or (1.25).

For instance, if the flue gas warms up the air the latter simplification is not possible because the order of magnitude of the heat transfer coefficient of both flue gas and air is the same.

The previous considerations also lead to the following considerations.

The designer is greatly interested in maximizing the value of U (as compatible with acceptable pressure drops). In fact, given equality of S and Δt, this allows the designer to increase the amount of heat transfer. If the heat to be transferred is predetermined instead, and the value of Δt is equally set in advance, it is possible to obtain the desired result with a reduced exchange surface and reduced costs.

If the values of the heat transfer coefficients of both fluids are of the same order of magnitude, it is in one's interest to increase both the heat transfer coefficient of the heating fluid and of the heated fluid.

But if the heat transfer coefficient of the heated fluid is much higher in comparison to that of the heating fluid, this creates an irrelevant advantage from the increase of the former. Therefore, significant results can be obtained solely by increasing the heat transfer coefficient of the heating fluid.

By the same token, there is no point in using finned tubes when the heat transfer coefficient of both fluids is of the same order of magnitude. This design solution is quite valid instead when the internal fluid has a high heat transfer coefficient while the external one is modest. In that case the small heat transfer coefficient is compensated by a great surface available to the external fluid.

As far as the temperatures of the wall, if we refer to (1.18) and indicate the temperature of the side licked by the heating fluid with t'_w and the temperature of the side icked by the heated fluid with t''_w, we have

$$t'_w = t' - \frac{U_i}{\alpha'}\left(t' - t''\right); \tag{1.26}$$

$$t''_w = t'' + \frac{U_i}{\alpha''}\frac{d_i}{d_o}\left(t' - t''\right). \tag{1.27}$$

Similarly, by referring to (1.20), we obtain

$$t'_w = t' - \frac{U_o}{\alpha'}\left(t' - t''\right); \tag{1.28}$$

$$t''_w = t'' + \frac{U_o}{\alpha''}\frac{d_o}{d_i}\left(t' - t''\right). \tag{1.29}$$

In general, note that the jump in temperature Δt between one of the two fluids and the wall, or through the wall, or through one of the layers that make up the wall is proportional to the relative thermal resistance given that

$$\Delta t = Rq. \tag{1.30}$$

By recalling the equation of global thermal resistance of the fluids/wall system, it follows that for each of the elements at play we have

$$\Delta t = RUS\left(t' - t''\right) \tag{1.31}$$

where R stands for thermal resistance of the element under scrutiny, while S stands for the reference surface of U, and it can either be the internal surface or the external surface of the wall, as long as the value of U is computed accordingly.

For example, if the heating fluid is on the outside and one wishes to know the difference in temperature between the fluid and the wall, if U refers to the external surface, i.e., U_o, based on (1.31) we obtain the following:

$$\Delta t = \frac{1}{\alpha' S_o} U_o S_o \left(t' - t''\right) = \frac{U_o}{\alpha'} \left(t' - t''\right). \tag{1.32}$$

Chapter 2
Steady Conduction

2.1 Introduction

Generally, the designer working on heat transfer concentrates on the computation of heat transfer from a heating fluid to a heated fluid through a wall separating the fluids.

The process occurs in three phases: first, the heat transfers from the heating fluid to the separation wall, then it passes through the wall, and finally from the wall to the heated fluid.

Each passage takes place according to laws that will be illustrated.

In this phase we will discuss heat transfer under steady conditions through a wall consisting of homogeneous material, and consider the case where the passage area of the heat has a constant surface (plane wall), as well as the case where this surface is variable (usually a tube).

As we shall see, the heat transfer is proportional to the thermal conductivity of the material, and we will analyze both the case when conductivity is constant and the one when it varies with the temperature, as is always the case.

2.2 Conduction through a Plane Wall

If we consider a plane wall crossed by heat in its perpendicular direction (x), and if there is no heat loss at the edges and the conductivity of the material of the wall is constant, based on Fourier's simple equation the heat transfer during the time unit q is equal to:

$$q = -kS\frac{dt}{dx} \tag{2.1}$$

where q is in W, k stand for constant thermal conductivity of the material in W/m K, S is the surface of the passage area of the heat in m^2, and t represents generic temperature, a function of x, in K; therefore, dt/dx is the gradient of temperature through the wall; it is negative given that the temperature decreases along the passage of heat.

D. Annaratone, *Engineering Heat Transfer*, DOI 10.1007/978-3-642-03932-4_2,

As a result,

$$dt = -\frac{q}{kS}dx; \tag{2.2}$$

and given that x_w stands for the thickness of the wall in m, by integration one obtains the following equation, given that t_1 and t_2 are the temperature of the warm side and the cold side, respectively:

$$t_1 - t_2 = \frac{qx_w}{kS}; \tag{2.3}$$

and

$$q = \frac{k}{x_w}S\,(t_1 - t_2)\,. \tag{2.4}$$

If k and S are constant, the gradient is constant, thus the behavior of the temperature is linear.

The value of thermal conductivity is quite variable if we consider the different solid materials that may be used.

There are materials with high conductivity that conduct heat well, and those with medium conductivity, and finally those with low conductivity that show considerable resistance to heat transfer and have consequently good insulating characteristics.

The use of materials with high conductivity, like metals, which is generally required by technical and structural constraints helps to facilitate the heat transfer through the wall, by reducing the impact of the wall itself on the conditions necessary for the heat to pass from fluid to the other to a minimum.

The use of materials with low conductivity usually originates from the desire to reduce the heat transfer to a minimum, i.e., normally heat loss outside. This leads to considerable energy saving.

The value of conductivity varies greatly even among metals which are highly conductive materials.

There are metals with especially high conductivity, such as copper ($k \approx 380$ W/m K), gold ($k \approx 290$ W/m K) or magnesium ($k \approx 160$ W/m K), and metals with particularly low conductivity, such as bismuth ($k \approx 8$ W/m K). As far as the others, thermal conductivity varies from 18 W/m K of antimony and 92 W/m K of cadmium. Steel has a thermal conductivity of about 48 W/m K. Even though it is not a metal, silicon carbide is within this range of values and is sometimes used to coat tubes in certain steam generators. Its thermal conductivity varies from 11 to 18 W/m K depending on the temperature.

Among materials with medium thermal conductivity we have concrete ($k \approx 1$ W/m K), window glass ($k = 0.5 - 1$ W/m K) and refractory bricks with an alumina base ($k \approx 3$ W/m K). As far as the latter, as for all porous materials, porosity is crucial. If porosity is high, thermal conductivity may be reduced even to a fourth of the typical allowance of the material. The apparent density of the material which depends on the number of pores and their size is helpful in judging the porosity of a material.

Many materials can be used for insulation. In principle, these are materials with conductivity below 0.5 W/m K. For instance, pumice has a thermal conductivity of 0.24 W/m K, whereas mineral wool has a conductivity of about 0.04 W/m K. This means that, e.g., given a layer of mineral wool of 100 mm of thickness, and a difference in temperature between internal and external side of 100°C, the heat loss is equal to 40 W/m^2, i.e., a much lower value.

Appendix A includes the values for the thermal conductivity of various materials.

Until now we considered thermal conductivity to be constant. In fact, this is not true because it varies with temperature; depending on the material, the variation may be modest or considerable.

As far as solid materials, like the ones we are discussing, the variation of conductivity with temperature may be considered linear, based on experimental data. Therefore,

$$k = a \pm bt \tag{2.5}$$

where the positive sign works for most materials with the exception of a few metals requiring the negative sign.

As far as carbon steel, we can assume that (with k in W/m K)

$$k = 53.19 - 2.49\frac{t}{100}. \tag{2.6}$$

As far as stainless austenitic steel we have instead

$$k = 14.56 + 1.45\frac{t}{100} \tag{2.7}$$

Based on (2.5) and recalling (2.1):

$$q = -(a \pm bt)\, S\frac{dt}{dx}; \tag{2.8}$$

then

$$\frac{q}{S}dx = -(a \pm bt)\, dt \tag{2.9}$$

By integration:

$$\frac{q}{S}x_w = a(t_1 - t_2) \pm \frac{b}{2}\left(t_1^2 - t_2^2\right) = (t_1 - t_2)\left[a \pm \frac{b}{2}(t_1 + t_2)\right]. \tag{2.10}$$

We introduce an average value of k and indicate it with k_m; based on (2.4):

$$\frac{q}{S} x_w = k_m (t_1 - t_2) ; \tag{2.11}$$

and then by comparison of (2.10) with (2.11):

$$k_m = a \pm \frac{b}{2} (t_1 + t_2) . \tag{2.12}$$

We establish that k_m stands for the average value of conductivities corresponding to temperatures t_1 and t_2; it is also equal to the conductivity corresponding to the average of these temperatures.

The latter consideration is particularly important. In fact, it suffices to consider the average temperature among the extreme ones of the wall, and to adopt the corresponding value of k based on this temperature as a constant value in the different equations.

Naturally, taking into account the variation of k with temperature, the pattern of the latter is no longer linear through the wall.

If conductivity increases with temperature, inside the wall the latter is equal to

$$t = -\frac{a}{b} + \sqrt{\frac{2qx}{bS} + \left(\frac{a}{b} + t_2\right)^2} \tag{2.13}$$

where t_2 stands for the temperature of the cold side and x stands for the distance of the section from this side.

On the other hand, if conductivity decreases with temperature, the latter is equal to:

$$t = \frac{a}{b} - \sqrt{\left(\frac{a}{b} - t_2\right)^2 - \frac{2qx}{bS}} . \tag{2.14}$$

2.3 Conduction through a Plane Multiwall

If a plane wall consists of different layers of materials with different conductivity, the computation of the heat transfer must be done as follows (Fig. 2.1).

Let us recall the well-known analogy existing between heat transfer and electric power transfer.

If we consider an electrical resistor, the law governing the amount of electricity transferred through it is as follows:

$$I = \frac{\Delta V}{R} \tag{2.15}$$

where I stand for the current intensity, ΔV the difference in potential between the extremities of the resistor and R its electric resistance.

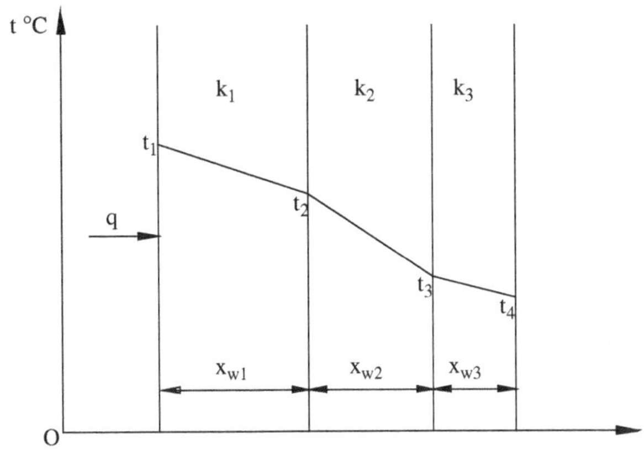

Fig. 2.1 Several walls in series

By analogy, if we substitute ΔV with the difference in temperature Δt, I with heat q, and electric resistance with thermal resistance we have

$$q = \frac{\Delta t}{R} \tag{2.16}$$

where R represents thermal resistance. Equation (2.16) corresponds to (2.3) if the following equation is to represent thermal resistance:

$$R = \frac{x_w}{kS}. \tag{2.17}$$

The analogy is not simply formal but a substantial one.

As far as electricity, the "engine" causing the passage of electricity is the difference in potential ΔV; similarly, the "engine" for the passage of the heat is the difference in temperature Δt; in the presence of ΔV electricity passes through the resistor identified by current intensity I, whereas in the presence of Δt heat q passes through the wall.

The entity of the intensity of either current or heat, ΔV or Δt being equal, is governed by the electrical resistance of the resistor or by the thermal resistance of the wall.

As far as electricity, if various resistors are placed in sequence their total resistance is equal to the sum of the resistances of the individual resistors.

Similarly, in our case if the wall is made of different layers that are in sequence with respect to the passage of heat, the thermal resistance of the wall is equal to the sum of the resistances of the single layers.

If R indicates this total resistance we have

$$R = \frac{x_{w1}}{k_1 S} + \frac{x_{w2}}{k_2 S} \cdots + \frac{x_{wn}}{k_n S} \tag{2.18}$$

where the meaning of $x_{w1}, x_{w2} \ldots x_{wn}$ and of $k_1, k_2 \ldots k_n$ is obvious (see Fig. 2.1).

Thus, if $t_1 - t_2$ indicate the difference in temperature between the warm and the cold side of the multilayer wall, we have

$$t_1 - t_2 = \frac{q}{S}\left(\frac{x_{w1}}{k_1} + \frac{x_{w2}}{k_2} + \cdots \frac{x_{wn}}{k_n}\right); \tag{2.19}$$

then

$$q = \frac{S}{\frac{x_{w1}}{k_1} + \frac{x_{w2}}{k_2} + \cdots \frac{x_{wn}}{k_n}}\left(t_1 - t_2\right). \tag{2.20}$$

At this point it is also possible to identify an equivalent conductivity which factors in the behavior of the multilayered wall globally with respect to heat transfer.

This conductivity amounts to the same transferred heat when considered constant across the entire wall.

If we indicate it with k_e:

$$q = \frac{k_e}{x_w} S\left(t_1 - t_2\right) \tag{2.21}$$

where x_w stands for the total thickness of the wall.

By comparing (2.21) with (2.20) we have:

$$k_e = \frac{x_w}{\frac{x_{w1}}{k_1} + \frac{x_{w2}}{k_2} + \cdots \frac{x_{wn}}{k_n}}. \tag{2.22}$$

The assumption with regard to multilayered walls is that the contact between layers is perfect.

If this is not so, as is often the case, there are thermal additional resistances due to the presence of a modest interspace between each layer.

A simple example suffices to fully comprehend the impact of this resistance.

We assume an interspace of air of just 1/10 mm; with air conductivity of 0.03 W/m K and a conductivity of steel equal to 48 W/m K, one can easily verify that the interspace in question has a thermal resistance identical to that of a steel wall with a thickness of 160 mm.

These additional resistances may be neglected only in the presence of insulating materials, i.e., materials characterized by high thermal resistance. Note that the conductivity of insulating materials is in the same order of magnitude of air, or even equal to 10–15 times that of air. The assumed interspace of air of 1/10 mm corresponds to a layer of insulation of up to 1–1.5 mm. Moreover, note that these additional resistances positively contribute to reach the desired goal, given that the aim is to insulate and reduce the heat transfer to a minimum.

On the other hand, they cannot be ignored if the goal is to transfer heat.

In fact, in that case the computation factors in a heat transfer which is greater than the actual one, or a difference in temperature between warm and cold side which is lower than the actual one for any given transferred heat. Therefore, the thermal resistances regarding these interspaces must be included in (2.18); (2.19) as well as (2.20) must be adjusted accordingly.

2.4 Conduction through a Cylindrical Wall

If the heat transfers through a cylindrical wall (the wall of a tube) surface S is variable.

Equation (2.2) must be rewritten considering the generic value of S as a function of the radius and substituting the radius at the abscissa x in (2.2).

Considering the unitary length of the cylinder; the generic surface through which the heat passes is equal to

$$S = 2\pi r. \tag{2.23}$$

Consequently, (2.2) must be written as follows:

$$dt = -\frac{q}{2\pi r}\frac{dr}{k}. \tag{2.24}$$

By integrating (2.24) we obtain:

$$t_1 - t_2 = \frac{q}{2\pi k}\log_e\frac{r_o}{r_i} = \frac{q}{2\pi k}\log_e\frac{d_o}{d_i} \tag{2.25}$$

where r_i and r_o are the inside and the outside radius of the cylinder, respectively, (see Fig. 2.2), whereas d_i and d_o are the inside and outside diameters. Heat q transferred during the time unit through the wall of the tube of unitary length is therefore equal to:

$$q = \frac{2\pi k}{\log_e\dfrac{d_o}{d_i}}(t_1 - t_2). \tag{2.26}$$

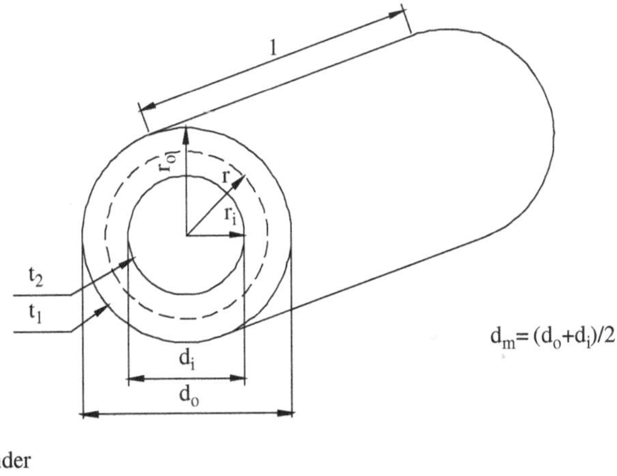

Fig. 2.2 Cylinder

Since the thermal gradient dt/dr is a function of the radius, the behavior of temperature is not linear.

If we want to consider the cylindrical wall like a plane wall with S ideally constant, we must introduce an average conventional surface S_m. Recalling (2.4) and (2.26) in order to derive an equation for S_m we must write that:

$$\frac{2\pi k}{\log_e \dfrac{d_o}{d_i}}(t_1 - t_2) = \frac{k}{x_w} S_m (t_1 - t_2); \tag{2.27}$$

then

$$S_m = \frac{2\pi x_w}{\log_e \dfrac{d_o}{d_i}} = \frac{\pi (d_o - d_i)}{\log_e \dfrac{d_o}{d_i}}. \tag{2.28}$$

By indicating the inside and outside surfaces with S_i and S_o, respectively, we have

$$S_i = \pi d_i; \quad S_o = \pi d_o. \tag{2.29}$$

Thus, from (2.28):

$$S_m = \frac{S_o - S_i}{\log_e \dfrac{S_o}{S_i}}. \tag{2.30}$$

Therefore, if we want to consider the tube like a plane wall, the mean logarithmic value of both external and internal surfaces must be introduced as the surface of reference. By analogy, in reference to the diameter the mean logarithmic value of both the outside and inside diameter must be considered as the diameter of reference.

The logarithmic average differs from the arithmetic average in a negligible way in certain instances, and in a considerable way in others. The latter are infrequent.

As a consequence, it is common to use the arithmetic average with a corrective factor that may or may not be considered.

In other words, for the cylinder we write that

$$t_1 - t_2 = \frac{q}{\pi d_m} \frac{x_w}{k} \phi \tag{2.31}$$

where d_m stands for the average diameter of the cylindrical wall (arithmetic mean value), x_w stands for its thickness, and ϕ is the corrective factor.

Table 2.1 Factor ϕ for cylindrical wall

$\frac{d_o}{d_i}$	ϕ	$\frac{d_o}{d_i}$	ϕ	$\frac{d_o}{d_i}$	ϕ
1.0	1.000	1.8	1.030	3.5	1.129
1.1	1.001	1.9	1.035	4.0	1.152
1.2	1.002	2.0	1.040	4.5	1.182
1.3	1.005	2.2	1.050	5.0	1.207
1.4	1.009	2.4	1.061	5.5	1.231
1.5	1.014	2.6	1.074	6.0	1.254
1.6	1.018	2.8	1.087	6.5	1.276
1.7	1.024	3.0	1.089	7.0	1.297

The equation of ϕ is obtained by imposing that

$$\frac{d_m}{\phi} = \frac{d_o - d_i}{2\phi} = \frac{d_o - d_i}{\log_e \frac{d_o}{d_i}}; \tag{2.32}$$

then

$$\phi = \frac{\left(\frac{d_o}{d_i} + 1\right) \log_e \frac{d_o}{d_i}}{2\left(\frac{d_o}{d_i} - 1\right)}. \tag{2.33}$$

The values of ϕ, as a function of d_o/d_i, are shown in Table. 2.1.

Clearly, the value of ϕ differs from unity in a negligible way if the values of the ratio d_o/d_i are small (roughly below 1.3).

It is the case of thin tubes destined to contain an internal fluid heated by an external fluid or heating an external fluid; in that case the coefficient ϕ may be ignored, and (2.31) becomes very similar to the analogous equation for plane walls. Note the fact, though, that the passage surface of the heat must be the one corresponding to the mean diameter.

The value of ϕ becomes significant only for high values of the ratio d_o/d_i.

For instance, this is the case with a cylindrical insulation around a tube.

As far as variability of k with temperature, as for plane walls it is possible to adopt a constant value of thermal conductivity, as long as it is assumed to be equal to the one corresponding to the mean temperature between the external and the internal.

2.5 Conduction through a Cylindrical Multiwall

As for plane walls, if we consider a cylindrical multiwall (Fig. 2.3) we must consider the thermal resistances of the different layers and compute the total thermal resistance.

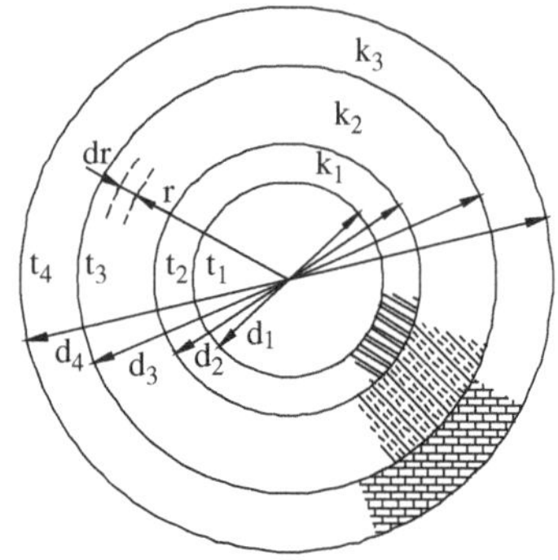

Fig. 2.3 Three concentric cylinders

With reference to a tube of unitary length, recalling (2.31) and the meaning of thermal resistance, for a cylindrical wall the latter is given by

$$R = \frac{x_w \phi}{\pi k d_m}. \tag{2.34}$$

Thus, considering the total thermal resistance of the multiwall:

$$R = \frac{x_{w1}\phi_1}{\pi k_1 d_{m1}} + \frac{x_{w2}\phi_2}{\pi k_2 d_{m2}} + \cdots \frac{x_{wn}\phi_n}{\pi k_n d_{mn}} \tag{2.35}$$

where $x_{w1}, x_{w2} \ldots x_{wn}$ and $k_1, k_2 \ldots k_n$ have an obvious meaning (Fig. 2.3), while $\phi_1, \phi_2 \ldots \phi_n$ represent the values of the corrective coefficient ϕ for the different layers, and $d_{m1}, d_{m2} \ldots d_{mn}$ are the mean diameters of the layers.

The heat q transferred through the wall by the time unit and by the length unit of the tube is therefore equal to:

$$q = \frac{t_1 - t_2}{\dfrac{x_{w1}\phi_1}{\pi k_1 d_{m1}} + \dfrac{x_{w2}\phi_2}{\pi k_2 d_{m2}} + \cdots \dfrac{x_{wn}\phi_n}{\pi k_n d_{mn}}}. \tag{2.36}$$

If the mean surface of the cylindrical wall is used as the surface of reference for q we write that

$$q = \frac{\pi d_m}{\dfrac{x_{w1}\phi_1}{k_1}\dfrac{d_m}{d_{m1}} + \dfrac{x_{w2}\phi_2}{k_2}\dfrac{d_m}{d_{m2}} + \cdots \dfrac{x_{wn}\phi_n}{k_n}\dfrac{d_m}{d_{mn}}} (t_1 - t_2). \tag{2.37}$$

As for the plane wall, we can consider an equivalent conductivity k_e for the entire wall; then it is possible to write that

$$q = \frac{k_e}{x_w \phi} \pi d_m (t_1 - t_2) \tag{2.38}$$

where x_w represents the total thickness of the wall, ϕ is the corrective factor for the entire wall (therefore a function of the ratio d_o/d_i of the wall), and d_m is the mean diameter of the entire wall.

A comparison between (2.38) and (2.37) leads to

$$k_e = \frac{x_w \phi}{\frac{x_{w1}\phi_1}{k_1}\frac{d_m}{d_{m1}} + \frac{x_{w2}\phi_2}{k_2}\frac{d_m}{d_{m2}} + \cdots \frac{x_{wn}\phi_n}{k_n}\frac{d_m}{d_{mn}}}. \tag{2.39}$$

Obviously, the same considerations made for plane walls as far as minute interspaces between the layers apply to cylindrical multiwalls, too. If thermal resistances relative to these interspaces must be considered, they must be added in (2.35) to obtain actual global thermal resistance. The equations following (2.35) will have to be modified accordingly.

2.6 Conduction through a Spherical Wall

If we consider heat transfer through a spherical wall, given that the surface of passage for the heat is equal to

$$S = 4\pi r^2, \tag{2.40}$$

we must write that

$$dt = -\frac{q}{4\pi r^2}\frac{dr}{k}. \tag{2.41}$$

By integrating (2.41) we obtain

$$t_1 - t_2 = \frac{q}{4\pi k}\left(\frac{1}{r_i} - \frac{1}{r_o}\right); \tag{2.42}$$

as usual, t_1 stands for the temperature on the warm side and t_2 on the cold side, while r_i and r_o are the inside and outside radius, respectively.

If we want to consider the spherical wall like a plane wall, as was previously done for the tube, it is a question of identifying the ideal surface for the passage of heat, assumed to be constant, to correctly compute the heat transfer.

By indicating this surface with S_m we have

$$\frac{q}{4\pi k}\left(\frac{1}{r_i} - \frac{1}{r_o}\right) = \frac{q x_w}{k S_m} = \frac{q\,(r_o - r_i)}{k S_m}; \tag{2.43}$$

then

$$S_m = 4\pi r_i r_o. \tag{2.44}$$

If S_i and S_o indicate the inside and outside surface, respectively, we know that

$$S_i = 4\pi r_i^2; \quad S_o = 4\pi r_o^2; \tag{2.45}$$

and through (2.44) we establish that

$$S_m = \sqrt{S_i S_o}. \tag{2.46}$$

In other words, the surface to consider to treat the spherical wall like a plane wall is the geometric average of the outside and inside surfaces.

By analogy with tubes, we consider the surface of the sphere corresponding to its mean radius, and by introducing a corrective factor ϕ we see whether it is possible to substitute the surface given by (2.44) with this one.

Therefore,

$$t_1 - t_2 = \frac{q}{S_m}\frac{x_w}{k}\phi \tag{2.47}$$

where S_m in this case is not the surface given by (2.44) but, as we said, the surface corresponding to the mean radius of the sphere.

Thus, we must have

$$4\pi\left(\frac{r_o + r_i}{2}\right)^2 \frac{1}{\phi} = \sqrt{S_o S_i} = 4\pi\, r_o r_i; \tag{2.48}$$

then

$$\phi = \frac{\left(\dfrac{r_o + r_i}{2}\right)^2}{r_o r_i} = \frac{r_o^2 + r_i^2 + 2r_o r_i}{4r_o r_i} = \frac{\left(\dfrac{r_o}{r_i}\right)^2 + 1 + 2\dfrac{r_o}{r_i}}{4\dfrac{r_o}{r_i}}. \tag{2.49}$$

Table. 2.2 shows the values of ϕ as a function of the ratio r_o/r_i. Clearly, the ratios between radii or diameters being equal, the values of the corrective factor are greater for the sphere with respect to the cylinder.

Nonetheless, for values of r_o/r_i below 1.2 the value of ϕ is so close to unity to become negligible. In that case the spherical wall may be treated like a plane wall, too, as long as the constant surface of reference is the one corresponding to the mean radius of the sphere.

Table 2.2 Corrective factor ϕ for spherical walls

$\frac{r_o}{r_i}$	ϕ	$\frac{r_o}{r_i}$	ϕ	$\frac{r_o}{r_i}$	ϕ
1.0	1.000	1.7	1.072	2.4	1.204
1.1	1.002	1.8	1.089	2.5	1.225
1.2	1.008	1.9	1.107	2.6	1.246
1.3	1.017	2.0	1.125	2.7	1.268
1.4	1.029	2.1	1.144	2.8	1.289
1.5	1.042	2.2	1.164	2.9	1.311
1.6	1.057	2.3	1.184	3.0	1.333

2.7 Conduction through Liquids and Gases

The laws governing conduction through a solid discussed above also apply to liquids, vapors and gases; the values of k for the latter are generally very small.

For instance, while conductivity for steel is known to be equal to about 48 W/m K, at low temperature for stagnant water it is equal to about 0.6 W/m K , whereas for air it is about 0.026 W/m K, still at low temperature.

In liquids conductivity generally decreases with an increase in temperature except in the case of water, as its conductivity increases in the beginning and then decreases. The conductivity of gas increases with temperature.

In the absence of experimental data, Smith suggests the following equation for the computation of thermal conductivity in liquids:

$$k = 0.004602 + 0.03676\left(c_p - 1.88\right)^3 + 0.519\left(\frac{\rho'}{M}\right)^{1/3} + 0.09\left(\frac{\mu}{\rho'}\right)^{1/9} \quad (2.50)$$

where k is in W/m K, c_p stands for isobaric specific heat in kJ/kgK, ρ' is the density relative to water, M is the mean molecular weight and μ stands for dynamic viscosity in kg/ms.

For example, for water at 50°C with $c_p = 4.18$ kJ/kgK, $\rho' = 1$, $M = 18$ and $\mu = 547.2 \times 10^{-6}$ kg/ms, we obtain $k = 0.688$ W/m K; the actual value of thermal conductivity is equal to about 0.66 W/m K; therefore, there is a error in excess of 4%.

As far as gas, note that Prandtl's dimensionless number (Pr) is given by

$$\mathrm{Pr} = \frac{c_p \mu}{k}. \quad (2.51)$$

In (2.51) c_p is in J/kg K, while the units of measure of k and μ are the ones already indicated.

According to Eucken, it is possible to write that

$$\mathrm{Pr} = \frac{4}{9 - \frac{5}{\gamma}} \quad (2.52)$$

where γ is the ratio between isobaric specific heat c_p and the specific heat at constant volume c_v.

Based on (2.52) and recalling (2.51), we obtain:

$$k = \frac{9 - \frac{5}{\gamma}}{4} c_p \mu = 0.25\left(9c_p - 5c_v\right)\mu \quad (2.53)$$

which makes it possible to compute k as a function of the specific heat values and of the viscosity.

Note that if gas is considered to be perfect gas, thus $c_p/c_v = 1.4$, Eq. (2.53) is reduced to the following:

$$k = 1.357 c_p \mu \tag{2.54}$$

with a constant value in Prandtl's number equal to Pr= 0.737.

For example, if we consider air at 0°C with $c_p = 1003.8$ J/kgK and $\mu = 17.07 \times 10^{-6}$ kg/ms, as far as thermal conductivity we obtain $k = 0.02325$ W/m K, and this is in great agreement with the actual value. If we consider air at 100°C instead, with $c_p = 1013.5$ J/kg K and $\mu = 21.63 \times 10^{-6}$ kg/ms we obtain $k = 0.02975$ W/m K, while the actual value is 0.02985 W/m K. Even in this case the approximation is excellent because the error only amounts to 0.3%.

Finally, let us consider flue gas at 500°C with a mass moisture percentage equal to 7%; with $c_p = 1182$ J/kgK and $\mu = 34.45 \times 10^{-6}$ kg/ms we obtain $k = 0.0541$ W/m K; the actual thermal conductivity is equal to 0.0552 W/m K; there is a modest error of 2%.

Note that in the case of transparent gas, such as air, the heat may be transferred by conduction, by convection and by radiation.

In most industrial exchangers, as well as in steam generators, the fluids receive and transfer heat as a result of characteristic convective motions. Therefore, they must be analyzed with this frame of mind, and we will do so by discussing convection; their conductivity is not the predominant element of the heat transfer, even though this contributes to determine the entity of the transferred heat by convection.

During discussion of the latter, the presentation of the phenomenon will be introduced by a section devoted to the physical characteristics of the fluids that contribute to assess the impact of the convective phenomenon. Besides isobaric specific heat and dynamic viscosity, these also include the thermal conductivity of the fluid, as anticipated earlier.

There will be more opportunities to discuss fluid conductivity, but for now we would like to anticipate a few approximate equations on the topic.

The thermal conductivity of water is basically independent from pressure; it may be computed with good approximation with respect to realistic values through the following equation:

$$k = 0.5755 + 0.1638 \frac{t}{100} - 0.05767 \left(\frac{t}{100}\right)^2 \tag{2.55}$$

with k in W/m K and t in °C.

Equation (2.55) is valid between 10°C and 300°C; the error is below or at the most equal to ± 1.3%.

The thermal conductivity of air may be computed according to the following:

$$k = 0.02326 + 0.06588 \frac{t}{100} \tag{2.56}$$

with t in °C. (2.56) is valid between 0°C and 300°C; the error is irrelevant.

For methane we can write that

$$k = 0.0306 + 0.01642\frac{t}{100} + 0.00065\left(\frac{t}{100}\right)^2 \tag{2.57}$$

with t in °C. (2.57) is valid between 0°C and 400°C; the error is below or equal to ± 1%.

Finally, for flue gas it is possible to write that

$$k = \left[21.924 - 0.0337m + (68.467 + 0.0966m)\frac{t}{1000} - (12.991 - 0.6229m)\left(\frac{t}{1000}\right)^2\right] \times 10^{-3} \tag{2.58}$$

where m stands for the mass moisture percentage of the gas and t is in °C. Equation (2.58) is valid for temperatures ranging from 50°C to 1200°C and for mass moisture under or equal to 12%.

Chapter 3
Transient Conduction

3.1 Introduction

In the 2nd chapter temperatures were considered independently from time. Under such conditions the heat transfer through the wall does not vary over time, and there is no storage of heat in the wall itself.

Let us consider Fig. 3.1 where the warm side of the wall is at temperature t_1 while the cold side is at temperature t_2; due to this temperature difference we know that a certain amount of heat transfers from the warm to the cold side.

Clearly, the heat entering the warm side is produced by an external source which transfers it to the wall; then the same heat is transferred from the cold wall to an external source.

Let us assume that the external source corresponding to the cold side suddenly transfers heat to it, thus increasing its temperature. For sake of simplicity, we assume that the temperature reaches t_1, i.e., the identical temperature of the other side.

Clearly, at this point the heat transfer through the wall stops because the temperature difference between the two sides that caused it no longer exists.

The temperature on the inside of the wall is lower compared to that of both sides, thus creating a heat flux towards the inside from both sides and progressively

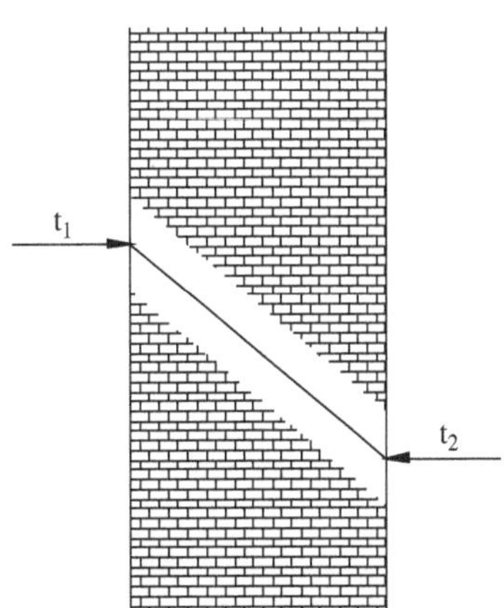

Fig. 3.1 Wall with steady conduction

D. Annaratone, *Engineering Heat Transfer*, DOI 10.1007/978-3-642-03932-4_3,

increasing the inside temperatures until the latter stabilize, depending on certain physical characteristics of the material. The example shows how the entire wall reaches temperature t_1, and the heat transfer through the wall stops.

Naturally the phenomenon takes place in the same way, even though to a different degree, even when the temperature increase of the cold side is lower than the one assumed earlier. In that case, when balance is reached the heat transfer through the wall corresponds to the new difference in temperature between the sides.

Based on our assumptions and as a result of the unbalances in temperature, there is an increase in heat content of the wall.

Of course, the opposite takes place if the temperature of the warm wall suddenly drops. In that case there is a slow decrease in temperature inside the wall at the expense of a heat flux moving towards both sides until balance is reached, and the heat transfer through the wall is in agreement with the new temperatures of the sides.

Evidently, if the temperature of the cold side is brought to be higher compared to the warm side, or if the temperature of the warm side is brought to be lower in comparison to the cold side, under steady conditions there is a change in heat transferring through the wall, and the direction of the heat flux is reversed.

As you can see, the situation shifts from initial balance of temperature to an unsteady state where temperatures vary over time, and finally to a new balance in agreement with the new temperatures of the sides.

If we consider a volume inside the wall that must receive heat from the adjacent areas during the unsteady state, this heat is proportional to the thermal conductivity of the material. The ensuing increase in temperature is proportional to the heat transfer, i.e., k, and is inversely proportional to the specific heat, as referred to the volume of the material. This specific heat (in $\mathrm{J/m^3\,K}$) that we indicate with c_{vol} is given by the product of the specific heat referred to mass (in J/kgK) by the density (in $\mathrm{kg/m^3}$), i.e.,

$$c_{vol} = c\rho. \tag{3.1}$$

Then we introduce thermal diffusivity a equal to

$$a = \frac{k}{c\rho}. \tag{3.2}$$

It is significant with regard to the duration of the transition.

The higher a, the faster the variations in temperature inside the wall with a consequent shorter duration of the unsteady state. Therefore, the latter is shorter in proportion to an increase of the value of k and a decrease in the value of product $c\rho$.

As we shall see, thermal diffusivity is included in Fourier's general law of thermal conduction.

3.2 General Law of Thermal Conduction

Let us consider the cubic element shown in Fig. 3.2 and assume that the element is crossed by heat only in the direction x.

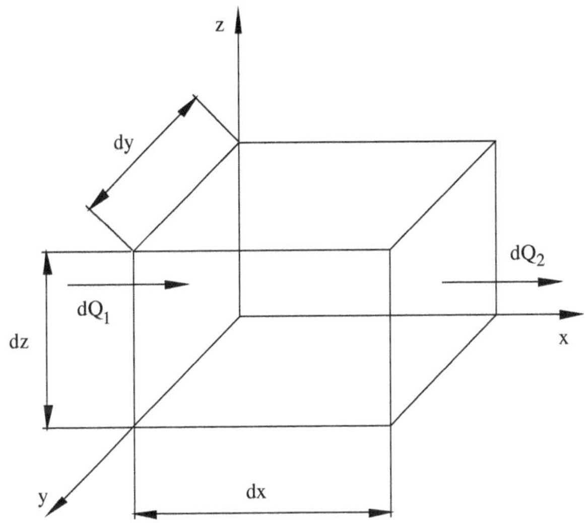

Fig. 3.2 Element with transient conduction

If conditions are unsteady, heat dQ_1 enters the cubic element and heat dQ_2 exits the element. The latter differs from dQ_1(under steady conditions both types of heat are, of course, identical); the heat dQ is stored in the cubic element and is equal to

$$dQ = dQ_1 - dQ_2. \tag{3.3}$$

If the left side of the element registers the thermal gradient $\partial t/\partial x$ (the derivative is partial given that t also depends on time), based on Fourier's well-known law, within time $d\theta$ and given that the surface crossed by the heat is equal to $dydz$,we have

$$dQ_1 = -kdydz\frac{\partial t}{\partial x}d\theta. \tag{3.4}$$

If we move to the other side of the cubic element, i.e., at distance dx from the previous one, the thermal gradient is given by

$$\frac{\partial t}{\partial x} + \frac{\partial}{\partial x}\frac{\partial t}{\partial x}dx = \frac{\partial t}{\partial x} + \frac{\partial^2 t}{\partial x^2}dx. \tag{3.5}$$

Therefore,

$$dQ_2 = -kdydzd\theta\left(\frac{\partial t}{dx} + \frac{\partial^2 t}{\partial x^2}dx\right). \tag{3.6}$$

Then, recalling (3.3)

$$dQ = kdydzd\theta\frac{\partial^2 t}{\partial x^2}dx. \tag{3.7}$$

This heat increases the temperature of the cube; within time $d\theta$ this increase is equal to

$$\frac{\partial t}{\partial \theta} d\theta. \tag{3.8}$$

The heat dQ is equal to the volumetric specific heat of the material (given, as we already saw, by the product of the specific heat referred to mass c by the density of material ρ) multiplied by the volume of the cubic element and by the noted increase in temperature. Therefore,

$$dQ = c\rho dxdydz \frac{\partial t}{\partial \theta} d\theta. \tag{3.9}$$

A comparison between (3.9) and (3.7) leads to

$$\frac{\partial t}{\partial \theta} = \frac{k}{c\rho} \frac{\partial^2 t}{\partial x^2}. \tag{3.10}$$

Equation (3.10) represents Fourier's general law of thermal conductivity.

As we can see, the partial derivative of temperature with respect to time goes up in sync with an increase in thermal diffusivity. Consequently, variations in temperature are faster and the duration of the unsteady state is reduced. This confirms the process described in Sect. 3.1.

The integration of this differential equation makes it possible in principle to compute the development of temperature t over time for every point of the wall.

As shown, (3.10) was derived assuming that the heat transfer occurs only in one direction (x); if this were not the case, the described procedure may be extended by considering the heat flux in direction y, as well as z.

In that case and under the assumption that thermal conductivity is the same in all three directions, the heat transferred to the cubic element, is equal to

$$dQ = kdydzd\theta \frac{\partial^2 t}{\partial x^2} dx + kdxdyd\theta \frac{\partial^2 t}{\partial z^2} dz + kdxdzd\theta \frac{\partial^2 t}{\partial y^2} dy; \tag{3.11}$$

this can also be written as follows:

$$dQ = kdxdydzd\theta \left(\frac{\partial^2 t}{\partial x^2} + \frac{\partial^2 t}{\partial y^2} + \frac{\partial^2 t}{\partial z^2} \right). \tag{3.12}$$

Recalling (3.9) we therefore obtain

$$\frac{\partial t}{\partial \theta} = \frac{k}{c\rho} \left(\frac{\partial^2 t}{\partial x^2} + \frac{\partial^2 t}{\partial y^2} + \frac{\partial^2 t}{\partial z^2} \right). \tag{3.13}$$

Based on (3.10) we obtain rather complicated equations for the computation of t. They are as follows:

$$t = A + Bx + Ce^{-p\theta} e^{qx}; \tag{3.14}$$

$$t = A + Bx + Ce^{-aq^2\theta} \cos(qx); \tag{3.15}$$

$$t = A + Bx + Ce^{-aq^2\theta} \sin(qx); \tag{3.16}$$

$$t = A + Bx + \frac{C}{\sqrt{\theta}} e^{\frac{(q-x)^2}{4a\theta}}; \tag{3.17}$$

$$t = A + Bx + C\frac{2}{\sqrt{\pi}} \int_{\eta=0}^{\eta=\frac{x}{2\sqrt{a\theta}}} e^{-\eta^2} d\eta; \tag{3.18}$$

where A, B, C, p, q are constants of integration that may be null, positive, negative or even imaginary; a stands for thermal diffusivity $(k/c\rho)$ included in (3.10) and in (3.13).

When examining equations from (3.14) to (3.18), if we compute the first partial derivative of t with respect to θ and the second partial derivative of t with respect to x, we will find, as is to be expected, that (3.10) is satisfied for all equations above.

Thus, all equations are valid in principle but during an attempt to apply them to a specific case of interest, it is established that some of them do not satisfy the borderline conditions of space and time.

Therefore, it is required to use the equation or the equations that satisfy these conditions.

The analysis of the equations in question clearly shows that the approach to solve the issues of unsteady state is anything but easy.

Only few physically elementary situations can be studied in a relatively simple way.

For instance, in reference to a sudden variation of the superficial temperature only few types of variation find a theoretical solution.

For other situations that apparently look quite simple but already show complexity as far as the theoretical computation, the available literature provides complex and difficult to use theoretical treaties, solutions requiring the awkward and not necessarily precise use of diagrams, or approximate solutions.

The interest in such procedures has faded after the introduction of automatic computation.

In fact, it is possible to substitute the resolution of differential equations with computation programs based on finite differences. The use of finite differences had already been hypothesized before the advent of computers, but obviously the ensuing manual computation was enormously burdensome in terms of time; only the adoption of bigger dimensions of the differences made it possible to reasonably adopt the method, but in that case the errors were great, and the computation turned out to be rough. The introduction of automatic computation opened up possibilities that were previously unthinkable and eliminated all the problems mentioned above. By exploiting the enormous potential of automatic computation today, it is possible to examine even very complex cases and to adopt dimensions of the differences so minute to make the results basically exact.

Therefore, we believe it is neither useful nor advisable to illustrate the manual procedures described above. By way of an example, we only illustrate a relatively simple type of theoretical computation; it will be an in-depth analysis because it will lead to qualitatively, generally valid considerations to address problems regarding unsteady state.

3.3 Surface Temperature Variation in Infinite Thickness Walls

We consider the side of an infinite thickness wall with a temperature equal to t_0 that is suddenly brought to temperature t_1.

In other words, the aim is to establish how this sudden increase in temperature on the side impacts the temperature inside the wall over time.

The wall is assumed to have infinite thickness to allow the theoretical computation but, as we shall see, the results are transferable to a wall of finite thickness too, provided certain conditions are set.

From a practical point of view with regard to the investigation, note that the imaginary situation can be related to the ground when its surface is hit by a sudden increase in temperature; in that case the infinite thickness of the wall corresponds to reality.

It can even be used in those instances where the temperature of the wall is increased to a value that remains constant over time, such as in the case of furnaces, even though the thickness is not infinite.

If we examine the equations from (3.14), (3.15), (3.16), (3.17) and (3.18), we see that the first four cannot be used because they disagree with the initial time conditions, as well as in correspondence of the side showing the variation in temperature. Only (3.18) satisfies those conditions.

Note that the integral in (3.18), i.e.,

$$\frac{2}{\sqrt{\pi}} \int_0^{\eta} e^{-\eta^2} d\eta \tag{3.19}$$

is none other than Gauss' *error integral*; its values may be taken from Table 3.1 or from Fig. 3.3; it is zero for $\eta = 0$ and it is equal to 1 for $\eta = \infty$.

At the beginning of the phenomenon, thus for $\theta = 0$, given that θ is time, and for $x = 0$, i.e., on the surface of the wall the value of η is zero, this means that the integral in question is zero.

Recalling (3.18) and the fact that the superficial temperature is increased to t_1 we have

$$t_1 = A. \tag{3.20}$$

Table 3.1 Error integral of Gauss

$\frac{x}{2\sqrt{a\theta}}$	$f()$	$\frac{x}{2\sqrt{a\theta}}$	$f()$	$\frac{x}{2\sqrt{a\theta}}$	$f()$	$\frac{x}{2\sqrt{a\theta}}$	$f()$
0.05	0.056	0.55	0.563	1.05	0.862	1.55	0.972
0.10	0.112	0.60	0.604	1.10	0.880	1.60	0.976
0.15	0.168	0.65	0.642	1.15	0.896	1.65	0.980
0.20	0.223	0.70	0.678	1.20	0.910	1.70	0.984
0.25	0.276	0.75	0.711	1.25	0.923	1.75	0.987
0.30	0.329	0.80	0.742	1.30	0.934	1.80	0.989
0.35	0.379	0.85	0.771	1.35	0.944	1.85	0.991
0.40	0.428	0.90	0.797	1.40	0.952	1.90	0.993
0.45	0.475	0.95	0.821	1.45	0.960	1.95	0.994
0.50	0.520	1.00	0.843	1.50	0.966	2.00	0.995

Considering the generic section of the wall with abscissa x in the beginning of the phenomenon, i.e., for $\theta = 0$; this leads to $\eta = \infty$, and as a result the integral is equal to one.

Based on (3.18) and recalling that the wall is at temperature t_0, we obtain

$$t = t_1 + Bx + C = t_0. \tag{3.21}$$

Equation (3.21) is in contrast with the fact that the entire wall is at temperature t_0; therefore, it is necessary to set $B = 0$; this leads to

$$C = t_0 - t_1. \tag{3.22}$$

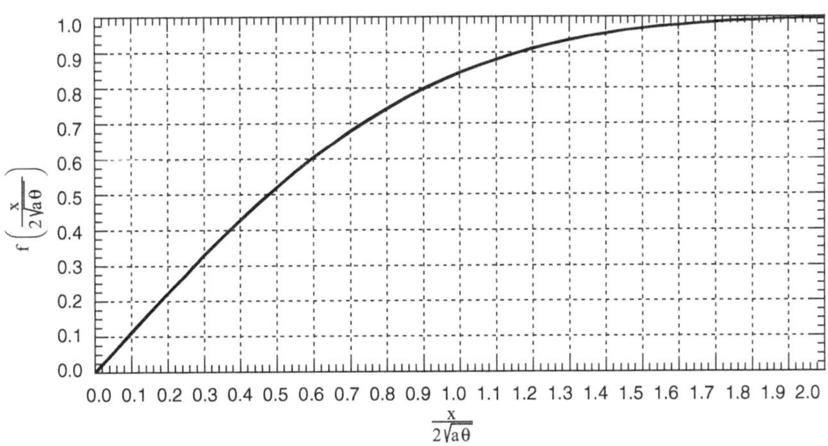

Fig. 3.3 Function $f\left(\frac{x}{2\sqrt{a\theta}}\right)$

Thus, based on (3.18)

$$t = t_1 + (t_0 - t_1) f\left(\frac{x}{2\sqrt{a\theta}}\right) \tag{3.23}$$

where $f(\ )$ indicates the error integral shown in Fig. 3.3; note that a stands for thermal diffusivity.

Thus, equation (3.23) makes it possible to compute the temperature in every section of the wall identified by the abscissa x and for any time indicated by θ.

Figure 3.3 establishes that if the error integral is equal to 1 only for $\eta = \infty$ from a mathematical point of view, for practical purposes it may already be considered equal to one for $\eta \geq 2$; this means that if this condition is satisfied, i.e., if the abscissa in question exceeds a certain value, the temperature in that area of the wall is still equal to t_0, in other words equal to the initial temperature of the wall. The new temperature t_1 on the surface of the wall has not impacted this area of the wall yet.

Note that for this to happen, by indicating with x_a the value of x that satisfies this condition, we must have

$$x_a \geq 4\sqrt{a\theta}. \tag{3.24}$$

During the initial phase of the phenomenon, i.e., for small values of θ, the value of x_a is small, too, and that means that only a thin layer is influenced by the new temperature on the surface. Over time while θ gradually increases the value of x_a increases, as well, so that a growing section of wall is more or less influenced by the new temperature t_1.

Now is a good time for a numerical example.

Assuming that the wall consists of a refractory material with the following characteristics: $k = 0.535$ W/mK; $c = 1025$ J/kgK; $\rho = 1650\ \mathrm{kg/m^3}$; thermal diffusivity is equal to $3.163 \times 10^{-7}\ \mathrm{m^2/s}$.

Time is set to be 15 min, i.e., 900 s; based on (3.24) we obtain $x_a = 0.067$ m; this means that after 15 min from the beginning of the phenomenon only 67 mm of the wall are impacted by the new temperature t_1; 5 h are necessary to extend the area impacted by the new temperature by 300 mm.

The behavior would be considerably different if the wall were made of steel.

In this case with $k = 45$ W/m K; $c = 500$ J/kg K; $\rho = 7850\ \mathrm{kg/m^3}$, i.e., with a thermal diffusivity equal to 1.146×10^{-5} after 15 min the area impacted by temperature t_1 would have a thickness of 408 mm and after 5 h of 1.82 m.

The characteristics of the phenomenon are well visible even computing the heat per surface unit crossing any section of the wall.

To that extent one must first compute the partial derivative $\partial t/\partial x$; note that the partial derivative with respect to x of function $f(\ )$ present in (3.23) is as follows:

$$\frac{\partial t}{\partial x} = \frac{2e^{-\frac{x^2}{4a\theta}}}{\sqrt{\pi}} \frac{1}{2\sqrt{a\theta}} = \frac{e^{-\frac{x^2}{4a\theta}}}{\sqrt{a\pi\theta}}. \tag{3.25}$$

Recalling that with reference to the units of time and surface $q = -k\partial t/\partial x$, by deriving (3.23) we obtain

$$q = k\,(t_1 - t_0)\,\frac{e^{-\frac{x^2}{4a\theta}}}{\sqrt{a\pi\theta}}. \tag{3.26}$$

For $x = 0$, i.e., in correspondence of the surface, (3.26) is reduced to the following:

$$q = k\frac{t_1 - t_0}{\sqrt{a\pi\theta}}. \tag{3.27}$$

Equation (3.27) shows that the heat crossing the surface decreases over time; (3.26) demonstrates that, given a value of θ, the transferred heat strongly decreases with an increase of x; thus, it is not surprising that the layers farthest away from the surface are only slightly impacted by the new temperature t_1 provided the situation does not persist over a considerable amount of time.

If q_s indicates heat crossing the surface with a suddenly increased temperature t_1, its value, as we saw, can be obtained through (3.27); at this point we introduce the ratio χ between the heat crossing the generic section of the wall, given by (3.26) and the heat q_s.

Based on (3.26) as well as (3.27) we obtain

$$\chi = \frac{q}{q_s} = e^{-\frac{x^2}{4a\theta}} = e^{-\left(\frac{x}{2\sqrt{a\theta}}\right)^2}. \tag{3.28}$$

The value of χ will be useful to our discussion about walls with finite thickness.

Finally, note that for the condition in which for $\theta = 0$ the temperature is the same on the entire wall and is equal to t_0, can be resolved by considering that such temperature is instead equal to

$$t = t_0 \pm bx. \tag{3.29}$$

Therefore, the temperature varies linearly along the wall, and this entails a constant amount of heat transferred over time and space passing through it. This passage of heat, occurring before the unsteady state caused by external factors that provoke it, must continue even during the unsteady state.

This situation is typical of a wall with finite thickness. Here it is suggested for the wall with infinite thickness, and in view of considering it even for a wall with finite thickness when certain conditions are in place, as we shall see later on.

Note that by reconsidering (3.18) once again, given that for $\theta = 0$ the error integral is equal to 1, for the generic abscissa x we obtain

$$t = t_0 \pm bx = A + Bx + C. \tag{3.30}$$

Still for $\theta = 0$ but for $x = 0$, given that the error integral is equal to 0, we have

$$A = t_1. \tag{3.31}$$

Based on (3.30) we have

$$t_0 \pm bx = t_1 + Bx + C. \tag{3.32}$$

This equation is satisfied by all values of x only if

$$B = \pm b \tag{3.33}$$

$$C = t_0 - t_1; \tag{3.34}$$

thus,

$$t = t_1 \pm bx + (t_0 - t_1) f\left(\frac{x}{2\sqrt{a\theta}}\right). \tag{3.35}$$

The thermal flux crossing the generic section of the wall is given by

$$q = k\left[\pm b + (t_0 - t_1)\,\frac{e^{-\frac{x^2}{4a\theta}}}{\sqrt{a\pi\theta}}\right]; \tag{3.36}$$

in correspondence of the surface (x=0) this is reduced to

$$q = k\left(\pm b + \frac{t_0 - t_1}{\sqrt{a\pi\theta}}\right). \tag{3.37}$$

Note that product $\pm kb$ represents the heat crossing the wall before the unsteady state.

3.4 Surface Temperature Variation in Finite Thickness Walls

Let us see if the conclusions drawn in the previous section about a wall of infinite thickness can be applied to a wall of finite thickness with a reasonable approximation to the actual phenomenon and under which conditions.

One possible and reasonable approach is through (3.24).

Based on previous statements and on the equation in question, and if we set a time frame for investigation θ_a and write that thickness x_w of the wall is such that

$$x_w \geq 4\sqrt{a\theta_a}, \tag{3.38}$$

we know that during the time in question the part of the wall beyond thickness x_w that is lost with respect to the wall of infinite thickness is irrelevant.

In other words, the conclusions drawn for the wall of infinite thickness are transferable to the wall with thickness x_w.

As we already pointed out during computation for (3.24), if we use the equal sign in (3.38) we obtain,

$$\frac{x_w}{2\sqrt{a\theta_a}} = 2. \tag{3.39}$$

If we consider the side of the wall opposite to the one brought to temperature t_1, which means $x = x_w$, based on (3.28) we obtain

$$\chi = e^{-4} = 0.0183. \tag{3.40}$$

This means that heat which is at the most equal to 1.8% of the heat crossing the other side is transferred outward through this side. Therefore, it is negligible.

This very restrictive condition always guarantees an actual correspondence between the behavior of the wall with thickness x_w and that of the wall with infinite thickness.

A less restrictive approach to the problem is also feasible; it consequently leads to the smallest possible thickness.

One accepts the fact that the second side of the wall is crossed by a considerable amount of heat, requiring at the same time that this heat corresponds to the one the wall is capable of transferring to the environment.

The condition according to which the environment is at same temperature t_0 as the wall before the unsteady state is analyzed.

This is a wall immersed in air at the same temperature of the air itself.

The heat that the wall, its side is at temperature t, can transfer to the outer environment at temperature t_0 within the units of time and surface is equal to

$$q = \alpha\,(t - t_0) \tag{3.41}$$

where α stands for the heat transfer coefficient of the external fluid (generally, it will be air).

Recalling (3.26) and given that the abscissa x is equal to thickness x_w

$$k\,(t_1 - t_0)\,\frac{e^{-\frac{x_w^2}{4a\theta}}}{\sqrt{a\pi\theta}} = \alpha\,(t - t_0). \tag{3.42}$$

Finally, recalling (3.23) and by indicating the time for analysis with θ_a

$$\alpha = \frac{ke^{-\frac{x_w^2}{4a\theta_a}}}{\sqrt{a\pi\theta_a}\left[1 - f\left(\frac{x_w}{2\sqrt{a\theta_a}}\right)\right]}. \tag{3.43}$$

We establish that the term $x_w/2\sqrt{a\theta_a}$ in which we are interested is a function of term $\alpha\sqrt{a\pi\theta_a}/k$; in turn, besides θ_a, this depends on the physical characteristics of the material the wall is made of (k,c,ρ) and from the ability of the surrounding fluid to receive heat.

For $x_w/2\sqrt{a\theta_a} \le 1.4$ one may assume with sufficient approximation for this type of analysis that

$$\frac{x_w}{2\sqrt{a\theta_a}} = 0.78\left(\frac{\alpha\sqrt{a\pi\theta_a}}{k} - 1\right). \tag{3.44}$$

Here is an example.

With reference to the refractory material already considered in Sect. 3.3, we have thermal diffusivity equal to 3.163×10^{-7} and $k = 0.535$ W/mK. If we consider a period of 3 h, i.e., 10800 s , and if we assume that the value of αin calm air is equal to $10\,\mathrm{W/m^2\,K}$, based on (3.44) we obtain

$$\frac{x_w}{2\sqrt{a\theta_a}} = 0.73 \tag{3.45}$$

matched by $x_w = 85$ mm.

We decided to analyze the actual behavior of the wall in question through a computational program using finite differences that we developed at the time.

We assumed that the wall has an initial temperature of 20°C, and that the air licking one of the sides has the same temperature with a heat transfer coefficient equal to $10\,\mathrm{W/m^2\,K}$.

We brought the temperature of the other side to 300°C; the different temperatures inside the wall as a function of time are shown in Fig. 3.4.

As you can see, after 3 h the temperatures are close to those under steady state; the unsteady state is almost finished. Conventionally, we consider the unsteady state to be basically over when the temperature of the side opposite to the one brought to 300°C show an increase in temperature equal to 95% of that which corresponds to reaching steady state. In our case the requirement is 16100 s; this corresponds to 4.5 h, i.e., 1.5 times the time of 10800 s considered for the analysis of that wall.

In order to evaluate if (3.23) allows the computation of the temperatures inside the wall in agreement with our digital computations, we examine the intermediate fiber of the wall ($x = 0.5x_w$) considering the same time units of the curves in Fig. 3.4.

Equation (3.23) typically underestimates the correct results. The errors slowly increase over time without ever exceeding 3%.

Under the same conditions, if we consider equation (3.39) we obtain a thickness of 233 mm.

This solution was analyzed through the same program based on finite differences. Considering that regardless of the considerable thickness, there is a slight increase in temperature beyond the original 20°C on the side opposite the one brought to 300°C because Gauss' error integral is not exactly equal to unity, we assumed that it

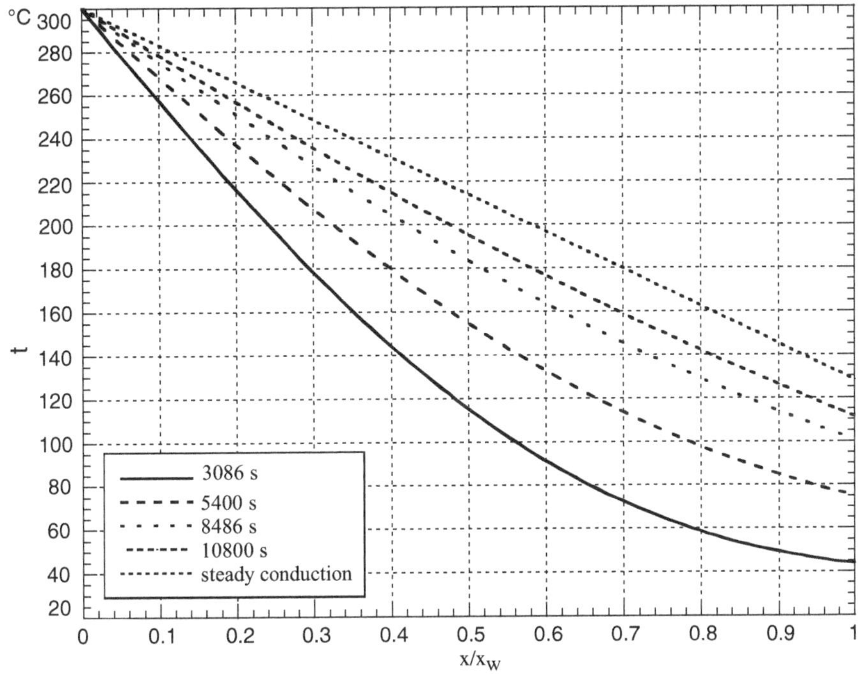

Fig. 3.4 Transient conduction in 85 mm thickness wall

was equally licked by a fluid at a temperature of 20°C and a heat transfer coefficient equal to 20 W/m^2 K. After 10800 s the temperature of this side has a temperature of 21.4°C.

The results are shown in Fig. 3.5; we see that after 3 h the heating process inside the wall is far from being over.

Even in this case we computed the temperatures through (3.23) and then we compared them to those obtained through automatic computation; the final temperature on the cold side is 21.4°C as calculated automatically; the errors do not exceed ±0.3%; therefore, we can state that the values obtained through the computer and those computed through (3.23) coincide.

This way we established that under certain conditions, like the ones that were discussed, (3.23) may very well be used even with walls of finite thickness.

To appreciate the impact of the different quantities on unsteady states we look at the wall of steel already discussed in Sect. 3.3; we keep the same values for α and θ_a taking into account that thermal diffusivity is equal to 1.146×10^{-5} and that $k = 45$ W/m K,

The term in parenthesis in (3.44) turns out to be negative; no solutions are possible from this point of view except with values of α or θ_a that are absolutely devoid of any sense of realism.

By applying (3.39) the outcome is a thickness equal to 1.4 m instead.

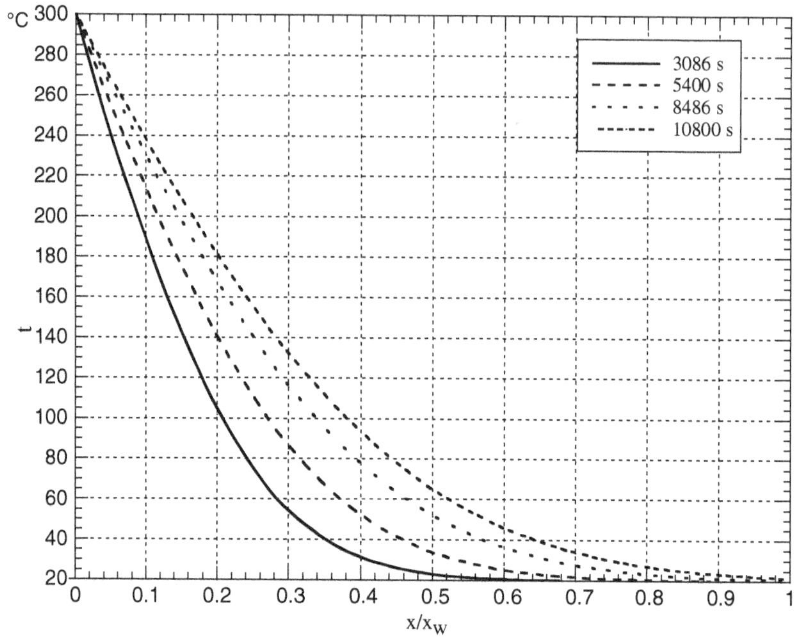

Fig. 3.5 Transient conduction in 233 mm thickness wall

Let us consider a wall of steel with a thickness of 50 mm at a temperature of 20°C. Subsequently, as was the case with the refractory material wall discussed earlier, we abruptly bring one side to 300°C, while the other is licked by a fluid at 20°C with a heat transfer coefficient of $10\,\mathrm{W/m^2\,K}$.

Note that as far as the refractory material wall with a thickness of 85 mm discussed earlier Fourier's number (fundamental in phenomena of unsteady state – see Sect. 3.6) is equal to 0.473. For the new case at hand we use a value similar of Fourier's number; this can be done by assuming a duration of the process of 100 s.

This low value for the time should not be surprising. We observe that the thermal diffusivity of the steel wall is 36 times its equivalent in the refractory material wall. Moreover, the layer is thinner, and in Fourier's number the thickness is to the square. Therefore, the duration of the unsteady state in the steel wall is much shorter.

The computation through the program mentioned earlier yields the temperatures shown in Fig. 3.6. Note that the behavior of the curves is qualitatively similar to those in Fig. 3.4 but the temperatures are very different. At the end of the process the temperature on the coldest side is higher because the heat transfer to the outside is greater.

The difference between the diagrams is mostly due to the fact that the heat transfer coefficient licking the cold wall is still at $10\,\mathrm{W/m^2\,K}$, the thermal diffusivity

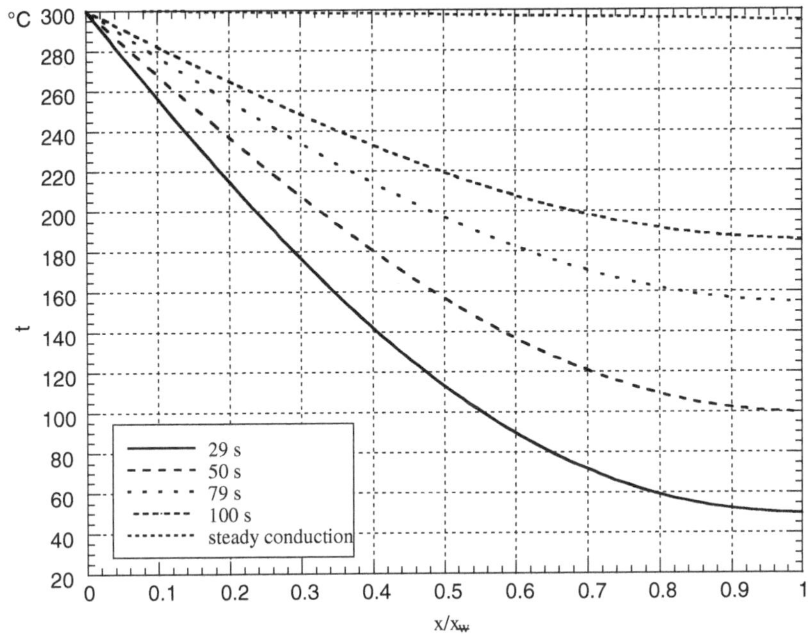

Fig. 3.6 Transient conduction in 50 mm thickness steel plate

of the steel wall is 36 times greater than that of the refractory material wall. If we increased the value of heat transfer coefficient by 4 times the ratio of the diffusivities, i.e. $\alpha = 1450\,\mathrm{W/m^2\,K}$, the resulting diagram of temperatures would basically coincide with that relative to the refractory material wall with a thickness of 85 mm. The unsteady state of the steel wall in Fig. 3.6 is not as advanced as that of the refractory wall. In fact, an increase in temperature on the cold side equal to 95% the increase in temperature corresponding to the steady state requires 277 s, i.e., 2.8 times the time of 100 s considered during the analysis of this wall. Note the very short duration, less than 5 min.

Let us compare the temperatures calculated via computer in correspondence of the middle of the wall ($x = 0.5x_w$) with those obtained through (3.23); we compute the errors that were made by using this equation as far as the increase in temperature with respect to the initial 20°. In the beginning the errors are irrelevant; then they slowly increase until they reach 18% at the end of the selected period of time. Evidently this is not acceptable, and it demonstrates that Eq. (3.23) cannot be used in this case.

This leads to the conclusion that the possibility to use the equations relative to walls of infinite thickness for walls of finite thickness are limited to materials with low conductivity and consequent limited thermal diffusivity.

3.5 Immersed Plane Wall in Fluid at Different Temperature

We analyze the behavior of a steel sheet with a thickness of 50 mm; initially at 20°C, it is immersed in a fluid at 300°C and with a heat transfer coefficient of 1000 W/m^2 K.

The evolution of temperatures of the sheet over time is shown in Fig. 3.7. As you can see, the process does not take long because of the high heat transfer coefficient of the surrounding fluid; after 4 min the warming up of the sheet amounts to 88% of its total; naturally, the heating process slows down as it gets closer and closer to steady state conditions, i.e., to a temperature of 300°C of the surrounding fluid. In theory, this condition can be reached in an infinite amount of time. In reality, the difference in temperature between fluid and sheet is reduced to about 7°C after 7 min with a heating percentage of 97.5%.

As far as the behavior of temperature inside the sheet, after 30 s the increase in temperature at the center of the sheet is half the increase in temperature at the edges. This difference diminishes over time; after 4 min it is reduced to about ten degrees.

We now assume to cool the same sheet down to 20°C from an initial temperature of 300°C through air. We also assume a heat transfer coefficient of the air equal to 20 W/m^2 K.

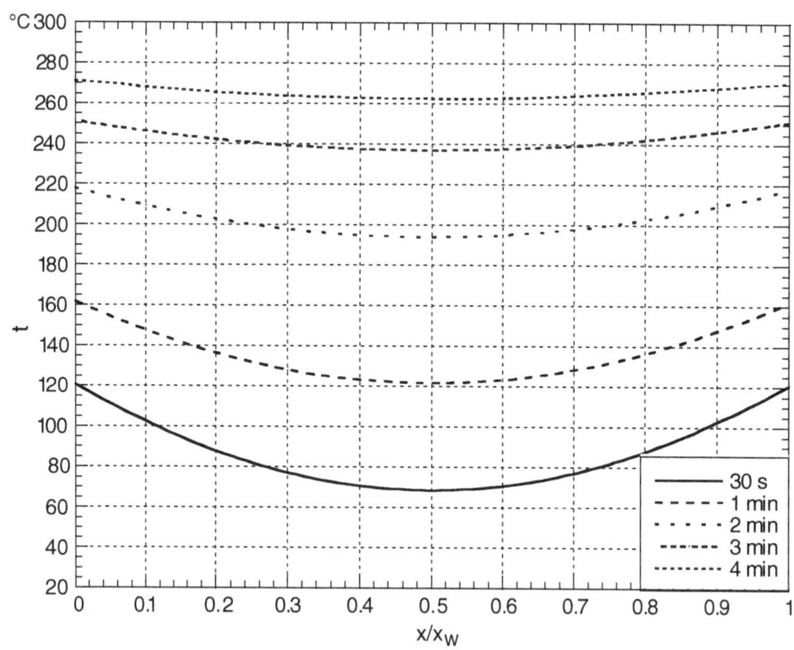

Fig. 3.7 Steel plate of 50 mm thickness heated to 300 °C

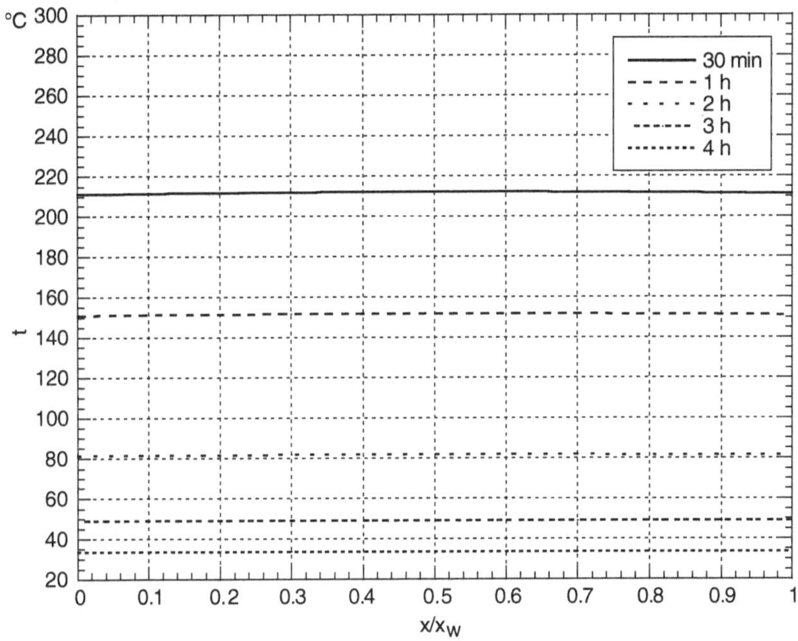

Fig. 3.8 Steel plate of 50 mm thickness cooled to 20°C

Figure 3.8 shows the resulting diagram.

As you can see, the process is particularly long because of the low value of the heat transfer coefficient of the air; 95% of the cooling is reached only after 4 h.; 53% of the cooling takes place within the first hour but three more hours are required to almost complete the cooling process.

The evolution of temperatures inside the sheet is significant. The considerable amount of time, as well as the thermal conductivity of the steel, contribute to keeping temperatures inside the sheet similar from point to point.

The behavior of temperatures inside the wall is quite different if thermal conductivity is low.

If we consider the refractory wall with a thickness of 85 mm already discussed in Sect. 3.3, including thermal conductivity of $0.535\ \mathrm{W/m^2\,K}$ that we would like to cool down by air to 20°C, as discussed in the previous example, we obtain the following diagram in Fig. 3.9.

Even in this case the process takes a long time; after 4 h 93% of the total cooling is accomplished.

Still, the most significant aspect of the process consists of behavior of temperatures inside the wall.

Note that after 20 min the decrease in temperature at the center of the wall with respect to the initial conditions is only 16%, compared to the decrease at the edges.

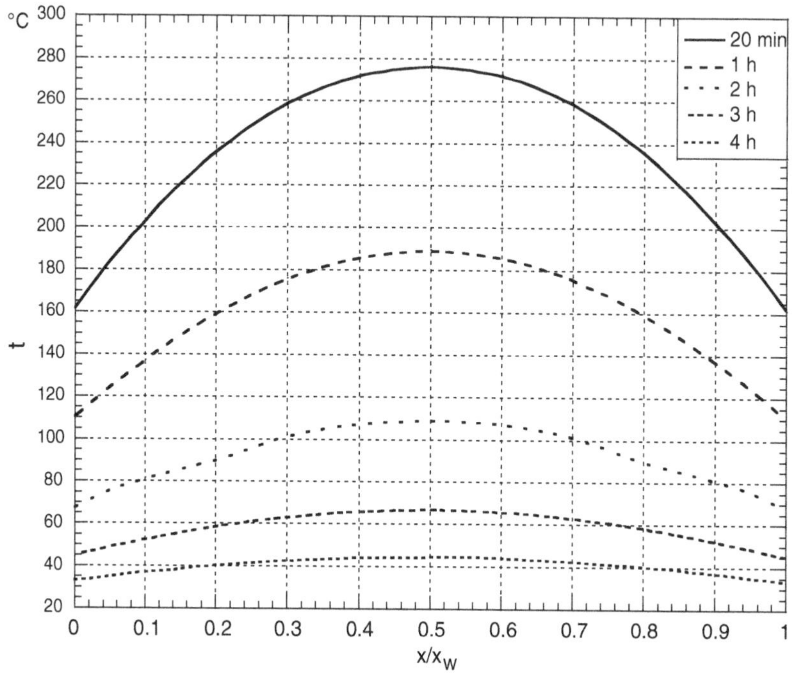

Fig. 3.9 Refractary plate of 85 mm thickness cooled to 20 °C

The difficulty encountered by the heat in transferring from the inside of the wall to the edges and then to the air is quite evident. This is due to the low value of the thermal conductivity of the material. Naturally, the impact lessens over time, yet after 4 h the temperatures at the center of the wall differ by about 12°C compared to those at the edges.

3.6 Transient Conduction in Tubes

Transient conduction in tubes is substantially quite similar to that of a plane wall except for the fact that the surfaces in contact with both internal and external fluid are different, and that the surface through which the heat passes inside the wall of the tube varies with the radius.

Here is an example.

Assuming a tube made of steel with an outside diameter of 40 mm and 5 mm thickness with an initial temperature of 20°C.

We introduce superheated steam at 400°C with a heat transfer coefficient of $1500\,\mathrm{W/m^2\,K}$ inside the tube; on the outside the tube is hit by flue gas at 1000°C with a heat transfer coefficient of $80\,\mathrm{W/m^2\,K}$.

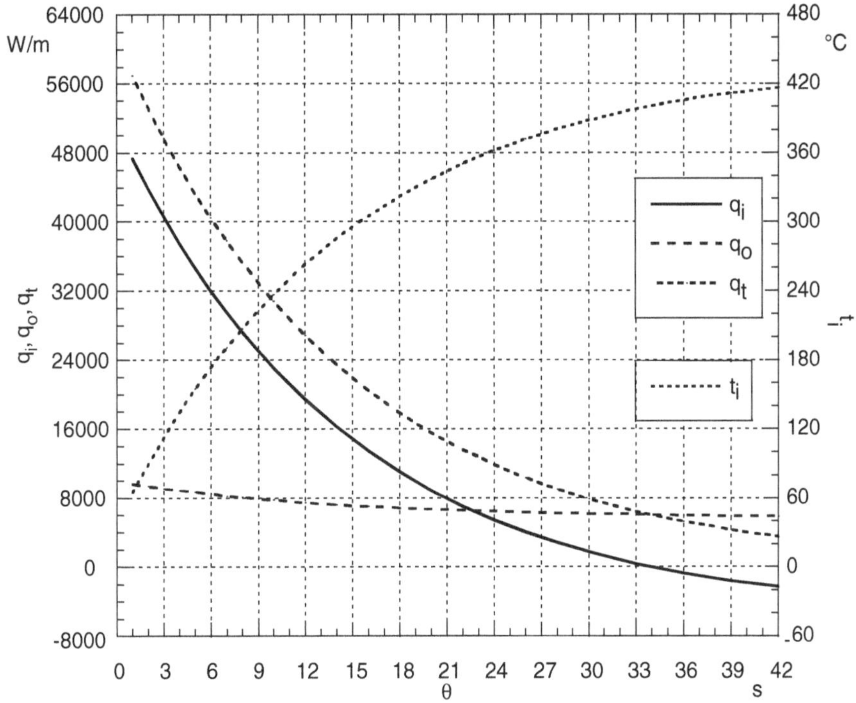

Fig. 3.10 Transient conduction in a tube

Unlike what was done for plane walls in the previous section, the behavior of temperatures inside the wall of the tube does not need to be displayed in a diagram. In fact, they vary insignificantly from point to point and this is partly due to their small thickness. In the beginning the temperature on the external fiber is lower than that on the internal fiber, but when they reach steady state the temperature on the external fiber becomes higher than that on the internal fiber, as it should be since flue gas transfers heat to the superheated steam in steady state. For this reason the diagram in Fig. 3.10 simply shows temperature t_i on the internal fiber as a function of time, because it is, in fact, the most significant one.In addition, we also included heat q_i as a function of time by length unit of the tube and by time unit transferred by the internal fluid to the tube, heat q_o transferred to the tube by the external fluid, and finally heat q_t which is transferred globally to the tube by both fluids.

The comments are as follows.

Heat q_o is lower than q_i for the first 20 s; even though the difference in temperature between flue gas and the tube is great, the heat transfer coefficient is low, thus the heat transfer is not considerable. Furthermore, even though it decreases over time as the temperature of the tube increases, the decrease is not big, given that the difference in temperature between flue gas and tube always remains big.

The behavior of q_i is quite different. Its value is initially very high, given the considerable difference in temperature between steam and tube, as well as the sizeable heat transfer coefficient. As the temperature of the tube increases, the difference in temperature between steam and tube, and the value of q_i drops. When the tube is brought to 400°C, i.e., the temperature of steam, the direction of the heat is reversed. The steam no longer transfers heat to the tube, but it is the tube that transfers heat to the steam instead. Of course, the tube does not cool down, but this heat originates from the flue gas through the wall of the tube.

As far as q_t, after 42 s the average temperature of the tube equals 418.45°C with an increase in temperature compared to initial conditions of 398.45°C. The mass of a meter of tube is equal to 4.316 kg. Considering that the specific heat equals 500 J/kg K, the heat transfer to the tube is $500 \times 4.316 \times 398.45 = 859855$ J/m, which corresponds to an average value of q_t equal to 20473 W/m.

In steady state the average temperature of the tube reaches 442.43°C matched by the heat transfer to the tube of 916028 J/m; after 42 s the heating of the tube is therefore equal to about 94% of its total.

3.7 Fourier's Number

In previous sections we often encountered either the term $x/\sqrt{a\theta}$ or $x_w/\sqrt{a\theta}$. These expressions are tied to a dimensionless number of great importance for the kinds of problems connected with heat transfer in unsteady state.

Recalling the equation on thermal diffusivity a, this is about Fourier's number given by

$$\mathrm{Fo} = \frac{a\theta}{{x_w}^2} = \frac{k\theta}{c\rho {x_w}^2}. \tag{3.46}$$

First of all, this is in fact a dimensionless number.

It is very important to qualitatively evaluate certain phenomena and to make comparisons.

As far as certain unsteady processes Fourier's number may remain constant, in the sense that modifications of the values of the quantities making it up do not vary its value.

For instance, this means that if the duration of the process has a certain value with a certain conductivity, if the latter is doubled, the duration of the process is cut in half.

The same is true for the specific heat which is proportional to time.

As the specific heat increases, with conditions being equal, the duration of the phenomenon increases in proportion, too.

Finally, note that thickness in Fourier's number is included squared; thus, if a process occurs in the presence of a given thickness, its doubling makes the duration of the phenomenon quadruple.

Chapter 4
Dimensional Analysis

4.1 Introduction

This chapter will focus on dimensional analysis to facilitate the study of heat transfer with respect to forced convection.

The first step in this direction is defining the dimensions of all quantities.

The dimensional equation in relation to each quantity which is significant with reference to a specific mechanical or thermal phenomenon, is the result of the combination of fundamental quantities at the base of the considered system.

Depending on the quantities chosen at the base of the system as fundamental (in principle they are arbitrary), the dimensional equation of a single quantity takes on a different structure.

The typical example is the system considering mass (system SI) as fundamental quantity, or that considering force instead (technical system); as a result, if this fundamental quantity is divided by the volume, the dimensional equation of the derived quantity (thus given by the fundamental quantity divided by the cubic length) represents density in the first case, and the specific weight in the second one. The dimensional equation for density is different in the second system.

The SI system we use considers that the fundamental mechanical quantities are mass (M), length (L) and time (θ); in the case of thermal phenomena it is necessary to introduce another fundamental quantity, i.e., temperature (T).

It would not be necessary to introduce heat as a fundamental quantity. In fact, based on the first principle of thermodynamics the dimensions of heat are those of work. Work is the product of force by length; in turn, force is the product of a mass by acceleration, i.e., the dimensions of heat with reference to M, L, θ are ML^2/θ^2.

In fact, even though this process is absolutely correct, it would actually be cumbersome to manage, given the fundamental role played by heat in thermal processes.

For this reason it is advisable to introduce heat (Q) as fifth fundamental quantity. This makes it possible to highlight heat in the quantities derived from it and to provide them with a simpler dimensional framework. In addition, the obtained formulation of the units of measure related to these quantities is easier and more straightforward.

D. Annaratone, *Engineering Heat Transfer*, DOI 10.1007/978-3-642-03932-4_4,

Table 4.1 Quantity dimensions and units of measure

Symbol	Quantity	Dimensions ML θ T	Dimensions ML θ QT	SI unit
L	Length	L	L	m
A	Area	L^2	L^2	m^2
V	Volume	L^3	L^3	m^3
M	Mass	M	M	kg
θ	Time	θ	θ	s
V	Velocity	L/θ	L/θ	m/s
g	Gravity acceleration	L/θ^2	L/θ^2	m/s^2
F	Force	ML/θ^2	ML/θ^2	N
P_v	Volumetric flow rate	L^3/θ	L^3/θ	m^3/s
P_m	Mass flow rate	M/θ	M/θ	kg/s
G	Mass velocity	$M/L^2\theta$	$M/L^2\theta$	kg/m^2s
ρ	Density	M/L^3	M/L^3	kg/m^3
v	Specific volume	L^3/M	L^3/M	m^3/kg
p	Pressure	$M/L\theta^2$	$M/L\theta_2$	Pa (bar)
μ	Dynamic viscosity	$M/L\theta$	$M/L\theta$	kg/ms
ν	Kinematic viscosity	L^2/θ	L^2/θ	m^2/s
Q	Heat	ML^2/θ^2	Q	J
T	Temperature	T	T	K (°C)
q	Heat in time unit	ML^2/θ^3	Q/θ	J/s =W
c	Specific heat	L^2/θ^2T	Q/MT	J/kg K
k	Thermal conductivity	ML/θ^3T	$Q/\theta LT$	W/m K
α	Heat transfer coefficient	M/θ^3T	$Q/\theta L^2T$	W/m^2K
U	Overall heat transfer coefficient	M/θ^3T	$Q/\theta L^2T$	W/m^2K
$k/c\rho$	Thermal diffusivity	L^2/θ	L^2/θ	m^2/s

Table 4.1 includes symbol, dimensional equation (for thermal quantities considering or not considering heat as a fundamental quantity), and unit of measure in the SI system for a series of both mechanical and thermal quantities.

Taking into account the dimensions of the different quantities it is possible to create dimensionless factors when the dimensions of the various quantities making up these factors are such that they become null.

These factors (called "numbers") are therefore pure numbers and independent from the adopted system of measure.

As we shall see, their use is fundamental to resolve numerous issues relative to heat transfer by convection.

Table 4.2 includes a few of these numbers (the most frequently used) with indication of the composing quantities. Clearly, these numbers are pure, i.e., dimensionless.

In Table 4.2 the different symbols have a well-known meaning except for β in Grashof's number, i.e. the cubic expansion coefficient with dimension T^{-1}.

If the factors influencing a given physical phenomenon are known, dimensional analysis makes it possible to use this knowledge to create correlations between these

Table 4.2 Dimensionless numbers of frequent use

Symbol	Name	
Re	Number of Reynolds	$Vd\rho/\mu$
Nu	Number of Nusselt	$\alpha d/k$
Pe	Number of Peclet	$Vd\rho c_p/k$
Pr	Number of Prandtl	$c_p\mu/k$
Gr	Number of Grashof	$d^3\beta\Delta t\rho^2 g/\mu^2$
Fo	Number of Fourier	$k\theta/c\rho s^2$

factors to program most rational experimentation and facilitate the interpretation of the results.

The method is particularly useful when a mathematical equation connecting the different quantities with one another is unknown, or when the mathematical development is very complex. In that case the grouping various quantities to form dimensionless numbers facilitates the most rational choice of research criteria.

Once the correlation between dimensionless numbers is established, the researcher will try to identify a mathematical equation to connect them to each other, but it will be possible to do so by varying any quantity among those forming the dimensionless number. This facilitates the task and provides a lot of leeway.

There are three methods to obtain significant dimensionless numbers for a given phenomenon:

(a) algebraic method, i.e., Rayleigh's classic method
(b) dimensionless numbers from differential equations
(c) considerations on similarity (geometric, kinematic and dynamic)

At the end of the Chapter a section will outline the theory of models and illustrate the usefulness of dimensionless numbers.

4.2 Three Methods to Find Dimensionless Groups

4.2.1 Algebraic Method

A specific problem is used to illustrate the method.

The goal is to establish the logical grouping of the different quantities influencing the value of the heat transfer coefficient for a fluid flowing inside a tube in turbulent flow which does not undergo any change of state but receives or provides heat.

The heat passes through the liquid film touching the wall, and must consequently be influenced by the thermal conductivity of the fluid k.

Since the thickness of the film depends on mass velocity G of the fluid, the diameter d and the dynamic viscosity μ must play a role. Moreover, heat q transferred in the time unit being equal, the temperature of the fluid at the center of the current is influenced by the isobaric specific heat c_p therefore intervening in the process.

Then we can write that

$$\alpha = f\left(G,d,c_p,\mu,k\right) \tag{4.1}$$

Within a reasonably sized field, the impact of the different quantities may be represented by a power the exponent of which and the proportionality factor are yet to be determined. Then, it is possible to write that

$$\alpha = CG^a d^b c_p^c \mu^d k^e \tag{4.2}$$

where C is a constant and a,b,c,d,e are dimensionless exponents.

Considering the dimensions of the different quantities, based on (4.2)

$$\frac{Q}{\theta L^2 T} = \left(\frac{M}{L^2\theta}\right)^a (L)^b \left(\frac{Q}{MT}\right)^c \left(\frac{M}{L\theta}\right)^d \left(\frac{Q}{\theta LT}\right)^e. \tag{4.3}$$

Considering the exponents of the fundamental quantities Q,M,L,θ,T, by sequence we must have:

$$\begin{array}{lll} \text{for} & Q & 1 = c + e \\ \text{for} & M & 0 = a - c + d \\ \text{for} & L & -2 = -2a + b - d - e \\ \text{for} & \theta & -1 = -a - d - e \\ \text{for} & T & -1 = -c - e \end{array} \tag{4.4}$$

The solution of the equation system (4.4) leads to the following results: $b = a - 1$; $d = c - a$; $e = 1 - c$. From (4.2) we obtain:

$$\alpha = C\frac{G^a d^a}{\mu^a}\frac{c_p^c \mu^c}{k^c}\frac{k}{d}. \tag{4.5}$$

Recalling the meaning of Nusselt's, Reynolds' and Prandtl's numbers (see Table 4.2), from (4.5) we obtain

$$\mathrm{Nu} = C\mathrm{Re}^a\,\mathrm{Pr}^c. \tag{4.6}$$

If we were talking about perfect gas Prandtl's number would be invariable; then we could simply write that

$$\mathrm{Nu} = C\mathrm{Re}^a. \tag{4.7}$$

By first approximation (4.7) may also be adopted for real gas.

As we shall see later on, in actual situations the value of exponent a present in previous equations ranges from 0.6 and 0.8; exponent c varies from 0 to 0.4; we establish the considerable impact of mass velocity G on heat transfer coefficient α, whereas the impact of the diameter is always rather limited.

The dimensional analysis leading to (4.5) and (4.6) is absolutely correct, once we are sure that this phenomenon is conditioned by the quantities under review, as in fact it is the case.

Still, the result must be considered with caution. In fact, even the ratio between two temperatures, between two values of specific heat, between two thermal conductivities, or between two viscosities is obviously a dimensionless number and may have a precise physical significance.

Note that the value of c_p,k,μ depends on the temperature of reference and that the considered typical temperatures are the temperature at the center of the current of the fluid (bulk temperature), and the temperature at the boundary layer (film temperature).

The researcher decides to refer to one of these temperatures and consequently assumes values of c_p,k,μ that correspond to this temperature; certain values of Re, Pr and Nu constitute the result.

Based on (4.6), each set of Re and Pr should be matched by a certain value of Nu, regardless of the ratio between the temperature in the center of the current and the temperature of the boundary layer.

If the researcher establishes that the value of Nu also depends on this ratio, Eq. (4.6) must be adjusted to take this into account.

This ratio may directly influence the value of Nu, or its influence may manifest itself through the corresponding values of the specific heat, of thermal conductivities or of dynamic viscosities.

Consequently, a corrective factor will have to be added into (4.6); it will represent the dimensionless ratio of the one of the quantities listed above raised to certain power.

On the other hand, such potential dependency is not conceptually in contrast with the analysis leading to (4.5) as well as (4.6)

For this reason we believe it is more correct to rewrite (4.6) as follows:

$$\mathrm{Nu} = C\mathrm{Re}^a\mathrm{Pr}^c J^e \tag{4.8}$$

where J is a dimensionless number given by the ratio of two equal quantities with a different value, and e is the exponent of this ratio.

4.2.2 *Use of Differential Equations*

It is possible to obtain the dimensionless numbers conditioning heat transfer by convection and their connection by using the differential equations regarding movement and thermal conduction of fluids.

Let us start by considering Stokes' equation relative to movement of fluids. With reference to direction x, it is as follows:

$$\frac{\partial V_x}{\partial \theta}+V_x\frac{\partial V_x}{\partial x}+V_y\frac{\partial V_y}{\partial y}+V_z\frac{\partial V_z}{\partial z}=-\frac{1}{\rho}\frac{\partial p}{\partial x}+\frac{1}{3}\frac{\mu}{\rho}\left(\frac{\partial^2 V_x}{\partial x^2}+\frac{\partial^2 V_y}{\partial x\partial y}+\frac{\partial^2 V_z}{\partial x\partial z}\right)+\frac{\mu}{\rho}\left(\frac{\partial^2 V_x}{\partial x^2}+\frac{\partial^2 V_x}{\partial y^2}+\frac{\partial^2 V_x}{\partial z^2}\right). \tag{4.9}$$

In (4.9) x,y,z are the coordinates of the system, V_x,V_y,V_z the velocity components, respectively, ρ stands for density, μ for dynamic viscosity, p for pressure and θ for time. After a permutation of the coordinates and the ensuing corresponding velocities we can write two additional equations relative to directions y and z that are quite similar to (4.9).

For two currents to be similar the different quantities at play in every point of the current must be connected to one another through a fixed ratio. This is in reference to length, velocity, time, pressure, density and viscosity.

By introducing even diameter d for the length beyond x,y,z we must write that

$$\text{For length}\quad \begin{aligned} x_2 &= j_d x_1\\ y_2 &= j_d y_1\\ z_2 &= j_d z_1\\ d_2 &= j_d d_1 \end{aligned} \tag{4.10}$$

$$\text{For velocity}\quad \begin{aligned} V_{x2} &= j_v V_{x1}\\ V_{y2} &= j_v V_{y1}\\ V_{z2} &= j_v V_{z1} \end{aligned} \tag{4.11}$$

$$\text{For time}\quad \theta_2 = j_\theta \theta_1 \tag{4.12}$$

$$\text{For pressure}\quad p_2 = j_p p_1 \tag{4.13}$$

$$\text{For density}\quad \rho_2 = j_\rho \rho_1 \tag{4.14}$$

$$\text{For viscosity}\quad \mu_2 = j_\mu \mu_1 \tag{4.15}$$

If we write (4.9) for current 1 we must insert the terms V_{x1},V_{y1},V_{z1} and so on in the equation. If we consider current 2 instead, we must insert the terms V_{x2},V_{y2},V_{z2} and so on. Recalling equations (4.10), (4.11), (4.12), (4.13), (4.14) and (4.15) and that based on differential calculus the derivative of the product of a constant for a variable is equal to the constant for the derivative of the variable, Eq. (4.9) written for current 2 is as follows:

$$\frac{j_v}{j_\theta}\frac{\partial V_{x1}}{\partial \theta_1}+\frac{j_v^2}{j_d}\left(V_{x1}\frac{\partial V_{x1}}{\partial x_1}+V_{y1}\frac{\partial V_{y1}}{\partial y_1}+V_{z1}\frac{\partial V_{z1}}{\partial z_1}\right)=-\frac{j_p}{j_\rho j_d}\frac{1}{\rho_1}\frac{\partial p_1}{\partial x_1}+\frac{1}{3}\frac{j_\mu}{j_\rho}\frac{j_v}{j_d^2}\frac{\mu}{\rho}\left(\frac{\partial^2 V_{x1}}{\partial x_1^2}+\frac{\partial^2 V_{y1}}{\partial_{x1}\partial_{y1}}+\frac{\partial^2 V_{z1}}{\partial x_1\partial z_1}\right)+\frac{j_\mu}{j_\rho}\frac{j_v}{j_d^2}\frac{\mu}{\rho}\left(\frac{\partial^2 V_{x1}}{\partial x_1^2}+\frac{\partial^2 V_{x1}}{\partial y_1^2}+\frac{\partial^2 V_{x1}}{\partial z_1^2}\right) \tag{4.16}$$

Given that (4.16) is the general equation in reference to current 2 which is similar to current 1 both equations must be exactly identical. For this to happen it suffices that all factors containing the constants j are identical. This is the only way to eliminate them in (4.16), thus reduced to (4.9) in reference to current 1.

Therefore,

$$\frac{j_v}{j_\theta} = \frac{j_v^2}{j_d} = \frac{j_p}{j_\rho j_d} = \frac{j_\mu}{j_\rho}\frac{j_v}{j_d^2}. \tag{4.17}$$

A comparison between the 2nd term with the fourth in (4.17) leads to

$$j_v = \frac{j_\mu}{j_\rho j_d}; \tag{4.18}$$

finally,

$$\frac{j_v j_\rho j_d}{j_\mu} = 1. \tag{4.19}$$

Recalling (4.10), (4.11), (4.12), (4.13), (4.14) and (4.15) we then have

$$\frac{V_{x2}\rho_2 d_2 \mu_1}{V_{x1}\rho_1 d_1 \mu_2} = 1; \tag{4.20}$$

then

$$\frac{V_{x2}\rho_2 d_2}{\mu_2} = \frac{V_{x1}\rho_1 d_1}{\mu_1}. \tag{4.21}$$

The components of velocity V_{x2} and V_{x1} but all components of velocity share a common ratio j_v; therefore, we can substitute component V_x with the mean velocity V in both tubes in (4.21) by writing that

$$\frac{V_2\rho_2 d_2}{\mu_2} = \frac{V_1\rho_1 d_1}{\mu_1}. \tag{4.22}$$

The terms in (4.22) are none other than Reynolds' numbers relative to current 1 and current 2.

Reynolds' numbers must be equal for the currents to be considered similar.

Until now we did not find a connection between heat transfer and the characteristics relative to the movement of fluids.

Thus, we consider other quantities regarding heat transfer, and for them we also assume that the value of each quantity for current 2 is equal to that of current 1 multiplied by a certain factor. So we assume that

$$\text{For temperatures} \quad t_2 = j_t t_1 \tag{4.23}$$

$$\text{For thermal conductivities} \quad k_2 = j_k k_1 \tag{4.24}$$

$$\text{For specific heat} \quad c_2 = j_c c_1 \tag{4.25}$$

Let us consider thermal gradient $\partial t/\partial L$ and observe that based on (4.10), as well as (4.23):

$$\frac{\partial t_2}{\partial L_2} = \frac{j_t}{j_d}\frac{\partial t_1}{\partial L_1}. \tag{4.26}$$

If t is the mean temperature of the current and t_w the temperature of the wall we also know that

$$t_2 - t_{w2} = j_t\left(t_1 - t_{w1}\right). \tag{4.27}$$

From (4.26) and (4.27) we obtain

$$\frac{\partial t_2}{\partial L_2} = \frac{t_2 - t_{w2}}{t_1 - t_{w1}}\frac{1}{j_d}\frac{\partial t_1}{\partial L_1}; \tag{4.28}$$

finally,

$$\frac{\partial t_2}{\partial L_2}\frac{d_2}{t_2 - t_{w2}} = \frac{\partial t_1}{\partial L_1}\frac{d_1}{t_1 - t_{w1}}. \tag{4.29}$$

If we consider that the gradient of temperature refers to the direction of the heat hitting the wall of the tube, and we assume the existence of a boundary layer of the fluid where its movement is laminar, the heat transfer to the tube through the boundary layer takes place according to the law of conduction. With reference to current 1 the heat transfer by time unit and surface unit is equal to

$$q_1 = k_1\frac{\partial t_1}{\partial L_1}. \tag{4.30}$$

On the other hand, we also know that by indicating the heat transfer coefficient of fluid 1 with α_1:

$$q_1 = \alpha_1\left(t_1 - t_{w1}\right). \tag{4.31}$$

From (4.30) and (4.31) we obtain

$$\frac{\alpha_1 d_1}{k_1} = \frac{\partial t_1}{\partial L_1}\frac{d_1}{t_1 - t_{w_1}}. \tag{4.32}$$

By analogy, for current 2 we can write that

$$\frac{\alpha_2 d_2}{k_2} = \frac{\partial t_2}{\partial L_2}\frac{d_2}{t_2 - t_{w2}}; \tag{4.33}$$

Finally, based on (4.29)

$$\frac{\alpha_2 d_2}{k_2} = \frac{\alpha_1 d_1}{k_1}. \tag{4.34}$$

This establishes that if currents are similar, Nusselt's number is the same.

Two similar currents are therefore characterized by equal numbers of Reynolds and Nusselt. This implies a connection among them and we are able to write that

$$\mathrm{Nu} = f\,(\mathrm{Re}) \tag{4.35}$$

Since experience has shown that α is proportional to V^m, given m as a constant, it is possible to write that

$$\mathrm{Nu} = C\mathrm{Re}^m. \tag{4.36}$$

As in equation (4.36) and in contrast to (4.6), Prandtl's number is missing, and this is not surprising because the presentation above never considered the specific heat of the fluid, even though, as experience demonstrates, plays a role in determining the value of α.

Equation (4.36) is therefore valid only if Prandtl's number is a constant; this is true only for perfect gas, as already pointed out in the previous section, but it is not true for real gas (it varies only slightly) and most of all for the other fluids.

Instead of pointing to Stokes' differential equation, in order to highlight the impact of specific heat we must refer to Nusselt's differential equation about heat transfer which is as follows:

$$c_p\left(\frac{\partial t}{\partial \theta} + V_x\frac{\partial t}{\partial x} + V_y\frac{\partial t}{\partial y} + V_z\frac{\partial t}{\partial z}\right) - \frac{1}{\rho}\left(\frac{\partial p}{\partial \theta} + V_x\frac{\partial p}{\partial x} + V_y\frac{\partial p}{\partial y} + V_z\frac{\partial p}{\partial z}\right) = \\ = \frac{k}{\rho}\left(\frac{\partial^2 t}{\partial x^2} + \frac{\partial^2 t}{\partial y^2} + \frac{\partial^2 t}{\partial z^2}\right) \tag{4.37}$$

Proceeding as for Stokes' differential equation leads to

$$\frac{j_c j_t}{j_\theta} = \frac{j_c j_v j_t}{j_d} = \frac{j_p}{j_\rho j_\theta} = \frac{j_v j_p}{j_\rho j_d} = \frac{j_k j_t}{j_\rho j_d^2}. \tag{4.38}$$

From (4.38) we also obtain

$$\frac{j_c j_v j_d j_\rho}{j_k} = 1. \tag{4.39}$$

Recalling the meaning of the different factors j, from (4.39):

$$\frac{V_2 \rho_2 d_2 c_{p2}}{k_2} = \frac{V_1 \rho_1 d_1 c_{p1}}{k_1} \tag{4.40}$$

Both terms for current 1 and current 2 are none other than Peclet's number relative to both currents; Peclet's number is given by the product of the number of Reynolds by Prandtl's number, as can easily be verified.

Therefore, if two currents are similar Peclet's number is the same

As before, we can write that:

$$\mathrm{Nu} = f(\mathrm{Pe}) = C\mathrm{Pe}^{\mathrm{m}} = C\mathrm{Re}^{m}\,\mathrm{Pr}^{m}\,. \tag{4.41}$$

As you can see, (4.41) includes Prandtl's number, but it has the same exponent as the number of Reynolds; this disagrees with experimental data showing how it is more correct to assign Pr a value of the exponent in general other than the exponent of Re.

If we assume this position which is coherent with experimentation, the result is as follows:

$$\mathrm{Nu} = C\mathrm{Re}^{m}\,\mathrm{Pr}^{n} \tag{4.42}$$

which is nothing else but (4.6).

We point out what was said about (4.6) with regard to the potential necessity to introduce another dimensionless factor consisting of the ratio between equal quantities, yet of different value, into the equation.

4.2.3 Geometric, Kinematical and Dynamical Similitude

The previous section established that with reference to Stokes' differential equation on fluid motion, for two currents to be similar the value of Reynolds' number must be equal.

It is possible to draw this conclusion directly based on similarity criteria.

If we consider two circular tubes with two fluids in turbulent flow, for the currents to be similar the conditions of geometric, kinematic and dynamic similarity must be fulfilled.

As far as geometric similarity, the following condition must be fulfilled having indicated with subscript 1 the quantities of the current 1 and with subscript 2 those of current 2.

$$\frac{dx_2}{dx_1} = \frac{dy_2}{dy_1} = \frac{dz_2}{dz_1} = \frac{d_2}{d_1}, \tag{4.43}$$

given that d_1 and d_2 are the diameters of both tubes.

As far as the kinematic similarity the velocity gradients in direction x of the flow and in orthogonal direction y must be in proportion to each other so that

$$\frac{\partial V_2/\partial y_2}{\partial V_1/\partial y_1} = \frac{\partial V_2/\partial x_2}{\partial V_1/\partial x_1}. \tag{4.44}$$

As far as dynamic similarity, the forces due to viscosity and acceleration must be in the same proportion; by indicating the former with dF_v and the latter with dF_a, we must have

$$\frac{dF_{v2}}{dF_{v1}} = \frac{dF_{a2}}{dF_{a1}}. \tag{4.45}$$

Force dF_v is equal to

$$dF_v = \mu dxdz \frac{\partial V}{\partial y}. \tag{4.46}$$

Force dF_a which is equal to elementary mass by acceleration is equal to

$$dF_a = \rho dxdydz \frac{\partial V}{\partial \theta} \tag{4.47}$$

where θ indicates time.

From (4.45) and recalling both (4.46) and (4.47):

$$\frac{\mu_2 dx_2 dz_2 \dfrac{\partial V_2}{\partial y_2}}{\mu_1 dx_1 dz_1 \dfrac{\partial V_1}{\partial y_1}} = \frac{\rho_2 dx_2 dy_2 dz_2 \dfrac{\partial V_2}{\partial \theta_2}}{\rho_1 dx_1 dy_1 dz_1 \dfrac{\partial V_1}{\partial \theta_1}}. \tag{4.48}$$

Note that

$$\frac{\partial V}{\partial \theta} = \frac{\partial x}{\partial \theta}\frac{\partial V}{\partial x} = V\frac{\partial V}{\partial x}; \tag{4.49}$$

then, from (4.48) we obtain

$$\frac{\mu_2 \dfrac{\partial V_2}{\partial y_2}}{\mu_1 \dfrac{\partial V_1}{\partial y_1}} = \frac{\rho_2 dy_2 V_2 \dfrac{\partial V_2}{\partial x_2}}{\rho_1 dy_1 V_1 \dfrac{\partial V_1}{\partial x_1}} \tag{4.50}$$

Finally, recalling both (4.43) and (4.44):

$$\frac{V_2 \rho_2 d_2}{\mu_2} = \frac{V_1 \rho_1 d_1}{\mu_1}. \tag{4.51}$$

Equation (4.51) tells us that if currents are similar the number of Reynolds is equal, thus confirming what already established otherwise.

4.3 Theory of Models

Dimensionless numbers may be used in the theory of models.

Let us assume that a certain situation can be described through a number n of dimensionless numbers; if we know $n-1$ values, the value of the missing one is set as a function of the known ones.

The assumption is to build a model by scale of a certain machine in order to study its behavior. The dimensions of the model are therefore in a certain ratio with respect to the real ones. Consequently, the dimensionless numbers where the dimensions appear would be different for the machine and its model. This makes it impossible to transfer the experimental results obtained through the model to the machine in question. In order to do so the value of another quantity contained in the dimensionless numbers above must be changed in such a way that the value of these numbers is the same for both machine and model.

A typical example involves the number of Reynolds. If the analyzed phenomenon depends on the number of Reynolds, and we build a model with a diameter a tenth of the actual one, a velocity ten times the real one must be adopted for the model because then the number of Reynolds does not change.

This method can be extended beyond a simple change of diameter and velocity. It is possible to change type of fluid, temperature and so on, as long as the existing relations between the various dimensionless numbers conditioning the phenomenon are taken into account.

For instance, let us take the case where we would like to determine the value of the heat transfer coefficient of flue gas at 500°C and with a mass moisture percentage of 5.94%, flowing through a tube with a diameter of 100 mm at a velocity of 30 m/s.

We have a tube with a diameter of 50 mm and air inside at 100°C.

Given that these are gases, the number of Nusselt solely depends on the number of Reynolds with a good degree of approximation, regardless of the number of Prandtl.

For gas $\rho = 0.46\,\text{kg/m}^3$, $\mu = 34.66 \times 10^{-6}\,\text{kg/ms}$, $k = 53.92 \times 10^{-3}\,\text{mK}$.

Thus, the number of Reynolds is equal to $30 \times 0.46 \times 0.1/34.66 \times 10^{-6} = 39815$.

For air $\rho = 0.946\ \text{kg/m}^3$, $\mu = 21.63 \times 10^{-6}\,\text{kg/ms}$, $k = 29.85 \times 10^{-3}$ W/mK.

For the number of Reynolds of air to be equal to 39815 the velocity must be equal to $39815 \times 21.63 \times 10^{-6}/0.946 \times 0.05 = 18.21$ m/s.

Under these conditions, the computation of the heat transfer coefficient of air leads to 58.09 W/m^{2}K with a value for the number of Nusselt equal to $58.09 \times 0.05/29.85 \times 10^{-6} = 97.30$. Therefore, the same number of Nusselt must be valid for flue gas, too. Consequently, the heat transfer coefficient is equal to

$$97.30 \times 53.92 \times 10^{-3}/0.1 = 52.26\ \text{W/m}^2\text{K}.$$

Note that if the heat transfer coefficient of air were not computed but based on experimental data instead, the described method shows that the experimentation done on air makes it possible to identify the behavior of flue gas without requiring specific experiments on them.

The heat transfer coefficient of gas can be computed directly and results equal to 53 W/m^2K; this value exceeds the value obtained through the described method by 1%.

Prandtl's number is equal to 0.7344 for air, whereas it is equal to 0.7533 for flue gas.

Given the impact of Prandtl's number on the heat transfer coefficient (it will be discussed in more detail later on in connection with convection), the biggest value for actual heat transfer coefficient of flue gas with respect to the one computed through the described method is fully justified by the higher value of Prandtl's number for flue gas versus air.

Chapter 5
Convection

5.1 Types of Motion

In the case of fluid inside a tube we may be faced with two types of current.

In the first case the threads of the fluid are parallel to the axis of the tube, and there are no components of velocity perpendicular to the axis. The velocity of the fluid has parabolic behavior and its maximum value along the axis of the tube. Therefore, the mean velocity is half the maximum velocity.

This type of motion is called laminar motion.

The current can be characterized almost in its entirety by vortexes instead. Thus, the components of velocity are present not only along the axis of the tube but also perpendicular to it. The velocity, except in a thin layer in contact with the wall of the tube, has a much less arched behavior compared to laminar motion, so that the mean velocity is equal to about 80% of maximum velocity.

This type of motion is called turbulent motion.

The onset of any two motions depends on the value of the number of Reynolds.

If the number of Reynolds is low there is laminar motion, if it is high there is turbulent motion.

Opinions are not unanimous on the number of Reynolds relative to both fields (but the suggested values are very close), especially because through adequate precautions it is possible to have laminar motion even for Re values which would imply turbulent motion.

There is a transitional zone between the values of Re which certainly means laminar motion and values of Re for which turbulent motion is certain.

For values of Re below 2000 the motion is certainly laminar, whereas for values of Re above 3000 the motion is certainly turbulent.

The behavior of a current in turbulent motion is completely different from that of a current in laminar motion, both as far as pressure drop (the latter will be discussed extensively in a dedicated chapter), and as far as heat transfer.

As far as pressure drops, note that in a current in laminar motion the viscosity of the fluid is crucial. This is why it is also called viscous motion. In fact, the pressure drop is proportional to viscosity. Moreover, it is proportional to velocity. The roughness of the tube wall has no impact on pressure drop.

D. Annaratone, *Engineering Heat Transfer*, DOI 10.1007/978-3-642-03932-4_5,

If the motion is turbulent instead, the roughness of the tube wall is most often crucial, and this is not the case for low values of Reynolds and limited roughness of the tube. In addition, the pressure drop is about proportional to the square of velocity.

As far as heat transfer, if the current has laminar motion there are no components of velocity transversal to the axis of the tube, as we pointed out earlier. Therefore, there is no convective motion towards the wall, and the heat transfer is entirely dependent on the thermal conductivity of the fluid. In addition, based on the law of conduction the heat transfer is inversely proportional to the thickness that the heat must cross. In our case the heat transfer to the tube will be inversely proportional to its diameter. The velocity of the fluid does not impact the heat transfer.

If the motion is turbulent the phenomenon is much more complex. The number of Reynolds is crucial because the velocity of the fluid is crucial. The diameter of the tube plays a role, and even in this case the heat transfer decreases with the increase in diameter, yet the influence of the latter is much less compared to laminar motion. Finally, both specific heat and dynamic viscosity are important together with the thermal conductivity of the fluid.

Then if we consider heat transfer from a fluid to a tube bank hit by the fluid itself even dimensional characteristics of the bank play a role (arrangement, pitches and diameter)

Finally, by increasing the velocity of the fluid we obtain a greater heat transfer, but the increase is matched by a corresponding increase in pressure drop.

Therefore, by improving the heat transfer we obtain a reduction of the surface necessary to reach certain results and a consequent reduction in plant costs, but this means that the increase in pressure drop implies a cost increase for the fan, together with an increase in running costs due to the greater energy requirements.

5.2 Physical Characteristics of Fluids

Heat transfer by convection is governed by the value of heat transfer coefficient α in W/m^2K. It represents the quantity of heat by time and surface unit that can be transferred if the difference in temperature is equal to 1 K (1°C).

As we shall see, the value of α depends on different quantities, including a few physical characteristics of the fluid in question. They are the isobaric specific heat (in J/kg K), the thermal conductivity (in W/m K) and the dynamic viscosity (in kg/ms).

Even density plays a role because it is included in the dimensionless number of Reynolds which is fundamental as far as the value of the heat transfer coefficient.

But if we introduce mass velocity instead of velocity in the number of Reynolds, the density disappears. The recommendation is to always refer to mass velocity, since the volumetric flow rate varies with temperature (and consequently the velocity of the fluid) but mass flow rate is invariable.

As we shall see, Prandtl's number is very important too. It is characteristic of the fluid in use because it contains only physical quantities of the fluid itself.

The values for density, specific heat, thermal conductivity and dynamic viscosity of various liquids and gases are included in Appendix A. For water and steam we refer to the publication "Properties of Water and Steam in SI-Units" by Springer Verlag in Heidelberg.

Here we will discuss how to compute the three cited quantities, as well as Prandtl's number for water, air and flue gas as a function of temperature through the equations developed by the author.

In general, the values are acceptable, and the errors never exceed ±2%. In fact, they are typically lower or much lower when using equations, such as

$$G = X + Yt + Zt^2 \tag{5.1}$$

where G indicates the quantity in question, t indicates temperature, and X, Y, Z are constants.

5.2.1 Water

5.2.1.1 Density

The density of water which is only slightly influenced by pressure may be computed through the following approximated equation:

$$\rho = 1006.68 - 20.07\frac{t}{100} - 25.15\left(\frac{t}{100}\right)^2 \tag{5.2}$$

where ρ is in kg/m^3 and temperature t in °C.; (5.2) is valid for temperatures between 10°C and 300°C and for pressures ranging from 5 to 100 bar.

Errors never exceed ±1%.

5.2.1.2 Specific Heat

The specific heat of water which is only slightly influenced by pressure may be computed through the following approximated equation:

$$c = 4219.58 - 187.25\frac{t}{100} + 172.17\left(\frac{t}{100}\right)^2 \tag{5.3}$$

with c in J/kg K and t in °C; (5.3) is valid for pressures ranging from 5 to 100 bar and for temperatures between 20 and 250°C.

Errors never exceed ±2%.

5.2.1.3 Thermal Conductivity

The thermal conductivity of water which is basically independent from pressure may be computed through the following approximated equation, which was already discussed in Chap. 2:

$$k = 0.5755 + 0.1638\frac{t}{100} - 0.05767\left(\frac{t}{100}\right)^2 \quad (5.4)$$

where k is in W/m K and temperature t is in °C.

Equation (5.4) is valid for temperatures between 10 and 300°C and for pressures between 5 and 100 bar.

The error caused by using (5.4) is at the most equal to ±1%.

5.2.1.4 Dynamic Viscosity

The dynamic viscosity is basically independent from pressure, but it varies greatly with variations in temperature. It is impossible to develop a simple equation to compute it within a reasonable margin of error.

Table 5.1 shows some values

5.2.1.5 Prandtl's Number

It is not possible to compute Prandtl's number with good approximation through a simple equation because of the behavior of the values, especially at low temperature.

Therefore, Table 5.2 indicates mean values of Pr between 5 and 100 bar. The differences between the values of Pr at different pressures are sensitive to low temperature only ($t \leq 100$°C); within this range by using the values in the Table the errors amount to roughly 1%. At higher temperatures the errors are smaller or irrelevant.

5.2.2 Air

5.2.2.1 Density

The density of air under normal conditions is equal to 1.293 kg/Nm^3.

Under different pressure and temperature conditions it is possible to adopt the law of perfect gas according to which density is directly proportional to absolute pressure and inversely proportional to absolute temperature.

Table 5.1 Dynamic viscosity in 10^{-6} kg/ms of water

t(°C)	μ	t(°C)	μ	t(°C)	μ
10	1300	110	253	210	128
20	1000	120	231	220	123
30	797	130	212	230	118
40	652	140	196	240	113
50	545	150	182	250	109
60	463	160	170	260	104
70	401	170	160	270	102
80	351	180	150	280	98.2
90	312	190	142	290	94.9
100	280	200	135	300	91.7

Table 5.2 Prandtl's number of water

t(°C)	Pr	t(°C)	Pr	t(°C)	Pr
10	9.16	110	1.55	210	0.879
20	6.86	120	1.42	220	0.861
30	5.33	130	1.31	230	0.845
40	4.26	140	1.21	240	0.837
50	3.51	150	1.14	250	0.835
60	2.94	160	1.07	260	0.840
70	2.52	170	1.02	270	0.845
80	2.19	180	0.969	280	0.865
90	1.93	190	0.933	290	0.895
100	1.72	200	0.902	300	0.940

Then

$$\rho = 1.293\frac{p}{1.013}\frac{273.15}{T} = 1.293\frac{269.64p}{T} = \frac{348.65p}{273.15+t} \tag{5.6}$$

with ρ in kg/m^3, p in bar, T in K, and t in °C.

5.2.2.2 Isobaric Specific Heat

Isobaric specific heat of air may be computed through the following approximated equation:

$$c_p = 1003.79 + 75.53\frac{t}{1000} + 216\left(\frac{t}{1000}\right)^2 \tag{5.7}$$

with c_p in J/kg K and temperature t in °C.

Equation (5.7) is valid for temperatures between 0°C and 300°C.

5.2.2.3 Thermal Conductivity

The thermal conductivity of air can be computed through the following approximated equation already discussed in Chap. 2.

$$k = 0.02326 + 0.06588\frac{t}{1000} \tag{5.8}$$

with k in W/m K and temperature t in °C.

Equation (5.8) is valid for t between 0°C and 300°C. The errors are irrelevant.

5.2.2.4 Dynamic Viscosity

The dynamic viscosity of air can be computed based on the following approximated equation:

$$\mu = \left[17.069 + 47.469\frac{t}{1000} - 18.708\left(\frac{t}{1000}\right)^2\right] \times 10^{-6} \tag{5.9}$$

with μ in kg/ms and temperature t in °C.

Equation (5.9) is valid for temperatures between 0°C and 300°C.

5.2.2.5 Prandtl's Number

Prandtl's number of air varies very little with variations in temperature. An increase from 0 to 300°C results in a reduction of about 2%.

The equation to compute it between 0 and 300°C including errors not to exceed ±0.25% is as follows:

$$\mathrm{Pr} = 0.7382 - 0.0554\frac{t}{1000}. \tag{5.10}$$

5.2.3 Flue Gas

5.2.3.1 Density

The density of flue gas under normal conditions (ρ_0) is equal to:

$$\rho_0 = \frac{G_m}{G_v} \tag{5.11}$$

where G_m and G_v are the amounts of gas per kg of fuel as mass and volume, respectively.

If n stands for the air index, i.e., the ratio between real air and theoretical air (stoichiometric) we have

$$G_m = nA_{tm} + 1 \tag{5.12}$$

given that A_{tm} stands for theoretical air in mass.

If we consider the fuel where gas is coming from, such as a mix of hydrocarbons, we have

$$G_v = nA_{tv} + 0.05558H = n\frac{A_{tm}}{1.293} + 0.05558H, \tag{5.13}$$

given that A_{tv} is the theoretical air in volume, and H is the mass percentage of hydrogen in the fuel.

Then

$$\rho_0 = 1.293\frac{nA_{tm} + 1}{nA_{tm} + 0.071865H}. \tag{5.14}$$

If the fuel is a mix of hydrocarbons, as we assumed, we get $C = 100 - H$, given that C is the percentage of carbon. In addition,

$$A_{tm} = 0.11484C + 0.34204H = 11.484 + 0.2272H. \tag{5.15}$$

Recalling (5.15) from (5.14), we obtain the following:

$$\rho_0 = 1.293 \frac{11.484n + 1 + 0.2272nH}{11.484n + (0.2272n + 0.071865)\,H}. \tag{5.16}$$

Table 5.3 is based on (5.16).

Note that Eq. (5.16) or Table 5.3 may be used even for fuels containing components other than carbon and hydrogen. For example, for fuel oil with $C = 84\%$, $H = 12\%, S = 4\%$ and $n = 1.2$, based on (5.16) we obtain $\rho_0 = 1.3029$ kg/Nm^3 , while real density is equal to 1.3031 kg/Nm^3. The difference is irrelevant.

Note that the density of flue gas may be less or more compared to air depending on the amount of hydrogen in the fuel. Therefore, it is less for natural gas and more for fuel oil.

As for air, flue gas may be considered like perfect gas. Therefore, their density at pressure different from the atmospheric, and at temperature other than 0°C, is proportional to absolute pressure and inversely proportional to absolute temperature.

5.2.3.2 Isobaric Specific Heat

The suggested approximated equation was obtained through a rather complex computational process that will not be discussed. It is included in "Steam Generators" by the author edited by Springer Verlag in Heidelberg.

The different components present in flue gas and their isobaric specific heat were examined to obtain this equation. In the end, it was possible to correlate the specific heat of gas only to that of air, which is well-known, and to the mass moisture percentage of gas besides, of course, temperature. The only variables in the equation are moisture and temperature.

Table 5.3 Flue gas density under normal conditions (kg/Nm^3)

H%	n			
	1.0	1.1	1.2	1.3
0	1.4056	1.3954	1.3868	1.3796
5	1.3568	1.3512	1.3464	1.3424
10	1.3181	1.3160	1.3141	1.3126
15	1.2867	1.2872	1.2877	1.2881
20	1.2606	1.2633	1.2656	1.2676
25	1.2387	1.2432	1.2470	1.2503

This is the equation in question:

$$c_p = 971.7 + 10.49m + (325.53 - 4.97m)\,\frac{t}{1000} - (76.59 - 6.07m)\left(\frac{t}{1000}\right)^2 \tag{5.17}$$

with c_p in J/kg K and t in °C, whereas m stands for the mass moisture percentage.

(5.17) may be used for temperatures between 50 and 1200°C and for moisture between 0 and 12%.

The behavior of c_p is shown in Fig. 5.1.

5.2.3.3 Thermal Conductivity

The computational process for thermal conductivity is similar to that used for specific heat. Even in this case the procedure used to obtain the equation is included in the cited publication of the author.

As for the specific heat, it was possible to make the value of thermal conductivity dependent only on temperature and on the mass moisture percentage of the gas.

The equation already discussed in Chap. 2 is as follows:

$$k = \begin{bmatrix} 21.924 - 0.0337m + (68.467 + 0.0966m)\,\frac{t}{1000} \\ - (12.991 - 0.6229m)\left(\frac{t}{1000}\right)^2 \end{bmatrix} \times 10^{-3} \tag{5.18}$$

with k in W/m K and t in °C ; m is the mass moisture percentage of gas.

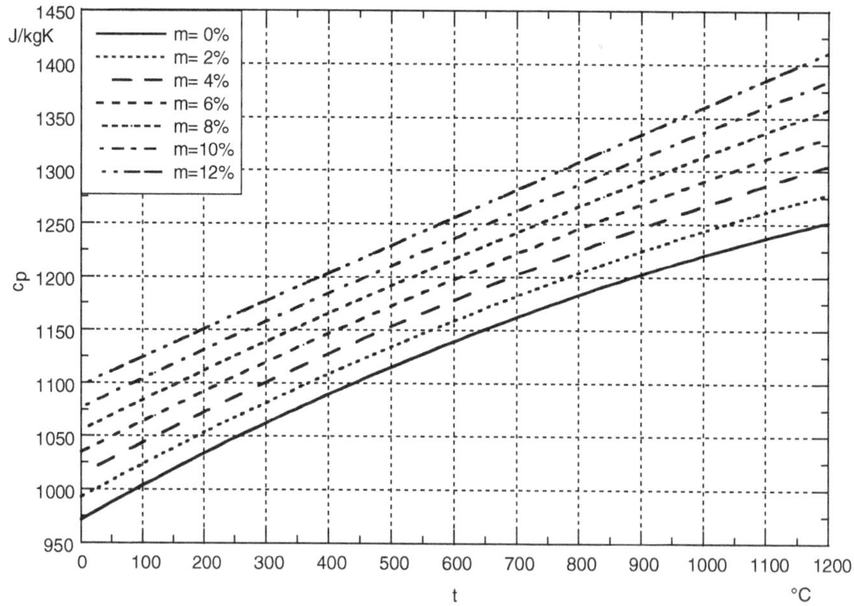

Fig. 5.1 Isobaric specific heat of flue gas

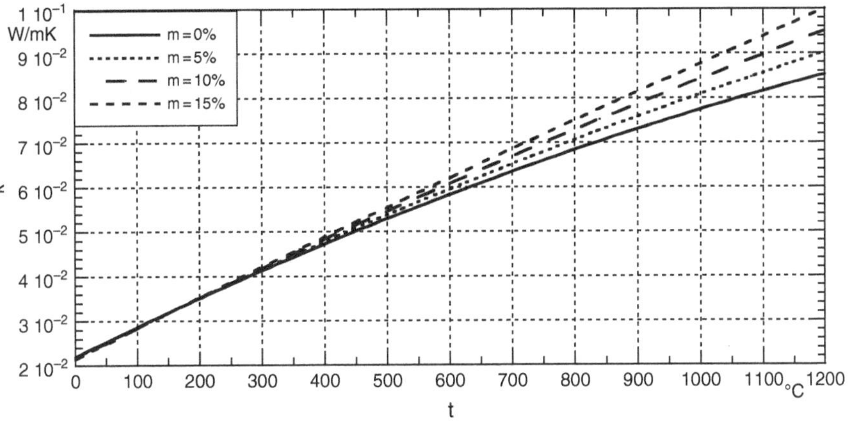

Fig. 5.2 Flue gas thermal conductivity

Equation (5.18) can be used for $t = 50 - 1200\,°\text{C}$ and for $m = 0 - 12\%$
The behavior of k is shown in Fig. 5.2.

5.2.3.4 Dynamic Viscosity

The computational process adopted for dynamic viscosity is similar to the one for specific heat. It is included in the cited publication.

As for the previous quantities, it was therefore possible to make the value of dynamic viscosity exclusively dependent on temperature and mass moisture percentage of the gas.

The equation in question is as follows:

$$\mu = \begin{bmatrix} 16.861 - 0.1106m + (43.449 - 0.111m)\,\frac{t}{1000} \\ \\ -\,(11.19 + 0.0985m)\left(\frac{t}{1000}\right)^2 \end{bmatrix} \times 10^{-6} \qquad (5.19)$$

where μ is in kg/ms and t is in °C, whereas m stands for the mass moisture percentage.

Equation (5.19) may be used for $t = 50 - 1200\,°\text{C}$ and for $m = 0 - 12\%$.

The behavior of μ is shown in Fig. 5.3.

5.2.3.5 Prandtl's Number

The following equation may be used to compute Prandtl's number:

$$\text{Pr} = 0.736 + 0.004m + (0.0377 - 0.0086m)\,\frac{t}{1000} \qquad (5.20)$$

with t in °C and the mass moisture m in percentage; the errors do not exceed $\pm 0.75\%$.

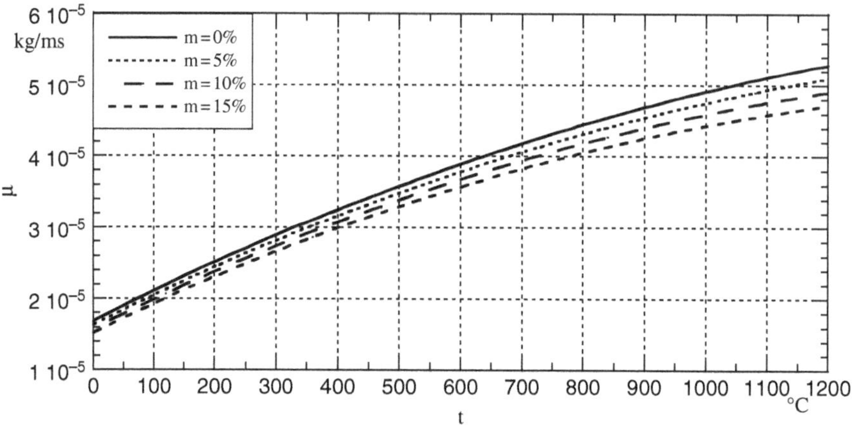

Fig. 5.3 Flue gas dynamic viscocity

5.3 Natural Convection

Let us consider a warm wall exposed to cold air. The air in contact with the wall warms up, and as a result its density decreases with respect to the surrounding air. Due to this reduction in density the air begins to move upward. Therefore, the warm air moves upward and is substituted by cold air on the bottom which in turn starts to become warmer.

Thus, the air naturally starts moving without the intervention of outside forces, and the heat is transferred to it either by conduction and convection.

Regardless of the presence of a conduction phenomenon, this type of heat transfer is called "*natural convection*".

As we shall see, the value of Nusselt's number which makes it possible to compute the heat transfer coefficient depends on the number of Grashof, as well as the number of Prandtl. This way Grashof's number substitutes the number of Reynolds which is crucial in the case of heat transfer with forced convection, as we established through the dimensional analysis.

The justification for Grashof's number as a fundamental element for heat transfer with natural convection originates from the following considerations.

As pointed out by Tribus, for geometric similarity between two currents and equal ratios between viscous forces and ascensional forces of both currents to take place, the following dimensionless number must be the same:

$$\frac{L\rho g\beta\Delta t}{\mu V}. \tag{5.21}$$

Yet, (5.21) contains the velocity of the fluid which is difficult to determine.

Based on dimensional analysis we know that equality in the number of Reynolds for two currents guarantees equality in the ratios between the forces of inertia and viscous forces. If we multiply the dimensionless number in question by the number of Reynolds we eliminate velocity and find the number of Grashof, thus guaranteeing the equality in the ratios of the forces of inertia, viscous forces and ascensional forces.

Grashof's number does not include velocity, and besides a length which identifies the dimensional characteristics of the element licked by the fluid, as well as the difference in temperature between wall and fluid, there are only characteristic quantities of the fluid itself.

They are the cubic expansion coefficient of the fluid, the density influencing the entity of the ascensional force causing the motion of the fluid and its viscosity which slows down the motion.

Note that this phenomenon can be associated with heat transfer by radiation. In that case, as we shall discuss in detail in the section on radiation, it is possible to factor in the simultaneity of both phenomena by introducing an ideal heat transfer coefficient that takes into account heat radiation. If we indicate it with α_r, while α_c indicates the heat transfer coefficient relative to the convective phenomenon, we may write that the total heat transfer by time and surface unit is given by

$$q = (\alpha_c + \alpha_r)\,(t_w - t_a) \tag{5.22}$$

where t_w is the wall temperature and t_a is the air temperature.

Based on suggestions by various scholars and researchers, we will be able to illustrate the characteristics of this phenomenon for the vertical plane wall, for horizontal cylinders, for the horizontal plane wall and for closed spaces. For ease of writing, the different equations will frequently if not exclusively include the symbols which refer to the numbers of Nusselt, Grashof and Prandtl instead of making the quantities which form them explicit. Equally remember that

$$\mathrm{Nu} = \frac{\alpha d}{k}; \tag{5.23}$$

$$\mathrm{Gr} = \frac{L^3 \rho^2 g \beta \Delta t}{\mu^2}; \tag{5.24}$$

$$\mathrm{Pr} = \frac{c_p \mu}{k} \tag{5.25}$$

where α is the heat transfer coefficient in W/m^{2}K, L is a characteristic length (the height of the vertical wall licked by the fluid, the outside diameter of a tube, the thickness of the interspace between two walls) in m, k stands for thermal conductivity in W/mK, ρ for density in kg/m^3, g for gravity acceleration in m/s^2, β for cubic expansion coefficient in °C^{-1}, Δt for the difference in temperature between wall and fluid in °C, μ for the dynamic viscosity in kg/ms, c_p for the isobaric specific heat in J/kg K.

5.3.1 Plane Vertical Wall and Vertical Tubes

Lorenz developed the first equation about plane vertical walls:

$$\mathrm{Nu} = 0.548\,\mathrm{Gr}^{0.25}\,\mathrm{Pr}^{0.25}. \tag{5.26}$$

As you can see, in (5.26) Grashof's, as well as Prandtl's number, have the same exponent which entails that α is influenced by viscosity μ because in Grashof's number μ is in the denominator to the square, whereas in Prandtl's number it is in the numerator at first power.

This is unquestionably true, as long as the motion of the ascensional current is laminar. When the motion becomes turbulent, the viscosity should cease to influence the value of α, and (5.26) would cease to correctly represent the phenomenon.

This lead Boussinesq to declare that Nusselt's number should necessarily depend on the following:

$$\mathrm{Gr}\,\mathrm{Pr}^2 = \frac{L^3\,g}{a^2}\beta\Delta t; \tag{5.27}$$

this is independent from viscosity.

In (5.27) a stands for thermal diffusivity; it was mentioned several times in connection with thermal transient states given by $a = k/c_p\rho$.

In fact, experimental data do not confirm such assumption because even in the case of turbulent motion a certain dependence of α from viscosity can be registered, as is the case with forced convection with turbulent flow, too.

Schmidt and Beckmann developed an equation based on experiments performed with air licking a wall not higher than 60 cm. It is as follows:

$$\mathrm{Nu} = 0.52\,\mathrm{Gr}^{0.25}\,\mathrm{Pr}^{0.25}. \tag{5.28}$$

Equation (5.28) appears to be very similar to Lorenz' equation except for the difference concerning the initial coefficient, but note that in (5.28) all quantities must refer to the mean temperature of the boundary layer which is conventionally assumed to be equal to the average between the wall temperature and the temperature of the fluid. This temperature is called film temperature.

Equation (5.28) sufficiently satisfies the characteristics of natural convection in the presence of air matched by a Prandtl number of about 0.73.

But many researchers pointed out that if Prandtl's number is substantially different from the number for air, the constant coefficient of 0.52 included in (5.28) is no longer acceptable. Moreover, its value varies as a function of Prandtl's number.

This lead Eckert to elaborate the following equation, by referring to the film temperature for the various quantities.

$$\mathrm{Nu} = 0.68\left(\frac{\mathrm{Pr}}{0.952+\mathrm{Pr}}\right)^{0.25}\mathrm{Gr}^{0.25}\mathrm{Pr}^{0.25}. \tag{5.28a}$$

If (5.28a) introduces Pr = 0.73 (for air) we obtain a coefficient of 0.55, which is only slightly greater than the coefficent in (5.28); on the other hand, if we assume, for instance, that Pr = 0.1 the coefficient amounts to 0.38, and that is quite a difference in comparison with the coefficient in (5.28).

Experiments performed by Weise and Saunders lead to the following equation with reference to laminar motion and to quantities referred to the film temperature:

$$\mathrm{Nu} = 0.59\mathrm{Gr}^{0.25}\mathrm{Pr}^{0.25}. \tag{5.29}$$

Note that the motion is laminar when the product GrPr stays between 10^4 and 10^9.

If the motion is turbulent instead because the product GrPr lies between 10^9 and 10^{12} the following equation is advisable:

$$\mathrm{Nu} = 0.13\,\mathrm{Gr}^{1/3}\mathrm{Pr}^{1/3}. \tag{5.30}$$

These equations agree with the following general equation about vertical tubes, where height L of the plane wall is substituted with the diameter of the tube.

$$\mathrm{Nu} = C\,\mathrm{Gr}^{\mathrm{n}}\mathrm{Pr}^{\mathrm{n}}. \tag{5.31}$$

In (5.31), if the product GrPr lies between 10^4 and 3.5×10^7 (laminar motion), $C = 0.55$ and n = 0.25; if GrPr lies between 3.5×10^7 and 10^{12} instead (turbulent motion), $C = 0.13$ and $n = 1/3$.

Observation of (5.29), (5.30) and (5.31) leads to interesting conclusions.

If motion is laminar ($n = 0.25$), the heat transfer coefficient is directly proportional to the fourth root of the temperature difference Δt, and inversely proportional to the fourth root of the length L. Thus, the impact of dimensions is quite modest.

If motion is turbulent ($n = 1/3$) the heat transfer coefficient is proportional to the cube root of Δt and independent from length L; therefore, the dimensions of the vertical wall or of the tube do not play a role.

Factoring in the above and the fact that the value of α only depends on the physical characteristics of the fluid, besides Δt and eventually L, it is possible to write the following general purpose equations:

For GrPr within 10^4 and 10^9 (laminar motion)

$$\alpha = A\left(\frac{\Delta t}{L}\right)^{0.25}; \tag{5.32}$$

For GrPr within 10^9 and 10^{12} (turbulent motion)

$$\alpha = B\Delta t^{1/3}, \tag{5.33}$$

where A and B depend on the type of fluid and on the film temperature.

For air which is the most likely fluid, at ordinary temperature as well as atmospheric pressure we can write that:

For GrPr within 10^4 and 10^9, in substitution for (5.29)

$$\alpha = \left(1.99 - 0.199 t_f^{0.25}\right)\left(\frac{\Delta t}{L}\right)^{0.25}; \tag{5.34}$$

For GrPr within 10^9 and 10^{12}, in substitution for (5.30)

$$\alpha = \left(2.42 - 0.359\, t_f^{0.25}\right) \Delta t^{1/3} \tag{5.35}$$

where t_f stands for the film temperature.

In (5.34) and in (5.35) α is in W/m^2K, t_f and Δt in °C and L in m.

If there is water instead, we can write that:

In substitution for (5.29):

$$\alpha = \left(29.58 + 22\sqrt{t_f}\right)\left(\frac{\Delta t}{L}\right)^{0.25}; \tag{5.36}$$

In substitution for (5.30):

$$\alpha = \left(49\sqrt{t_f} - 22.3\right) \Delta t^{1/3}. \tag{5.37}$$

Based on studies by Schmidt and Beckmann, it is also possible to suggest the following:

$$\alpha = C\left(\frac{\Delta t}{T_0 L}\right)^{0.25}. \tag{5.38}$$

In (5.38) T_o is the absolute temperature of air in K. According to Schmidt and Beckmann, coefficient C should be 5.58, but Schack considers the value of α to be underestimated this way and suggests a coefficient of 7 instead, and this seems exaggerated.

Equation (5.38) uses the absolute temperature of air correctly, given that an increase of this temperature is followed by an increase of the film temperature, the value of α decreases, but there is no clear correlation between air temperature and film temperature, because the latter is also influenced by the wall temperature. Therefore, reference to (5.34) is more correct.

5.3.2 Horizontal Cylinders

Based on theory, Hermann suggested the following assumption for a cylinder licked by biatomic gas and for Grashof numbers greater than 10^4:

$$\mathrm{Nu} = 0.37\,\mathrm{Gr}^{0.25}. \tag{5.39}$$

Since biatomic gases have a Prandtl number equal to about 0.74, and assuming that Prandtl's number impacts Nu through its fourth root, as we have already see for the plane wall, (5.38) may be generalized as follows:

$$\mathrm{Nu} = 0.4\,\mathrm{Gr}^{0.25}\mathrm{Pr}^{0.25}. \tag{5.40}$$

In fact, based on a series of esperiments on various types of gas, as well as on water, it was established that (5.40) leads to Nu values that are lower than actual ones by 25–30%.

As a result, the following equation was recommended. It is valid for values of the product GrPr between 10^3 and 10^9, i.e.,

$$\mathrm{Nu} = 0.53\,\mathrm{Gr}^{0.25}\mathrm{Pr}^{0.25}. \tag{5.41}$$

Both in Nusselt's and in Grashof's number the characteristic length is the outside diameter of the cylinder. The film temperature is always used as reference.

Note that if we considered half the circumference instead of the outside diameter of the cylinder, in other words if we assumed that $L = \pi d_o/2$, given that d_o is the outside diameter, the coefficient in equation (5.41) would become 0.59, i.e., the same value in (5.29) relative to the vertical plane wall.

As in the case of vertical plane walls, for horizontal cylinders it is equally possible to indicate simplified equations, as far as air under normal conditions and under atmospheric pressure.

They are as follows.

For GrPr between 10^3 and 10^9 (laminar motion) (typical case for tubes):

$$\alpha = \left(1.79 - 0.179 t_f^{0.25}\right)\left(\frac{\Delta t}{d_o}\right)^{0.25}; \tag{5.42}$$

For GrPr between 10^9 and 10^{12} (turbulent motion):

$$\alpha = \left(2.16 - 0.32 t_f^{0.25}\right)\Delta t^{1/3} \tag{5.43}$$

where t_f is the film temperature.

As usual, α is in W/m^{2}K, t_f and Δt in °C and d_o in m.

Based on his experience, Wamsler suggests the following equation for horizontal tubes and quiet air:

$$\alpha = 1.1\frac{\Delta t^{0.233}}{d_o^{0.3}}. \tag{5.44}$$

A comparison of the values obtained from this equation with those relative to (5.42) for $\Delta t = 20 - 100°\mathrm{C}$, $d_o = 50 - 200$ mm and $t_f = 40 - 70°\mathrm{C}$ shows that (5.44) leads to lower values of α the differences are equal to 10–25%.

If we consider water at 20°C and under absolute pressure of 1 bar instead, and if we assume that the motion is laminar

$$\alpha = \left(26.57 + 19.77\sqrt{t_f}\right)\left(\frac{\Delta t}{d_o}\right)^{0.25}. \qquad (5.45)$$

Note that based on Grigull's experiments, Schack developed the following equation relative to water with laminar motion:

$$\alpha = \left(18.6 + 20.7\sqrt{t_f}\right)\left(\frac{\Delta t}{d_o}\right)^{0.25}. \qquad (5.46)$$

As you can see, (5.46) is quite similar to (5.45). For the values of t_f to be encountered in reality, the values obtained by both equations differ only slightly, at the most by 2%.

Motion may become turbulent only for high values in diameter and temperature difference between tube and water.

We already pointed out that, under certain conditions, heat by natural convection may be combined with heat by radiation.

For instance, this is the case of horizontal tubes at a temperature greater than the surrounding air temperature.

With reference to experiments by Heilman and McMillan, and considering room temperature at 25°C, the ideal heat transfer coefficient by radiation may be computed with acceptable approximation through the following equation:

$$\alpha_r = \left[5.5 + 2.3\frac{\Delta t}{100} + 1.25\left(\frac{\Delta t}{100}\right)^2\right] d_o^{0.1}. \qquad (5.47)$$

In (5.47) α_r is in W/m^{2}K, Δt in °C and the outside diameter of the tube d_o in m. Equation (5.47) can be used for diameters ranging from 50 to 300 mm and for values of Δt between 25 and 400°C. Heat transfer by natural convection prevails only for very small differences in temperature between tube and environment. In all other instances heat transfer by radiation is prevalent (and for high values of Δt it prevails by far). Thus, it is clear why we talk about heat loss by radiation in the context of heat loss towards the environment in reference to a coating at a temperature greater than the environment (for example the coating of a steam generator), even though there is heat loss by natural convection.

5.3.3 Horizontal Plane Plates

There is a distinction between heat moving upward or downward.

The former case occurs if the warmer plate transfers heat to the surrounding fluid through the upper side, or if the colder plate receives heat from the surrounding fluid through the bottom side.

In that case, according to experiments by Fishenden and Saunders on a square plate we can write that:

For laminar motion (the product GrPr varies between 10^5 and 2×10^7)

$$\mathrm{Nu} = 0.54\mathrm{Gr}^{0.25}\mathrm{Pr}^{0.25}; \tag{5.48}$$

For turbulent motion (the product GrPr varies between 2×10^7 and 3×10^{10})

$$\mathrm{Nu} = 0.14\mathrm{Gr}^{1/3}\mathrm{Pr}^{1/3}. \tag{5.49}$$

In (5.48) and (5.49) the characteristic dimension L in Nu and Gr is shown from the side of the plate. The value of the various quantities must be referred to the mean temperature of the boundary layer called film temperature.

But if the warmer plate transfers heat to the environment through the bottom side, or if it receives heat through the upper side when it is colder, and considering laminar motion:

$$\mathrm{Nu} = 0.27\mathrm{Gr}^{0.25}\mathrm{Pr}^{0.25}. \tag{5.50}$$

For air in substitution of (5.48)

$$\alpha = \left(1.82 - 0.182t_f^{0.25}\right)\left(\frac{\Delta t}{L}\right)^{0.25}; \tag{5.51}$$

Instead of (5.49):

$$\alpha = \left(2.61 - 0.39t_f^{0.25}\right)\Delta t^{1/3}. \tag{5.52}$$

Finally, instead of (5.50) we can write that

$$\alpha = \left(0.913 - 0.0913t_f^{0.25}\right)\left(\frac{\Delta t}{L}\right)^{0.25}. \tag{5.53}$$

For water (5.48) can be substituted by

$$\alpha = \left(27.07 + 20.14\sqrt{t_f}\right)\left(\frac{\Delta t}{L}\right)^{0.25}. \tag{5.54}$$

Equation (5.49) can be substituted by

$$\alpha = \left(52.82\sqrt{t_f} - 24\right) \Delta t^{1/3}. \tag{5.55}$$

Finally, (5.50) can be substituted by

$$\alpha = \left(13.53 + 10.07\sqrt{t_f}\right) \left(\frac{\Delta t}{L}\right)^{0.25}. \tag{5.56}$$

In all equations above α is expressed in W/m^2K, t_f and Δt in °C and L in m.

5.3.4 Interspace Between two Plane Walls

We consider two plane walls with height L at different temperatures where Δt stands for the difference between their temperatures. They are separated by an interspace with thickness x consisting of air. There is heat transfer between the two walls consisting of heat by radiation that will be ignored in this case, and depending on the heat transferred by conduction through air or by natural convection.

If the heat transfer is by conduction nothing new can be added. If it is by natural convection instead, research has, as usual, correlated the value in Nusselt's number to Grashof's number by considering the thickness of the interspace as the characteristic length of the phenomenon.

In this case

$$\mathrm{Nu} = \frac{\alpha x}{k}. \tag{5.57}$$

Similarly,

$$\mathrm{Gr} = \frac{x^3 \rho^2 g \beta \Delta t}{\mu^2}. \tag{5.58}$$

The temperature of reference for the various quantities is the mean temperature of the boundary layer called film temperature.

Let us assume two vertical walls.

If Grashof's number, computed as discussed, is less that 2×10^3 there is no convective motion in the interspace and conduction by air should be considered.

If Grashof's number is greater instead, we have natural convection, and it is possible to use the following equation based on research by Mull, Reiher, Schmidt and Nusselt:

$$\mathrm{Nu} = \frac{C}{(L/x)^{1/9}} \mathrm{Gr}^{\mathrm{n}} \mathrm{Pr}^{\mathrm{n}}. \tag{5.59}$$

If Gr ranges from 2.1×10^3 to 2×10^4, $C = 0.20$, and $n = 0.25$; the motion is laminar.

If Gr ranges from 2.1×10^5 to 1.1×10^7, the motion is turbulent and $C = 0.071$, while $n = 1/3$. In this case, for a given value of L/x the value of α is independent from x.

If the two plane walls are horizontal and the heat flows upward, if Gr is less than 10^3 there is no convective motion and the heat transfer takes place by conduction. If the value of Gr is greater, based on research by Mull and Reiher, it is possible to use the following equation:

$$\mathrm{Nu} = C\,\mathrm{Gr}^{\mathrm{n}}\mathrm{Pr}^{\mathrm{n}}. \tag{5.60}$$

In this case, for Gr between 10^4 and 3.2×10^5 $C = 0.21$ and $n = 0.25$, whereas for Gr between 3.2×10^5 and 10^7 , $C = 0.075$ and $n = 1/3$.

5.4 Forced Convection Inside the Tubes

If we consider forced convection relative to a fluid inside a tube the first question should be about whether we are facing laminar or turbulent motion.

It was already noted earlier that if the number of Reynolds is lower or equal to 2000, we are certainly confronted with laminar motion.

Recalling the significance of Re, it must be:

$$Vd \leq 2000\frac{\mu}{\rho}, \tag{5.61}$$

where velocity V is in m/s, the diameter is in m, the dynamic viscosity μ is in kg/ms and density ρ is in kg/m^3.

Considering two characteristic fluids, water and air, we look at where (5.61) is leading to.

For water between 20 and 300°C, equation (5.61) leads to the following borderline values of product Vd

$$Vd = 2.53 \times 10^{-4} - 2 \times 10^{-3}, \tag{5.62}$$

where the highest value refers to water at 20°C, for which the value of dynamic viscosity is the highest.

For instance, if we consider a tube with a diameter of 20 mm, based on (5.62):

$$V = 0.012 - 0.1\,\mathrm{m/s}. \tag{5.63}$$

These are very small velocity values, yet they cannot be ignored when considering a velocity of 0.1 m/s.

Of course, if the diameter increases the values of V decrease even further.

Based on this simple investigation we can conclude that laminar motion of water is actually possible at low temperatures, small diameters and low velocities.

Now we consider air between 0 and 300°C.

Based on (5.61) the borderline values of product Vd are given by

$$Vd = 0.0264 - 0.0960 \tag{5.64}$$

where the highest value refers to 300°C.

Still considering a tube with a diameter of 20 mm we obtain:

$$V = 1.32 - 4.8\,\text{m/s}. \tag{5.65}$$

If we consider mass velocity G instead of velocity V we obtain:

$$G = 1.706 - 2.95\,\text{kg/m}^2\text{s}. \tag{5.66}$$

As you can see, by referring to G, the range narrows, and the temperature has a much lesser impact, even though it goes from 0 to 300°C.

These are very low velocities but they cannot be ruled out completely, given that laminar motion can occur only for small diameters (unlikely for air) and low velocities.

We should also consider a particularly viscous fluid, for instance a mineral oil with a petroliferous origin; fluids of this kind are used, for example, in boilers with diathermic fluid.

If we consider an oil at 200°C with $\mu = 3 \times 10^{-3}$ kg/ms and $\rho = 780$ kg/m^3, from (5.61) we obtain $Vd \leq 7.7 \times 10^{-3}$; with $d = 0.02$ m we have $V \leq 0.385$ m/s; these are unlikely values as far as the diameter of the tube, as well as the velocity of the oil.

Based on these considerations and taking into account industrial realities, we can conclude that motion is generally turbulent. Chances for it to be laminar are limited to a few rare and unlikely situations.

In any case, to check whether motion is definitely turbulent (Re $\geq$ 3000), it suffices to check mass velocity which must be as follows:

$$G \geq \frac{3000\mu}{d}. \tag{5.67}$$

Therefore, our attention will focus mostly on turbulent motion, and devote only a few comments to laminar motion in Sect. 5.7.

If we examine a current of fluid in turbulent motion, we see that a thin layer of fluid with laminar motion builds up in contact with the wall of the tube (boundary layer). There is a transition area towards the center of the current, and finally there is a central area with vortexes. Thus, the motion is typical of turbulent motion.

In the boundary layer (also called film) the heat transfer takes place by conduction. In other words, the value of thermal conductivity of the fluid is crucial. It is not surprising that the value of the heat transfer coefficient also depends on thermal conductivity even though we are faced with turbulent motion, as we shall see later on; in fact, it is explicitly included in Nusselt's number.

The temperature through the boundary layer varies between that of the wall and that in the central area of the current. From now on we will call the latter temperature bulk temperature and indicate it with t_b.

The mean temperature of the boundary layer is conventionally assumed to be equal the average between the wall temperature and the bulk temperature, and is called film temperature. It will be indicated by t_f.

During development of the equations to compute the heat transfer coefficient, researchers referred to bulk temperature or film temperature. Therefore, when examining the different possibilities, we consider which of two temperatures the equation is referring to. This is not crucial if the difference between wall temperature and bulk temperature of the fluid is small, as is the case for water and partly even for superheated steam. This depends on the high value of the heat transfer coefficient of water.

But in the case of fluids with low heat transfer coefficients, such as gases, this difference can be great, so that the bulk temperature may differ considerably from film temperature.

During dimensional analysis we established that the value of Nusselt's number, therefore the value of the heat transfer coefficient depends on the number of Reynolds. This dependency is fully confirmed by experimental data.

It is customary to formalize the dependency of Nu from Re through a power of the number of Reynolds. Moreover, note that velocity V or mass velocity G is included in the number of Reynolds. We always prefer the reference to mass velocity for the simple reason that the value of V varies with temperature, given that the volumetric flow rate varies, whereas the value of G does not change.

In addition, we see that diameter d (in this case it is the inside diameter d_i) is included both in Nusselt's as well as Reynolds' number.

Therefore, if the exponent of the number of Reynolds is n, we may write that

$$\alpha \equiv \frac{G^n}{d_i^{1-n}}. \tag{5.68}$$

The dimensional analysis also demonstrated that beyond G and d_i, α depends on the values of the isobaric specific heat c_p, on thermal conductivity k and on dynamic viscosity μ.

The value of these three quantities depends on the type of fluid, on the temperature and, under certain circumstances, on pressure.

Therefore,

$$\alpha = f\,(t,p)\,\frac{G^n}{d_i^{1-n}}, \tag{5.69}$$

where $f\,(t,p)$ stands for a function of temperature and pressure.

Generally, maintaining that thermal conductivity is included in Nusselt's number, and that viscosity is included in the number of Reynolds, the dependence of α from the values of c_p, k, μ also happens through Prandtl's number, and this brings to mind the outcome of the dimensional analysis. This, on the other hand, is not always true, as we shall see later on.

In any case, it is almost always possible to express the value of α through an equation like (5.69), by adopting approximate, yet acceptable expressions for $f(t,p)$.

In conclusion, note that during the dimensional analysis we already pointed out that it may be necessary to make Nusselt's number dependent on a dimensionless term represented by two different values of a same quantity, for instance on the values of the specific heat at bulk temperature and at film temperature.

(5.69) may therefore generically be rewritten as follows:

$$\alpha = f(t,p)\frac{G^n}{d_i^{1-n}}J^m; \tag{5.70}$$

J stands for the dimensionless term discussed above and m is a constant.

At this point we will examine a series of fluids in turbulent motion inside the tubes.

5.4.1 Water

The following equation by Dittus and Boelter is one of the most well-known ones for water:

$$\mathrm{Nu} = 0.023\,\mathrm{Re}^{0.8}\mathrm{Pr}^{0.4}. \tag{5.71}$$

In (5.71) the temperature of reference is the bulk temperature.

Recalling the significance of Nu, Re and Pr, based on (5.71) we obtain

$$\alpha = 0.023\frac{c_{pb}^{0.4}k_b^{0.6}}{\mu_b^{0.4}}\frac{G^{0.8}}{d_i^{0.2}}; \tag{5.72}$$

we adopted the subscript b for all quantities to indicate that the temperature of reference is the bulk temperature.

Equation (5.72) is none other than (5.69), taking into account that c_p, k, μ are a function of temperature and pressure.

In the case of water which is of interest in this case, the impact of pressure is negligible, and it is possible to consider bulk temperature only.

An examination of the values of the three quantities of interest for values of t_b ranging from 20 to 300°C leads to

$$K_{w1} = 5.80 + 9.19\frac{t_b}{100} - 1.395\left(\frac{t_b}{100}\right)^2 \approx 0.023\frac{c_{pb}^{0.4}k_b^{0.6}}{\mu_b^{0.4}} \tag{5.73}$$

with t_b in °C.

Equation (5.72) is reduced to the following:

$$\alpha = K_{w1}\frac{G^{0.8}}{d_i^{0.2}} \tag{5.74}$$

with α in W/m^{2}K and K_{w1} given by (5.73).

According to Colburn, we must adopt the following equation:

$$\mathrm{Nu} = 0.023\mathrm{Re}^{0.8}\mathrm{Pr}^{1/3}\frac{c_{pb}}{c_{pf}} \tag{5.75}$$

where c_{pb} and c_{pf} stand for the isobaric specific heats with respect to bulk temperature and film temperature. The temperature of reference is that of film.

From (5.75) and recalling significance of Nu, Re and Pr:

$$\alpha = 0.023\frac{c_{pf}^{1/3}k_f^{2/3}}{\mu_f^{0.4666}}\frac{G^{0.8}}{d_i^{0.2}}\frac{c_{pb}}{c_{pf}}. \tag{5.76}$$

Equation (5.76) is none other than (5.70) taking into account that the three quantities of interest are a function of temperature.

By analogy with (5.72), we may write that

$$K_{w2} = 4.735 + 9.98\frac{t_f}{100} - 1{,}515\left(\frac{t_f}{100}\right)^2 \approx 0.023\frac{c_{pf}^{1/3}k_f^{2/3}}{\mu_f^{0.46666}} \tag{5.77}$$

with t_f in °C.

Then (5.76) may be written as follows:

$$\alpha = K_{w2}\frac{G^{0.8}}{d_i^{0.2}}\frac{c_{pb}}{c_{pf}} \tag{5.78}$$

with α in W/m^{2}K and K_{w2} derived from (5.77).

Figure 5.4 shows the values of K_{w1} and K_{w2}.

A comparison between (5.78) and (5.74) is not easy because the temperatures of reference are different.

As we already saw, for water the film temperature is quite similar to the bulk temperature. For instance, if there is thermal flux equal to 60 kW/m^2 (this is a high value) and a heat transfer coefficient equal to 3000 W/m^{2}K, the difference between the wall temperature and the bulk temperature is equal to 60000/3000 = 20°C. Therefore, the film temperature differs from the bulk temperature by 10°C.

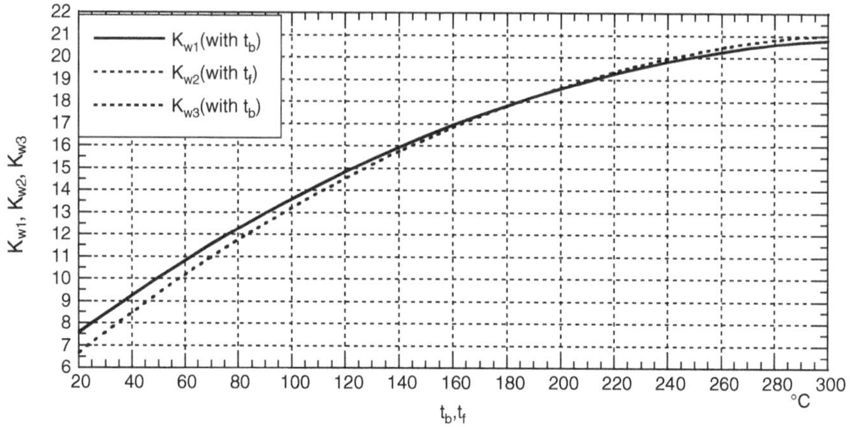

Fig. 5.4 Factors K_{W1}, K_{W2}, K_{W3}

At any rate, note that when considering the heating of water the film temperature is greater than the bulk temperature, and this leads to a greater computational value of α; vice versa, the ratio c_{pb}/c_{pf} leads to a reduction of the computational value of α.

Still, these are not the substantial differences between (5.75) and (5.71). The fundamental difference is the different exponent of Prandtl's number. This has little or no impact at high temperatures, where Prandtl's number is close to unity, but this is not the case at low temperatures, when Prandtl's number is high due to the high value of dynamic viscosity (see Table 5.2); if we consider the ratio $\Pr^{0.4}/\Pr^{1/3} = \Pr^{0.06666}$, at 30°C it is equal to 1.12, at 100°C it is equal to 1.037, and at 150°C it is equal to 1.008.

In conclusion, the biggest differences between (5.75) and (5.71), and consequently between (5.78) and (5.74), are registered at low temperatures, as shown in Fig. 5.4 by the values of K_{w1} and K_{w2}.

Sieder and Tate developed the following equation:

$$\mathrm{Nu} = 0.023\mathrm{Re}^{0.8}\Pr^{1/3}\left(\frac{\mu_b}{\mu_w}\right)^{0.14}; \qquad (5.79)$$

μ_w stands for dynamic viscosity at wall temperature; the temperature of reference is the bulk temperature, as for Eq. (5.71).

Proceeding as usual we obtain

$$\alpha = 0.023\frac{c_{pb}^{1/3}k_b^{2/3}}{\mu_b^{0.46666}}\frac{G^{0.8}}{d_i^{0.2}}\left(\frac{\mu_b}{\mu_w}\right)^{0.14}. \qquad (5.80)$$

Therefore, recalling (5.77), we have

$$K_{w3} = 4.735 + 9.98\frac{t_b}{100} - 1.515\left(\frac{t_b}{100}\right)^2 \approx 0.023\frac{c_{pb}^{1/3}k_b^{2/3}}{\mu_b^{0.46666}}; \tag{5.81}$$

then,

$$\alpha = K_{w3}\frac{G^{0.8}}{d_i^{0.2}}\left(\frac{\mu_b}{\mu_w}\right)^{0.14} \tag{5.82}$$

with α in W/m^{2}K and K_{w3} derived from (5.81).

Taking into account that in this case the temperature of reference is identical to the one in (5.74), the considerations made when comparing (5.75) with (5.71) are confirmed. As far as the impact of the ratio μ_b/μ_w, note that it leads to an increase of the computational value of a, and to the fact that it is more relevant at low temperatures (see Table 5.1). Thus, its presence reduces the influence of the smaller exponent of Pr with respect to (5.71).

Another suggested equation with reference to bulk temperature is as follows:

$$\mathrm{Nu} = 0.024\mathrm{Re}^{0.8}\mathrm{Pr}^{0.33}. \tag{5.83}$$

Besides the slight change concerning the coefficient from 0.023 to 0.024, (5.83) is comparable to (5.71) but with a different exponent for Pr. The same considerations relative to (5.75) and (5.79) may be made for low temperatures of water.

Equation (5.83) may be substituted by

$$\alpha = K_{w4}\frac{G^{0.8}}{d_i^{0.2}} \tag{5.84}$$

with α in W/m^{2}K, given that

$$K_{w4} = 4.94 + 10.41\frac{t_b}{100} - 1.58\left(\frac{t_b}{100}\right)^2. \tag{5.85}$$

Babcock and Wilcox suggest the following instead:

$$\mathrm{Nu} = 0.023\mathrm{Re}^{0.8}\mathrm{Pr}^{0.4}\left(\frac{T_b}{T_f}\right)^{0.8} \tag{5.86}$$

with reference to bulk temperature; T_b and T_f are the absolute temperatures of the fluid at bulk as well as film temperature.

This is a variation on (5.71) by introducing the ratio T_b/T_f.

Given that t_f is greater than t_b, the variation leads to a small reduction of the computational value of α.

Of course, (5.86) may be substituted by

$$\alpha = K_{w1}\frac{G^{0.8}}{d_i^{0.2}}\left(\frac{T_b}{T_f}\right)^{0.8} \tag{5.87}$$

where α is in W/m^{2}K, whereas K_{w1} is given by (5.73).

The following equation by Hansen is not part of the common structure among the five illustrated equations, given that

$$\mathrm{Nu} = 0.037\left(\mathrm{Re}^{0.75} - 180\right)\mathrm{Pr}^{0.42}\left[1 + \left(\frac{d_i}{L}\right)^{2/3}\right]\left(\frac{\mu_b}{\mu_w}\right)^{0.14}. \tag{5.88}$$

In (5.88) L stands for the length of the tube, μ_b and μ_w represent the dynamic viscosities with respect to bulk as well as wall temperature. The temperature of reference is bulk temperature.

Equation (5.88) factors in the phenomena occuring at the extremities of the tube. Obviously, they only impact relatively short tubes. According to this equation, if the tube is sufficiently long (roughly $L \geq 300d_i$), the phenomenon may be labeled as negligible, and then

$$\mathrm{Nu} = 0.037\left(\mathrm{Re}^{0.75} - 180\right)\mathrm{Pr}^{0.42}\left(\frac{\mu_b}{\mu_w}\right)^{0.14}. \tag{5.89}$$

Given the structure of (5.89) it is impossible to introduce the simplifications shown for the other equations.

Based on the considerations above, we believe that (5.74) or (5.84) may be preferable for the computation of α for their simplicity and the easier reference to bulk temperature.

5.4.2 Superheated Steam

The equations discussed in reference to water may also be used for superheated steam, since they are designed for fluids where the difference in temperature between wall and fluid is small. This is the case for superheated steam, as well, because it generally has a high heat transfer coefficient like water, even though it is not comparable to that of water itself.

Nonetheless the process to obtain simplified equations factoring in the heat transfer coefficient is much more complex, given that pressure is crucial in the case of superheated steam. Therefore, in our case the function *f(t,p)* considered in (5.69) or in (5.70) is a function of both t and p, whereas for water it could be reduced to a simple function of t.

For this reason, we developed an equation where α is included in an explicit way just for one equation that is not part of those discussed for water.

Besides the equations in reference to water discussed above, the following equations by McAdams may be used for superheated steam

$$\mathrm{Nu} = 0.0214\left(1 + \frac{2.3d_i}{L}\right)\mathrm{Re}^{0.8}\,\mathrm{Pr}^{1/3} \tag{5.90}$$

with reference to bulk temperature.

Clearly, Eq. (5.90) is structurally similar to a few equations that were discussed for water with $\mathrm{Re}^{0.8}$ and $\mathrm{Pr}^{1/3}$. Note that the coefficient 0.0214 is below 0,023, i.e., the one generally used for water. If $L \geq 120d_i$ the term in parenthesis may be ignored.

While examining the results obtained by various scholars on research related to different types of gas including superheated steam, Schack reached the conclusion that it was possible to elaborate a general equation for all types of gas, as well as superheated steam, i.e.,

$$\alpha = 0.0254c_{1pb}^{0.81}k_b^{0.19}\frac{V_0^{0.75}}{d_i^{0.25}}. \tag{5.91}$$

In (5.91) α is in W/m^{2}K, c_{1pb} is the isobaric specific heat referred to volume and bulk temperature in J/Nm^{3}K, k_b is the thermal conductivity referred to bulk temperature in W/m^{2}K, and d_i is the inside diameter of the tube in m. Moreover, V_0 (in m/s) is the velocity under normal conditions, i.e., the velocity picked up by the fluid if brought to 0°C and to atmospheric pressure.

Equation (5.91) is not based on a classic equation, where Nusselt's number is a function of the number of Reynolds and Prandtl, as is usually the case for water.

This would be the case if the exponent of c_{1pb} were 0.75 and that of k were 0.25.

In fact, if V stands for velocity under actual working conditions, c_{pb} for specific heat referred to mass unity, ρ_0 for density under normal conditions, and ρ for density under working conditions:

$$c_{1pb} = c_{pb}\rho_0; \tag{5.92}$$

$$V_0\rho_0 = V\rho. \tag{5.93}$$

If the exponents of $c_{1\mathrm{pb}}$ and k_b were 0.75 and 0.25, (5.91) may be written as follows with C being the coefficient to specify:

$$\alpha = Cc_{pb}^{0.75}k_b^{0.25}\frac{V^{0.75}\rho^{0.25}}{d_i^{0.25}}; \tag{5.94}$$

or this way

$$\alpha = C\left(\frac{V\rho d_i}{\mu_b}\right)^{0.75}\left(\frac{c_{pb}\mu_b}{k_b}\right)^{0.75}\frac{k_b}{d_i}; \tag{5.95}$$

then

$$\mathrm{Nu} = \mathrm{CRe}^{0.75}\mathrm{Pr}^{0.75} = \mathrm{CPe}^{0.75} \tag{5.96}$$

given that Pe stands for Peclet's number.

On the other hand, c_{p1} and k are a function of the temperature; we may write

$$0.0254c_{1pb}^{0.81}k^{0.19} = A + Bt_b + Ct_b^2 \tag{5.97}$$

where A,B and C are constants.

Calculating the values of A,B and C for the superheated steam we obtain the following equation showing only V_0,t_b and d_i:

$$\alpha = \left(4.42 + 0.302\frac{t_b}{100}\right)\frac{V_0^{0.75}}{d_i^{0.25}}. \tag{5.98}$$

Clearly, velocity V_0 has no physical significance for steam, but this is the logic behind Schack's equation. Therefore, V_0 is a conventional velocity computed by considering steam as perfect gas capable of staying that way until 0°C.

Thus, if V stands for the velocity of the steam under actual running conditions

$$V = V_0\frac{273 + t_b}{273}\frac{1.013}{p}, \tag{5.99}$$

with p in bar and keeping in mind that atmospheric pressure is equal to 1.013 bar.

Therefore, (5.98) can also be written like this:

$$\alpha = \frac{294 + 0.201t_b}{(273 + t_b)^{0.75}}\frac{(pV)^{0.75}}{d_i^{0.25}}. \tag{5.100}$$

Knowing the absolute pressure, the velocity, the temperature and the inside diameter of the tubes, (5.100) helps to compute the heat transfer coefficient α.

If v is the specific volume of the steam

$$V = Gv, \tag{5.101}$$

given that G is, as usual, the mass velocity.

From (5.100)

$$\alpha = \frac{294 + 0.201t_b}{(273 + t_b)^{0.75}}(pv)^{0.75}\frac{G^{0.75}}{d_i^{0.25}}. \tag{5.102}$$

Assuming that

$$K_s = \frac{294 + 0.201 t_b}{(273 + t_b)^{0.75}} (pv)^{0.75}, \tag{5.103}$$

we establish that this factor is a function of pressure and temperature because the specific volume is a function of these quantities. Based on the values of v, we obtain the following approximated equation of K_s:

$$K_s = 5.069 - 0.0529p + (4.467 + 0.169p)\frac{t_b}{1000} - (1.268 + 0.143p)\left(\frac{t_b}{1000}\right)^2 \tag{5.104}$$

valid for $p = 10 - 100$ bar and $t_b = 180 - 550°C$.

The greatest error made by using (5.104) instead of (5.103) is equal to about 1.5% plus or minus, and this is acceptable because of the modest influence of the heat transfer coefficient of steam in the computation of the overall heat transfer coefficient.

The heat transfer coefficient of the superheated steam can be computed much more easily based on the following equation:

$$\alpha = K_s \frac{G^{0.75}}{d_i^{0.25}}, \tag{5.105}$$

where α is expressed in W/m^{2}K, G in kg/m^{2}s and d_i in m; factor K_s can be computed through (5.104).

We would like to provide some information regarding the velocities to adopt with superheated steam.

The assumption is that $\alpha = 1200$ W/m^{2}K. Note that exceedingly low values of α cause an increase in temperature of the tube which may reach values that may endanger its conservation.

If pressure is 100 bar, temperature 500°C and the inside diameter 28 mm, based on (5.104) the result is $K_s = 6.57$; given that $d_i^{0.25} = 0.409$, from (5.105) we obtain $G^{0.75} = 74.70$, and then $G = 314{,}6$ kg/m^{2}s.

On the other hand, if $p = 25$ bar, $t_b = 400°C$ and $d_i = 32$ mm, this would lead to $K_s = 6.488$; given that $d_i^{0.25} = 0.423$, from (5.105) we obtain $G^{0.75} = 78.72$ and $G = 337.4$ kg/m^{2}s.

In both cases the value of mass velocity is almost the same but the velocity is quite different. In the first instance, with a mass volume of 0.03276 m^3/kg the velocity is $314.6 \times 0.03276 = 10.31$ m/s. In the second instance, with a mass volume of 0.12 m^3/kg we obtain $V = 337.4 \times 0.12 = 40.49$ m/s, and this is an extremely high value.

This basic example shows that superheaters at low pressure are difficult to implement under safety conditions and with acceptable steam velocity.

5.4.3 Mineral Oils

Based to some extent on the recommendations of producers of mineral oils, the heat transfer coefficient relative to heat transfer between tube wall and fluid can be computed based on the following equations.

Well-known Eq. (5.71) is among the recommended ones, i.e.,

$$\mathrm{Nu} = 0{,}023\mathrm{Re}^{0.8}\,\mathrm{Pr}^{0.4}\,. \tag{5.106}$$

Equation (5.83) is equally worth recommending:

$$\mathrm{Nu} = 0.024\mathrm{Re}^{0.8}\,\mathrm{Pr}^{0.33}\,. \tag{5.107}$$

It is valid for very small values of the ratio between inside diameter and tube length, and for values closed to unity of the ratio between viscosity of the fluid at bulk temperature and at wall temperature (in other words, for not exceedingly high thermal flux).

In order to factor in the different values of viscosity of the fluid at bulk temperature, and at wall temperature, and at times the difference is considerable, we recommend to use (5.79):

$$\mathrm{Nu} = 0{,}023\mathrm{Re}^{0.8}\,\mathrm{Pr}^{1/3}\left(\frac{\mu_b}{\mu_w}\right). \tag{5.108}$$

All these equations refer to the bulk temperature of the fluid. In (5.108) μ_w stands for viscosity at wall temperature. Based on (5.108)

$$\alpha = 0.023\frac{c_{pb}^{1/3}k_b^{2/3}}{\mu_b^{0.3267}\mu_w^{0.14}}\frac{G^{0.8}}{d_i^{0.2}}. \tag{5.109}$$

Note that the specific heat varies only slightly when moving from one type of mineral oil to another. Thermal conductivity does not vary in any substantial way from one mineral oil to another, even though it varies more in comparison with the specific heat. The dynamic viscosity, on the other hand, may vary considerably depending on the mineral oil.

Therefore, it is impossible to simplify (5.109) with an equation equally applicable to all mineral oils and that considers bulk temperature only, yet taking into account the intervention of wall temperature.

It can be done, though, if we highlight viscosity. This leads to the following equation applicable by first approximation to any mineral oil:

$$\alpha = \left[7.15 + 3.81\frac{t_b}{1000} - 5.93\left(\frac{t_b}{1000}\right)^2\right]\frac{G^{0.8}}{100\mu_b^{0.3267}\mu_w^{0.14}d_i^{0.2}}. \tag{5.110}$$

Equation (5.110) is applicable for $t_b = 20 - 320°\mathrm{C}$.

A comparison between (5.106) and (5.110) is based on the assumption that: $G = 2000\ \mathrm{kg/m^2s}$, $d_i = 50\,\mathrm{mm} = 0.05\mathrm{m}$, $t_b = 150°\mathrm{C}$, $t_w = 250°\mathrm{C}$.

Then: $c_{pb} = 2400$ J/kg K, $k_b = 0.121$ W/m K, $\mu_b = 3 \times 10^{-3}$ kg/ms, $\mu_w = 1 \times 10^{-3}$ kg/ms, Re = 33333, Pr = 59.5.

From (5.106) we obtain $\alpha = 1184.7\ \mathrm{W/m^2K}$; from (5.110) $\alpha = 1059.6\ \mathrm{W/m^2K}$ instead.

Clearly, in this case (5.110) produces a value of the heat transfer coefficient below the one obtained through (5.106).

5.4.4 Air

The Dittus and Bölter equation already discussed in relation to water (5.71) may be considered for air, as well. Here it is:

$$\mathrm{Nu} = 0.023\,\mathrm{Re}^{0.8}\,\mathrm{Pr}^{0.4} \tag{5.111}$$

It is best, though, to keep the results of other researchers into account, too, because we pointed out earlier that this equation, as much as (5.75) and (5.79), can be used only if the temperature difference between fluid and wall is modest. It is not our case, so it would not seem appropriate to refer to it.

Evans and Sarjant experimented with gas and air and then summarized it as follows:

$$\mathrm{Nu} = 0.020\,\mathrm{Re}^{0.8}. \tag{5.112}$$

Experiments on warm air by Zellnick and Churchill led to the following:

$$\mathrm{Nu} = 0.0231\,\mathrm{Re}^{0.8}\,\mathrm{Pr}^{1/3}. \tag{5.113}$$

Bulk temperature must be considered both in (5.112) and (5.113). A comparison between (5.112) and (5.111) for Pr=0.735 (mean value of air) shows a difference in the value of Nu equal to 1.7% which is still modest.

If we compare (5.113) with (5.111) the difference amounts to 2.5% instead.

Finally, the results obtained from experiments with gas by Desmon and Sams show good correspondence with (5.111) if the reference is to film temperature.

From (5.111) we obtain

$$\alpha = 0.023\mathrm{Re}^{0.8}\,\mathrm{Pr}^{0.4}\,\frac{k_f}{d_i}; \tag{5.114}$$

it may be written as follows:

$$\alpha = 0.023\frac{c_{pf}^{0.4}k_f^{0.6}}{\mu_f^{0.4}}\,\frac{G^{0.8}}{d_i^{0.2}}. \tag{5.115}$$

The values of c_{pf}, k_f, μ_f can be computed through equations (5.7), (5.8) and (5.9), and we establish that their value depends only on temperature. Based on the value of the three quantities, it is therefore possible to obtain the following approximate equation for the heat transfer coefficient of air:

$$\alpha = (3.087 + 0.186\frac{t_f}{100})\frac{G^{0.8}}{d_i^{0.2}} \tag{5.116}$$

with α in W/m^{2}K, G in kg/m^{2}s, d_i in m and t_f in °C.

Equation (5.116) is valid for $t_f = 0 - 300$°C. The errors caused by using (5.116) instead of (5.115) are irrelevant.

In relation to superheated steam we already mentioned that Schack developed a series of equations for the computation of α relative to different kinds of gas by deriving them from general Eq. (5.91). The equations are as follows

$$\alpha = f(t_b)\frac{V_0^{0.75}}{d_i^{0.25}}, \tag{5.117}$$

given that $f(t_b)$ is a function of bulk temperature t_b and V_0 velocity under normal conditions.

We prefer to refer to mass velocity G, as usual. Given that $G = V_0\rho_0$, ρ_0 stands for density under normal conditions, and this is well-known for many types of gas, we divide Schack's term $f(t_b)$ by $\rho^{0.75}$ and replace $V_0{}^{0.75}$ with $G^{0.75}$ obtaining the equations to be discussed later on.

Schack recommended to reduce the values of α obtained through his equations by 5% when we consider the turbulent current to have calmed down, i.e., when the phenomena at the ends of the tube are gone. We followed this criterion, too.

For air we obtain

$$\alpha = \left[3.23 + 0.182\frac{t_b}{100} - 0.006\left(\frac{t_b}{100}\right)^2\right]\frac{G^{0.75}}{d_i^{0.25}}. \tag{5.118}$$

As we said, if we consider a range of temperatures between 0 and 300°C the term within round brackets can be eliminated by modifying (5.118) like this:

$$\alpha = \left(3.23 + 0.173\frac{t_b}{100}\right)\frac{G^{0.75}}{d_i^{0.25}}. \tag{5.119}$$

A general comparison between (5.119) and (5.116) is impossible given the different reference temperature and the different exponents of G and d_i. Here are a few examples.

Assuming that $t_b = 50$°C, $t_f = 100$°C, $G = 16$ kg/m^{2}s and $d_i = 40$ mm, from (5.116) we get $\alpha = 57.26$ W/m^{2}K, whereas from (5.119) we get $\alpha = 59.33$ W/m^{2}K. The difference amounts to 3.6%.

Let us now assume that the temperatures do not change but that $G = 8$ kg/m^{2}s and $d_i = 60$ mm; from (5.116) we get $\alpha = 30.33$ W/m^{2}K, whereas from (5.119) we get $\alpha = 31.88$ W/m^{2}K. The difference amounts to 5.1%.

Equation (5.119) leads to values of α that are greater than those to be obtained through (5.116).

In the end, (5.116) is more cautionary and preferable in our opinion.

5.4.5 Different Kinds of Gas

At this point, we consider the different equations that can be derived from Schack's equations for different kinds of gas.

Carbon dioxide (CO_2)

$$\alpha = \left[2.529 + 0.464\frac{t_b}{100} - 0.0212\left(\frac{t_b}{100}\right)^2\right]\frac{G^{0.75}}{d_i^{0.25}}; \tag{5.120}$$

Hydrogen (H_2)

$$\alpha = \left(35.70 + 0.876\frac{t_b}{100}\right)\frac{G^{0.75}}{d_i^{0.25}}; \tag{5.121}$$

Methane (CH_4)

$$\alpha = \left[5.998 + 1.418\frac{t_b}{100} - 0.027\left(\frac{t_b}{100}\right)^2\right]\frac{G^{0.75}}{d_i^{0.25}}; \tag{5.122}$$

Ethylene (C_2H_4)

$$\alpha = \left[4.159 + 1.317\frac{t_b}{100} - 0.065\left(\frac{t_b}{100}\right)^2\right]\frac{G^{0.75}}{d_i^{0.25}}; \tag{5.123}$$

Acetylene (C_2H_2)

$$\alpha = \left[4.52 + 0.99\frac{t_b}{100} - 0.054\left(\frac{t_b}{100}\right)^2\right]\frac{G^{0.75}}{d_i^{0.25}}; \tag{5.124}$$

According to Schack's criterion it is possible to develop the equation for the computation of α for two additional kinds of gas:

Oxygen (O_2)

$$\alpha = \left[3.02 + 0.24\frac{t_b}{100} - 0.00803\left(\frac{t_b}{100}\right)^2\right]\frac{G^{0.75}}{d_i^{0.25}}; \tag{5.125}$$

Carbon monoxide (CO)

$$\alpha = \left[3.29 + 0.204\frac{t_b}{100} - 0.00545\left(\frac{t_b}{100}\right)^2\right]\frac{G^{0.75}}{d_i^{0.25}}. \tag{5.126}$$

Figure 5.5 shows the values of *f(t)* for the different kinds of gas we examined with the exception of hydrogen.

As far as flue gas, we refer to (5.114) as we do for air; consequently, in (5.115). The values of c_{pf}, k_f, μ_f can be derived from (5.17), (5.18) and (5.19). We establish that they depend on temperature and on mass moisture percentage of the gas. Therefore, it is possible to obtain the following approximate equation to be used to compute the term containing the three cited quantities:

$$K_g = 3.00 + 0.018m + (2.161 + 0.0117m)\frac{t_f}{1000} - (0.658 - 0.0257m)\left(\frac{t_f}{1000}\right)^2 \approx 0.023\frac{c_{pf}^{0.4}k_f^{0.6}}{\mu_f^{0.4}} \tag{5.127}$$

valid for $t_f = 50 - 1200°C$ and m=0–12%; the biggest error with this approximate equation amounts to about ± 0.6% which is acceptable.

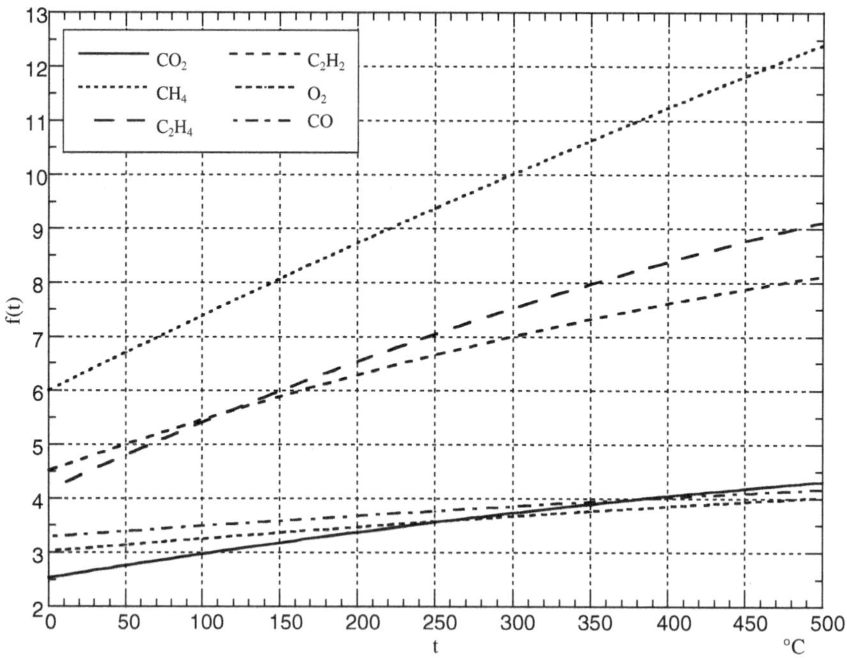

Fig. 5.5 Factor f(t) for several gases

Thus, the equation for the computation of the heat transfer coefficient of flue gas is as follows:

$$\alpha = K_g \frac{G^{0.8}}{d_i^{0.2}} \tag{5.128}$$

with α in W/m^2K, d_i in m and factor K_g computed through (5.127).

5.5 Heat Transfer in the Initial Section

When a fluid starts flowing in a tube the thickness of the boundary layer, where there is laminar motion and the heat transfer is by conduction, increases until it completely fills up the tube.

In this area the heat transfer coefficient is very high at the entrance. Then it slowly decreases along the length of the tube. The velocity constantly changes until it reaches eased turbulence state. At this point the heat transfer coefficient becomes constant, since there are no more variations along the tube.

The distance of the location where the eased turbulence occurs with respect to the entrance, depends up to a certain degree on the value of the number of Reynolds.

According to Latzko's theory, this distance indicated by L_c is obtained through the following equation, assuming that the tube has a flaring inlet:

$$L_c = 0.693 d_i \mathrm{Re}^{0.25}. \tag{5.129}$$

We establish, for instance, that with Re = 30000, L_c equals 9.1 diameters, whereas with Re = 60000 it equals 10.8 diameters. We establish that the number of Reynolds, according to this theory, impacts the value of L_c, yet only slightly.

If we consider a tube length under or equal to L_c, Latzko suggests an equation for the computation of the mean heat transfer coefficient in that section of the tube. By indicating this heat transfer coefficient with α_m, the one relative to eased turbulence with α_∞, and the length of the considered tube with L, the equation is as follows:

$$\alpha_m = 1.11 \alpha_\infty \left[\frac{\mathrm{Re}^{0.2}}{(L/d_i)^{0.8}} \right]^{0.275}. \tag{5.130}$$

For instance, if we consider Re = 40000 and $L/d_i = 8$ we obtain $\alpha_m = 1.26\alpha_\infty$.

If we consider a tube length greater then L_c instead, Latzko suggests the following equation:

$$\alpha_m = \alpha_\infty \left(1 + \frac{C}{L/d_i} \right) \tag{5.131}$$

where the dimensionless factor C is equal to

$$C = 0.144 \mathrm{Re}^{0.25}. \tag{5.132}$$

For instance, if we consider Re $= 40000$, as in the previous example, and we consider a length $L = 20d_i$ we obtain $C = 2.036$ and $\alpha_m = 1.10\alpha_\infty$.

The value of C is considerably influenced by the type of inlet. Figure 5.6 shows a few types of possible inlets, as illustrated by Schack. They are numbered and we will refer to them later on. The values of C obtained through experiments are shown in Table 5.4.

Note that if the inlet has a sharp edge, $C = 2.3$; this is a value we already encountered in (5.90).

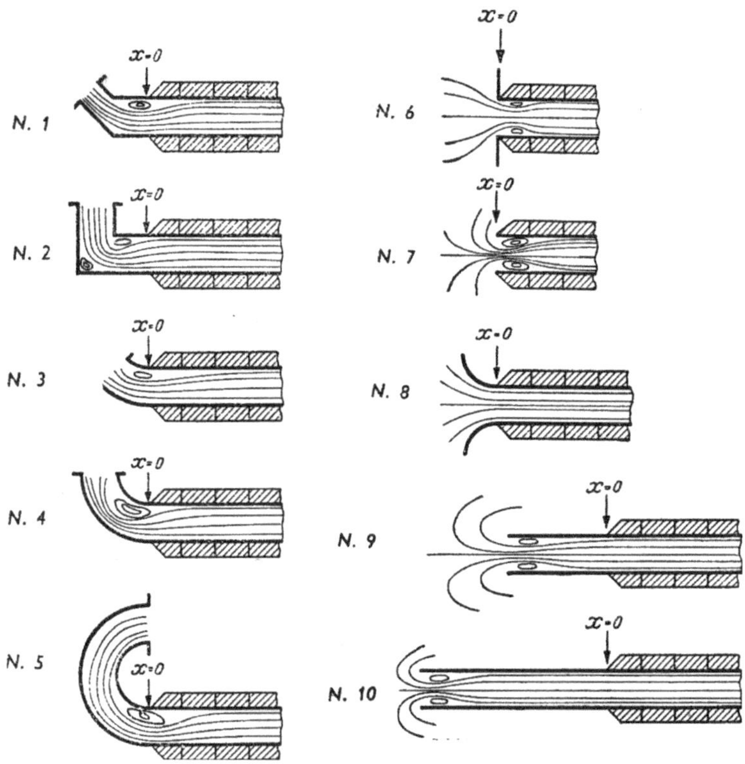

Fig. 5.6 Inlet type

Table 5.4 Values of C (Re=26000–56000 and $L/d_i \geq 5$)

Nr	Inlet type	C
1	Inlet at 45°	ab. 5
2	Inlet at 90°	ab. 7
8	Flaring inlet	1.4
9	Short calm section $L_c/d_i = 2.8$	ab. 3
10	Long calm section $L_c/d_i = 11.2$	1.4

Still, in reference to a sharp edge, the following equation is advisable, as well:

$$\alpha_m = \alpha_\infty \left[1 + \left(\frac{d_i}{L} \right)^{0.7} \right]. \tag{5.133}$$

Note that with $L/d_i = 20$ $\alpha_m = 1.123\alpha_\infty$, a value corresponding to the one previously found for Re = 40000. Variations of the value of Re are not followed by big variations.

Figure 5.7 shows experimental diagrams relative to a few types of inlet. The x-axis shows the ratio between the distance from the entrance and the inside diameter, the y-axis the values of α.

Another procedure consists of identifying the behavior of α starting from the inlet and deriving from this behavior the mean value of the heat transfer coefficient in a certain section.

Based on experimental data, Schack identified the following equation to compute α in the area adjacent to the inlet; by indicating the distance from the inlet with x and a factor that we will specify with c:

$$\alpha = \alpha_\infty \left(1 + \frac{cd_i}{x} \right). \tag{5.134}$$

From our perspective we believe it is useful to compute the mean value of α (α_m) derived as a result in the examined area, starting from (5.134); the phenomenon basically starts for $x = 0.5d_i$, and then we integrate (5.134) from $0.5d_i$ to L and obtain

$$\alpha_m = \frac{\alpha_\infty}{L - 0.5d_i} \int_{0.5d_i}^{L} \left(1 + \frac{cd_i}{x} \right) dx; \tag{5.135}$$

then

$$\alpha_m = \frac{\alpha_\infty}{L - 0.5d_i} \left(L - 0.5d_i + cd_i \log_e \frac{L}{0.5d_i} \right); \tag{5.136}$$

finally,

$$\alpha_m = \alpha_\infty \left(1 + \frac{cd_i}{L} \frac{\log_e \frac{2}{d_i/L}}{1 - \frac{d_i}{2L}} \right). \tag{5.137}$$

Assuming that

$$K = \log_e \frac{2}{d_i/L} \tag{5.138}$$

and neglecting the term $(1 - d_i/2L)$ in (5.137) because it is always close to unity we obtain

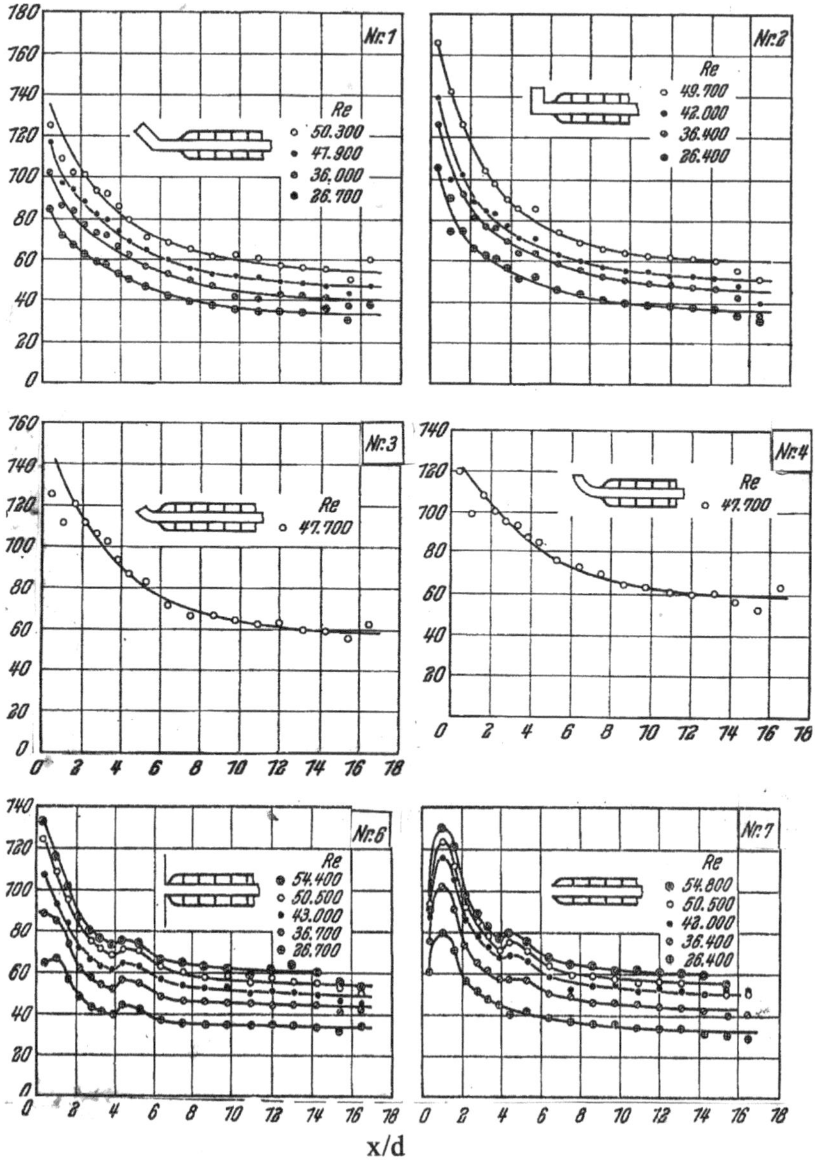

Fig. 5.7 Heat transfer coefficient at the inlet section

$$\alpha_m = \alpha_\infty \left(1 + \frac{Kcd_i}{L}\right). \tag{5.139}$$

Based on the experimental data (see Fig. 5.7) the phenomenon occurring at the inlet is almost completely over after about 10 diameters; this is also the result obtained through (5.129).

By first approximation we can adopt this length to obtain the value of the mean heat transfer coefficient α_m in the examined area.

With $L = 10d_i$ from (5.138) we obtain $K = 3$; coefficient c relative to the local heat transfer coefficient must be multiplied by 3 to obtain the mean heat transfer coefficient in the section of the tube that is 10 diameters long where the phenomenon occurs and mostly ends; with $L/d_i = 10$ (5.139) is simply reduced to the following:

$$\alpha_m = \alpha_\infty(1 + 0.3c). \tag{5.140}$$

Equation (5.140) makes it possible to compute the mean value of α in the section of the tube that is 10 diameters long, i.e., the most significant one.

At any rate, if we wanted to extend the examination to a section of the tube with a length of a random number of diameters, generally we can write that

$$\alpha_m = \alpha_\infty\,(1 + Ac)\,, \tag{5.141}$$

where A is a function of the ratio L/d_i, as established through (5.138) and (5.139); it is shown in Fig. 5.8.

The values of c obtained experimentally for various situations are included in Table 5.5.

Finally, note that contrary to (5.131) and (5.132) showing a higher ratio between the heat transfer coefficient at the inlet, and the one for eased turbulence for high values of Re, there are instances where this ratio is higher for low values of Re.

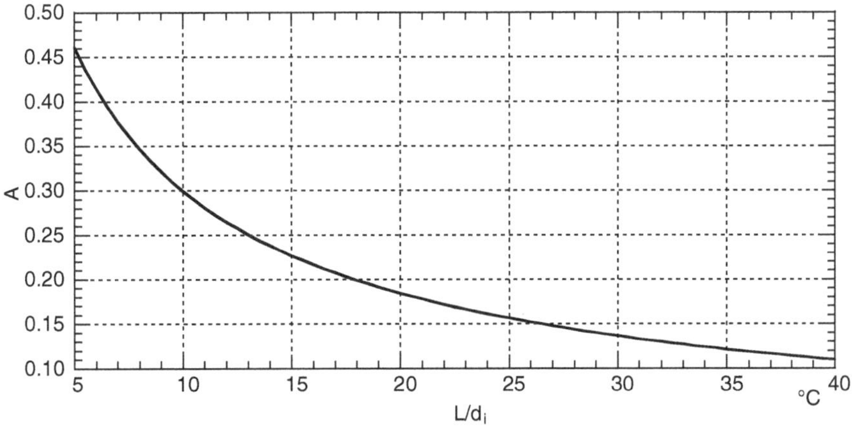

Fig. 5.8 Factor A

Table 5.5 Values of c for various inlet types (see Fig. 5.6)

Nr	Inlet type	c
1	Angle at 45°	1.8
2	Angle at 90°	2.0
3	Arch at 45°	1.7
4	Arch at 90°	1.3
5	Arch at 180°	2.1
6	Sharp corners with plate	0.9
7	Sharp corners without plate	1.2
8	Flaring inlet	0.4
9	Short calm section	1.0
10	Long calm section	0.7

5.6 Special Instances

5.6.1 Annular Interspace

If we consider two concentric cylinders with a fluid flowing inside the internal cylinder, a second fluid flowing in the annular interspace between the two cylinders and a third fluid outside the biggest cylinder, we want to determine the value of the heat transfer coefficient relative to the fluid in the interspace; it is relevant to both heat transfer towards the outside wall of the inside cylinder and heat transfer towards the inner wall of the outer cylinder.

To that extent, as for pressure drop, we must refer to the hydraulic diameter.

By indicating the outer diameter of the inner cylinder with d_{1o} and the inner diameter of the outer cylinder with d_{2i}, the area for the passage of fluid with A and the wet perimeter with P, the hydraulic diameter d_{hy}, as we know, is given by

$$d_{hy} = \frac{4A}{P} = \frac{\pi\left(d_{2i}^2 - d_{1o}^2\right)}{\pi\left(d_{2i} - d_{1o}\right)} = d_{2i} - d_{1o}. \tag{5.142}$$

As far as the heat transfer to the inner surface of the outer tube, we generally use the following equation:

$$\mathrm{Nu} = 0.023\phi \mathrm{Re}^{0.8}\,\mathrm{Pr}^{1/3} \tag{5.143}$$

with reference to bulk temperature. Of course, in both Nusselt's and Reynolds' number we must consider the hydraulic diameter. Factor ϕ is as corrective factor recommended by Monrad and Pelton to be assumed as follows:

$$\phi = 1.17\left(\frac{\mu_b}{\mu_w}\right)^{0.14} \tag{5.144}$$

where μ_b is the dynamic viscosity at bulk temperature, while μ_w is the viscosity at wall temperature.

In fact, considering various research fonts sources, the coefficient 1.17 can be eliminated, and (5.143) is reduced to

$$\mathrm{Nu} = 0.023\mathrm{Re}^{0.8}\,\mathrm{Pr}^{1/3}\left(\frac{\mu_b}{\mu_w}\right)^{0.14} \tag{5.145}$$

This is Eq. (5.79) that was already considered at the time.

If the fluid in question is water we may also write that

$$\alpha = K_{w3}\frac{G^{0.8}}{d_{id}^{0.2}}\left(\frac{\mu_b}{\mu_w}\right)^{0.14} \tag{5.146}$$

with K_{w3} derived from (5.81).

It is still possible to use (5.143) for the heat transfer to the outer surface of the inner tube; as far as ϕ, Wiegard suggested

$$\phi = \left(\frac{d_{2i}}{d_{1o}}\right)^{0.45} \tag{5.147}$$

valid for $d_{2i}\,/\,d_{1o}$ ranging from 1 to 10.

Research on both water and air has confirmed that even in this case it is correct to set $\phi = (\mu_b/\mu_w)^{0.14}$; this means that (5.145) and, in the case of water, even (5.146) may be used for heat transfer to the outer wall of the inner tube.

5.6.2 Plane Wall

For plane walls the computation of the heat transfer coefficient is essentially of interest for the heat transfer from the wall to the air licking it, as long as it is not by natural convection, as discussed earlier, but by forced convection instead; this situation takes place when the wall is licked by wind.

If Nusselt's and Reynolds' numbers are used, as customary, it is important to keep in mind that the characteristic dimension (for tubes it is the diameter) in this case is the length of the examined wall.

If the number of Reynolds is greater than 500000, it is possible to use the following equation resembling (5.75), i.e.,

$$\mathrm{Nu} = 0.036\mathrm{Re}^{0.8}\,\mathrm{Pr}^{1/3}\,\frac{c_{pb}}{c_{pf}} \tag{5.148}$$

where the values of the different quantities must be assumed with reference to the film temperature, and where c_{pb} and c_{pf} represent the isobaric specific heat of air under atmospheric pressure at room temperature and at film temperature.

Note that with a film temperature of 60°C, a wall length of 1 m Re = 500000 is reached with a wind velocity of 9.35 m/s which is equal to 33 km/h. This is a modest velocity that can certainly be increased. Note that greater wall lengths would result in even greater values of Re. Experimentation leading to (5.148), though, is on plates of modest dimensions. The use of this equation therefore requires caution when the dimensions of the wall are huge.

With Re = 500000 $t_b = 20°\text{C}$ and $t_f = 60°\text{C}$ from (5.148) we obtain Nu = 1174 and $\alpha = 32\ \text{W/m}^2\text{K}$.

Juerges' experiments lead him to suggest the use of the following equations. They refer to the entire range of velocity and consider the condition of the surface having a considerable influence on the value of the heat transfer coefficient.

If V indicates the wind velocity the equations are as follows:

For $V \leq 5$ m/s

Smooth surface

$$\alpha = 5.47 + 3.95\,V \tag{5.149}$$

Rough surface

$$\alpha = 6.16 + 4.19\,V \tag{5.150}$$

For $V \geq 5$ m/s

Smooth surface

$$\alpha = 7.12\,V^{0.78} \tag{5.151}$$

Rough surface

$$\alpha = 7.52\,V^{0.78} \tag{5.152}$$

where α is in $\text{W/m}^2\text{K}$.

The experimentation leading to the equations above was done with air at 20°C. If temperature t is different, the velocity to be introduced is the real one multiplied by the ratio 293/(273+ t).

We see that with $V = 9.35$ m/s from (5.152) we obtain $\alpha = 43\ \text{W/m}^2\text{K}$.

This value is considerably greater than the one computer earlier with (5.148). A simple example shows that the equations above lead to overestimated values of the heat transfer coefficient, as other authors already pointed out.

Interestingly, the wall length is not included, even though it is part of (5.148). In fact, if α is explicit the heat transfer coefficient becomes inversely proportional to $L^{0.2}$.

Longer surfaces compared to the ones used in experiments, where the range of temperatures tends to build itself or exists already, as is the case in long tubes, the heat transfer coefficient must have a lower value.

Based on the fact that experimentation focused on plates of 0.5 m of length, one would instinctively want to introduce a corrective factor given by $(0.5/\,L)^{0.2}$. In that case instead of $\alpha = 43$ W/m^{2}K, we would get $\alpha = 37.4$ W/m^{2}K, i.e., a result close to the 32 W/m^{2}K obtained with (5.148).

Probably this may be the right way for wind with velocity parallel to the wall, as we hypothesized so far.

Nonetheless, given their simplicity and the possibility to consider any wind velocity, we believe that these equations, regardless of the outlined uncertainties, are preferable to (5.148) and can be adopted as they are. This standpoint leads to the following consideration.

Until now we assumed that the wind moves parallel to the wall. In fact, this situation can be seen as exceptional because the velocity of the wind is more likely to have a perpendicular component to the wall, or even that the wind hits the wall from upfront.

In that case, the value of the heat transfer coefficient increases considerably. The values obtained from (5.149), (5.150), (5.151) and (5.152) become even cautionary. It is impossible to provide precise information about that increase. The heat transfer coefficient may even be 3–4 times greater than the values relative to parallel wind when it hits from upfront.

In addition, the heat transfer to the environment does not exclusively depend on the convective motion of the wind but also on wall radiation towards the environment, and this cannot be ignored. It will be discussed in more detail later on.

5.7 Laminar Motion in the Tubes

We already pointed out the fact that laminar motion in the tubes is to be considered exceptional because the motion is usually turbulent.

The velocities in question are such that the number of Reynolds is always greater than 3000 even in the case of highly viscous fluids.

Still, given that laminar motion cannot be ruled out completely, we provide a few indications as to how to proceed.

Nusselt developed a theory in this field leading to a complicated equation to show that the heat transfer coefficient greatly decreases in the first section of the tube, before it stabilizes at a minimum value once it reaches a certain distance from the inlet.

This distance, according to Nusselt, is given by

$$L = 0.2\frac{c_{pv} V_m r_i^2}{k}. \tag{5.153}$$

In (5.153) L represents this distance in m, c_{pv} is the specific heat but referred to the volume in J/m^{3}K, V_m is the mean velocity in the section of the tube in m/s, r_{i} is the radius of the tube in m, and k is the thermal conductivity in W/m K. Equation (5.153) can also be written in a way more customary to us:

$$L = 0.05\frac{c_p \rho V_m d_i^2}{k} \tag{5.154}$$

where c_p is the specific heat under constant pressure in J/kg K, ρ is the density in kg/m^3 and d_i is the inside diameter.

For instance, if we consider water at 1 bar and 50°C flowing in a tube with a diameter of 50 mm at a velocity of 0.02 m/s, the motion is certainly laminar because the number of Reynolds is equal to 1800.

Under these conditions, from (5.154) we obtain $L = 16$m; it equals 320 diameters.

Note that this length of the tube is necessary to reach the final minimum value of α only if the fluid has a uniform distribution of temperature at the entrance. Otherwise, and this is more likely, the minimum value of α is reached immediately regardless of phenomena taking place at the inlet, even though they are located in a limited section.

Nusselt found the following simple relation for the number carrying his name once the current was stabilized:

$$\text{Nu} = 5.15\,; \tag{5.155}$$

then

$$\alpha = 5.15\frac{k}{d_i}. \tag{5.156}$$

With regard to the previous example it would be $\alpha = 66.2$ W/m^2K.

If we consider a vertical tube equation (5.156) which ignores ascensional phenomena, as well as disturbances at the inlet, leads to values of α that are too small.

Colburn developed an equation to take them into account that includes both the number of Grashof, characteristic of natural convection, and the ratio d_i/L between the inner diameter of the tube and its length.

It is as follows:

$$\alpha = 1.65\frac{k}{d_i}\left(\frac{Qc_{pv}}{kL}\varphi\right)^{1/3}. \tag{5.157}$$

In (5.157) Q is the volumetric flow rate in m^3/s of the fluid (=VA), given that A is the area of the passage section ($= \pi d_i^2/4$); c_{pv} stands for the specific heat referred to the volume in J/m^3K, and it is equal to $c_p\rho$, given that c_p is the specific heat in J/kg K and ρ the density in kg/m^3; k stands for thermal conductivity in W/m K; φ is a function of the number of Grashof and it is equal to:

$$\varphi = 1 + 0.015\text{Gr}^{1/3}. \tag{5.158}$$

We see that (5.157) may be written as follows:

$$\alpha = 1.65\frac{k}{d_i}\left(\frac{A}{d_i L}\frac{V\rho d_i}{\mu}\frac{c_p\mu}{k}\varphi\right)^{1/3}; \qquad (5.159)$$

or like this:

$$\alpha = 1.52\frac{k}{d_i}\left(\frac{d_i}{L}\mathrm{Re}\,\mathrm{Pr}\,\varphi\right)^{1/3}; \qquad (5.160)$$

or even as follows:

$$\mathrm{Nu} = 1.52\left(\frac{d_i}{L}\mathrm{Re}\,\mathrm{Pr}\,\varphi\right)^{1/3}. \qquad (5.161)$$

Both (5.160) and (5.161) show the dependency of α from Gr and from ratio d_i/L.

If we consider our previous example and assume that the difference in temperature between the wall of the tube and water equals 50°C, that $L = 5$ m while $\beta=4.94\times10^{-4}$ 1/K, the number of Grashof is 10^7 , and that leads to $\varphi=4.23$; as you see, the impact of the number of Grashof is considerable.

Based on (5.161) we obtain Nu = 9.82 and finally α=126 W/m^{2}K; this value is almost double the one obtained with (5.156).

In terms of the diameter of the tube and its length, a doubling of the diameter for L being constant, considerably increases the value of φ, whereas the values of Re and d_i/L double. The number of Nusselt almost doubles and the value of α decreases very little, in clear contrast with (5.156) where d_i is crucial. Note that in this example there would be no more laminar motion.

A doubling of L leads to a decrease of α of about 20% instead.

The example shows what tests have pointed out, i.e., on one hand, the good correspondence of experimental data with (5.160), on the other hand, the fact that (5.156) cannot be used because it leads to unrealistic values of α. Equation (5.157) or its derivation (5.160) can definitely be applied.

5.8 Forced Convection Outside a Tube Bank

5.8.1 Introduction

Industrial processes are frequently characterized by the presence of tube banks hit by gas on the outside.

Generally, the gas hits the bank transversally, i.e., its velocity is perpendicular to the axis of the tubes.

The gas moving in parallel to the tubes represents a structural solution, generally leading to a reduction in heat transfer, and for this reason it is not recommended.

There are no difficulties, though, in computing the heat transfer coefficient even in this case. In fact, the gas moves through channels consisting of the spaces between the tubes. One must calculate the area of passage section A and the wet perimeter P (Fig. 5.9). Then, based on these one must calculate the hydraulic diameter given by

$$d_{hy} = \frac{4A}{P} \tag{5.162}$$

Based on the value of d_{hy} it is possible to determine the heat transfer coefficient according to the criteria discussed in Sect. 5.4, provided this done in reference to the film temperature.

Generally, the tube bank is hit transversally (Figs. 5.10 and 5.11).

The conditions of the current and the dependance of the heat transfer coefficient on the values of significant quantities are, of course, completely different from those in the case of motion inside the tubes.

A discharge vortex forms behind every tube and hits the following tube. In other words, there is vertical motion getting increasingly stronger as the gas hits the tubes after that. Therefore, the number of hit tubes impacts the mean heat transfer coefficient in the bank, as well as the amount of heat transfer. As we shall see, its influence stops if the number of tube rows exceeds 10.

The number of Reynolds plays a fundamental role on the value of the heat transfer coefficient, but its value impacts even the characteristics of vortexes. Thus, this process influences even the value of α.

Finally, the arrangement of the tubes is quite important, specifically the type of arrangement, either in-line tubes or staggered tubes (see Figs. 5.10 and 5.11). Moreover, even the ratio between transversal and longitudinal pitches and the diameter influence the value of α.

Much research was done on this topic and the results are mostly in agreement.

In 1933 Colburn suggested the following equation for in-line tubes (Fig. 5.10)

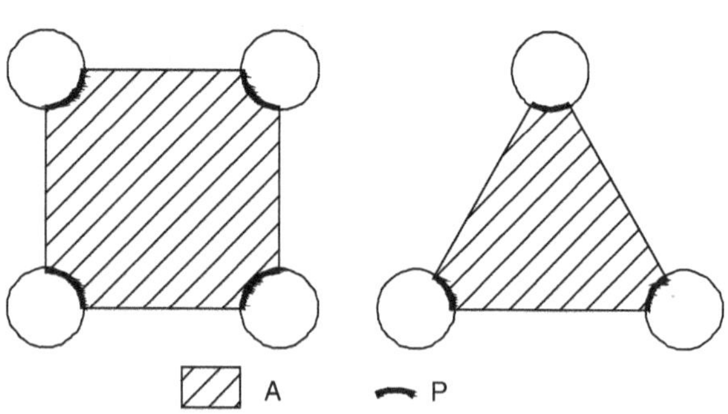

Fig. 5.9

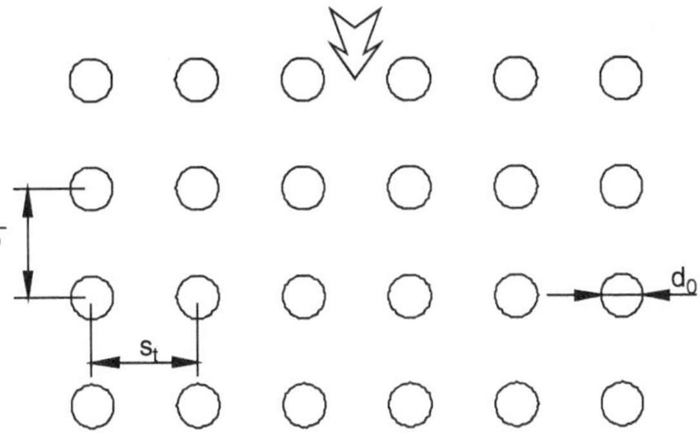

Fig. 5.10 Arrangement with in line tubes

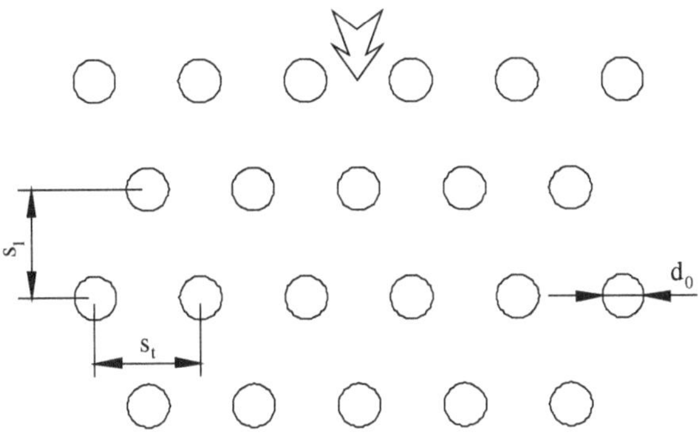

Fig. 5.11 Arrangement with staggered tubes

$$\mathrm{Nu} = 0.26\mathrm{Re}^{0.6}\,\mathrm{Pr}^{1/3}\,, \tag{5.163}$$

and this for staggered tubes (Fig. 5.11)

$$\mathrm{Nu} = 0.33\mathrm{Re}^{0.6}\,\mathrm{Pr}^{1/3}\,. \tag{5.164}$$

Equations (5.163) and (5.164) are in reference to film temperature.

Note that in both (5.163) and (5.164) the different type of arrangement yields a different value of the fixed coefficient, but subsequent research proved this to be simplistic.

Fundamental research was done later on by Pierson and Huge at Babcock and Wilcox.

Based on the results of this research, and after comparison with the results of other researchers, Grimison developed the following equation which is valid for a number of at least 10 rows crossed by gas:

$$\mathrm{Nu} = 0.284 f_a \mathrm{Re}^{0.61}\,. \tag{5.165}$$

The reference is to film temperature.

In (5.165) f_a is an arrangement factor the value of which depends on Re, the ratio between the transverse pitch s_t and the outside diameter of the tubes d_o, and the ratio between longitudinal pitch s_l and the diameter (Figs. 5.10 and 5.11). In addition, it depends on the type of arrangement (in the case of in-line and staggered tubes). Equation (5.165) is modified as follows to factor in the influence of the number of rows (when below 10):

$$\mathrm{Nu} = 0.284 f_d f_a \mathrm{Re}^{0.61}, \tag{5.166}$$

where f_d does, in fact, take this influence into account; the value of depth factor f_d can be taken from Table 5.6.

The values of f_a can be obtained from Figs. 5.12 and 5.13.

Grimison's equation responds quite well to the reality of the phenomenon.

In any case, note that Grimison's equation does not include Prandtl's number in contrast to Colburn's equations (5.163) and (5.164), and to the majority of equations in Sect. 5.4.

This number is included in an equation suggested later on by Babcock and Wilcox, i.e.,

$$\mathrm{Nu} = 0.287 f_d f_a \mathrm{Re}^{0.61}\,\mathrm{Pr}^{1/3} \tag{5.167}$$

in reference to film temperature like (5.165) and (5.166). Equation (5.167) represents a refinement compared to Grimison's equation, and we believe it to be preferable.

Highlighting the heat transfer coefficient leads from (5.166) to the following:

$$\alpha = 0.284 f_d f_a \mathrm{Re}^{0.61} \frac{k}{d_o}. \tag{5.168}$$

Similarly, from (5.167) we obtain the following:

$$\alpha = 0.287 f_d f_a \mathrm{Re}^{0.61}\,\mathrm{Pr}^{1/3} \frac{k}{d_o}. \tag{5.169}$$

The value for the number of Prandtl varies for air, between 0 and 300°C, from 0.74 to 0.72, so that the difference between the values of the two equations amounts to 9–10%.

For flue gas, between 50 and 1200°C, the difference is equal to 7–10%.

Table 5.6 Depth factor f_d

Number of rows	1	2	3	4	5	6	7	8	9
f_d	0.70	0.82	0.87	0.91	0.93	0.95	0.97	0.98	0.99

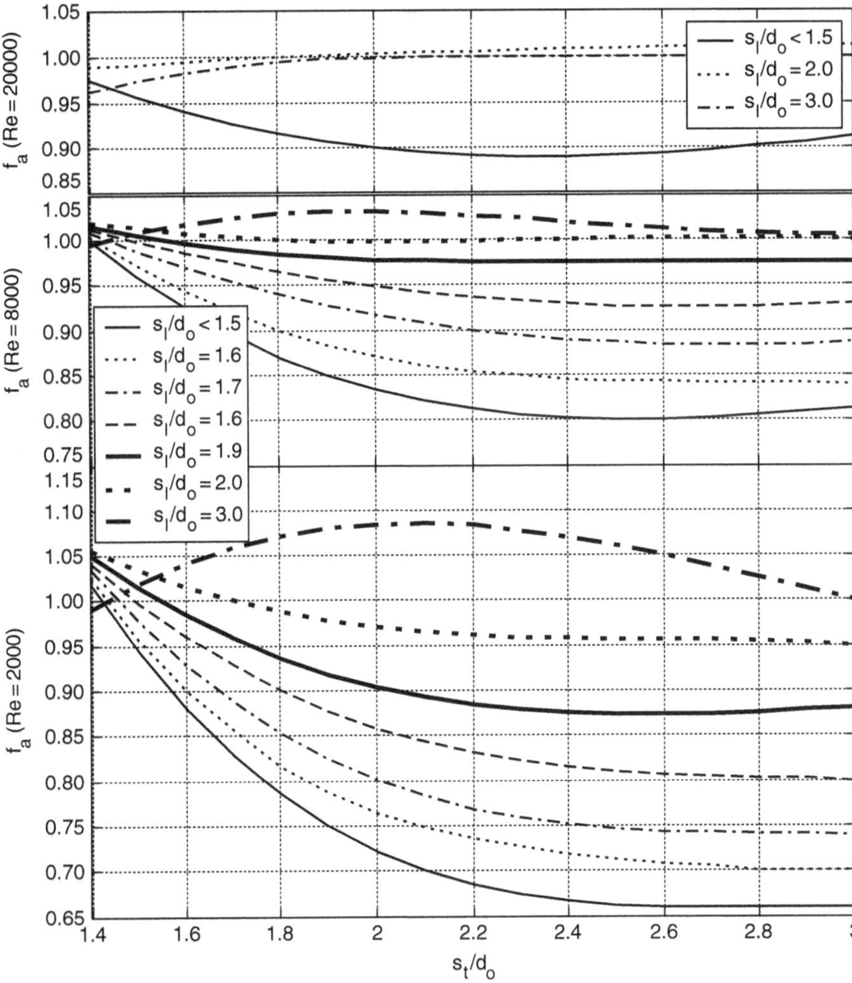

Fig. 5.12 Arrangement factor f_a for in line tubes

Equation (5.169) can be written as follows:

$$\alpha = 0.287 f_d f_a \frac{G^{0.61}}{d_o^{0.39}} \frac{\left(c_p k^2\right)^{1/3}}{\mu^{0.2767}}; \tag{5.170}$$

or

$$\alpha = K f_d f_a \frac{G^{0.61}}{d_0^{0.39}} \tag{5.171}$$

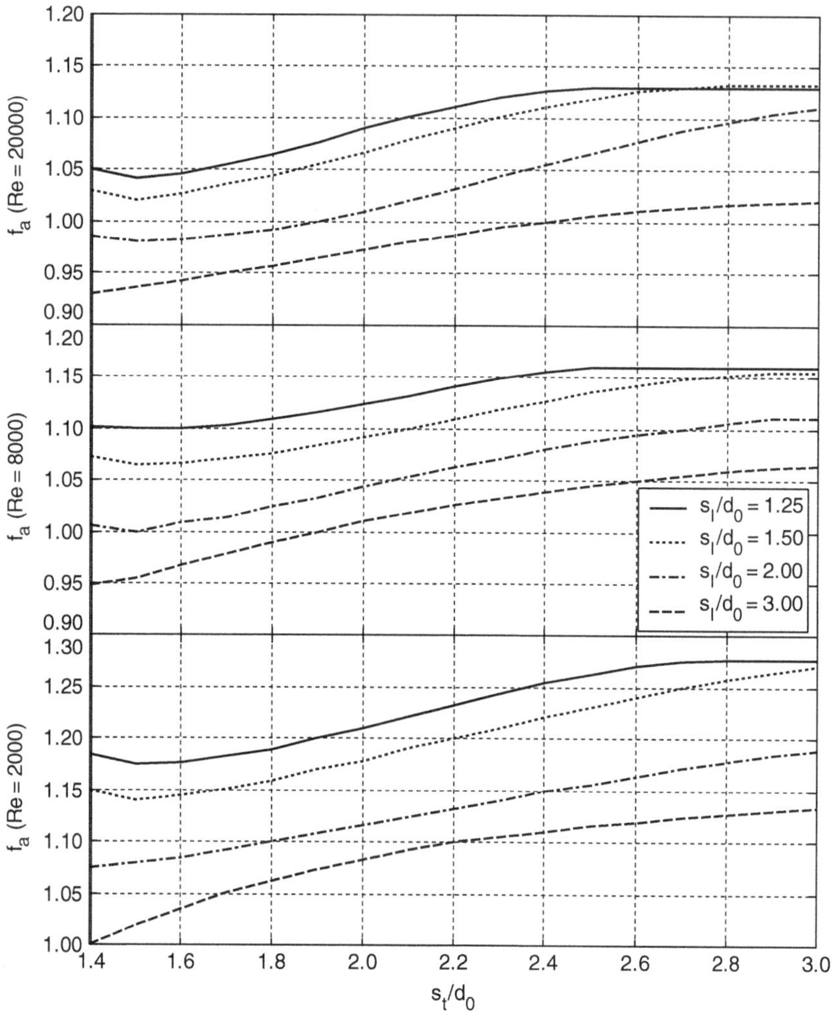

Fig. 5.13 Arrangement factor f_a for staggered tubes

with

$$K = 0.287\frac{(c_p k^2)^{1/3}}{\mu^{0.2767}}. \tag{5.172}$$

Factor K contains only physical characteristics of the fluid at hand, therefore its value depends on the type of fluid and on its temperature (the assumption must be film temperature).

Thus, it is possible to examine different types of gas and indicate for each one the function of t_f representing K. Then 5.171) is valid for all types of examined gas.

5.8.2 Air

For air the values of c_c, k, μ are obtained based on (5.7), (5.8) and (5.9).

An examination of the values leads to the following approximate equation of α

$$\alpha = \left[4.884 + 0.545\frac{t_f}{100} - 0.012\left(\frac{t_f}{100}\right)^2\right] f_d f_a \frac{G^{0.61}}{d_o^{0.39}}. \tag{5.173}$$

Equation (5.173) is valid for $t_f = 0\text{-}300°C$. The errors deriving from using (5.173) instead of (5.170) are irrelevant.

Note that for $t_f = 0°C$ the function of t_f included in (5.173) is 4.884, whereas for $t_f = 300°C$ it is 6.411. It increases with temperature and, mass velocity being equal, the value of α increases.

This characteristic is common to all types of gas which transfer more heat to the tubes, or receive more heat from them if temperature is high.

The function of t_f for air is shown in Fig. 5.14.

5.8.3 Various Types of Gas and Superheated Steam

The following equations for the computation of α are obtained after examination of the different types of gas.

Carbon dioxide (CO_2)

$$\alpha = \left[3.427 + 1.188\frac{t_f}{100} - 0.0836\left(\frac{t_f}{100}\right)^2\right] f_d f_a \frac{G^{0.61}}{d_o^{0.39}}; \tag{5.174}$$

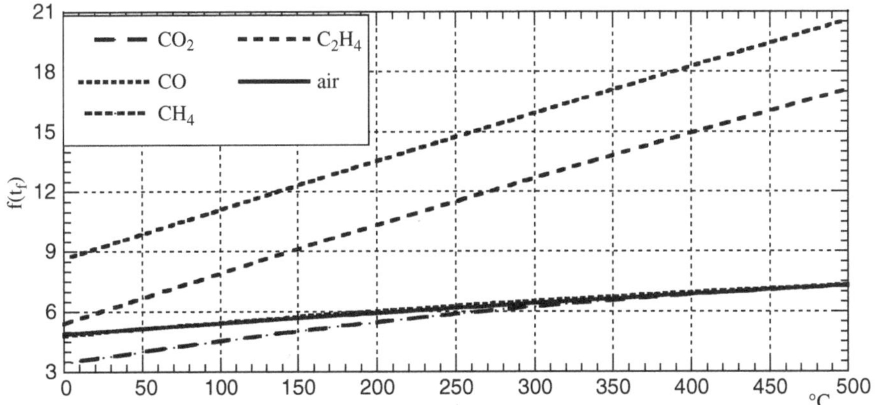

Fig. 5.14 Function $f(t_f)$ for air and several gases

Carbon monoxide (CO)

$$\alpha = \left[4.808 + 0.675\frac{t_f}{100} - 0.0344\left(\frac{t_f}{100}\right)^2\right] f_d f_a \frac{G^{0.61}}{d_o^{0.39}}; \tag{5.175}$$

Methane (CH_4)

$$\alpha = \left[8.64 + 2.48\frac{t_f}{100} - 0.018\left(\frac{t_f}{100}\right)^2\right] f_d f_a \frac{G^{0.61}}{d_o^{0.39}}; \tag{5.176}$$

Ethylene (C_2H_4)

$$\alpha = \left[5.412 + 2.54\frac{t_f}{100} - 0.04\left(\frac{t_f}{100}\right)^2\right] f_d f_a \frac{G^{0.61}}{d_o^{0.39}}; \tag{5.177}$$

Hydrogen (H_2)

$$\alpha = \left[53.83 + 4.93\frac{t_f}{100} - 0.218\left(\frac{t_f}{100}\right)^2\right] f_d f_a \frac{G^{0.61}}{d_o^{0.39}}. \tag{5.178}$$

Equations (5.174), (5.175), (5.176), (5.177) and (5.178) are valid for t_f=0-500°C Their function $f(t_f)$, with the exception of hydrogen, is shown in Fig. 5.14.

As far as flue gas, remember that the values of c_p, k, μ can be obtained through (5.17), (5.18) and (5.19). Clearly, the value of the three quantities depends on the temperature and the mass moisture percentage of the gas m. For this reason, even the value of factor K in (5.171) depends on temperature and moisture.

By indicating it with K_g we obtain

$$K_g = 4.752 + 0.0204m + (5.553 + 0.0294m)\frac{t_f}{1000} - (1.614 - 0.0479m)\left(\frac{t_f}{1000}\right)^2. \tag{5.179}$$

Once K_g is known, the value of the heat transfer coefficient can be obtained through (5.171).

(5.179) is valid for t_f = 50-1200°C and for $m = 0 - 12\%$.

As far as the superheated steam, the value of K obviously depends on temperature and pressure; the values of the quantities involved lead to the following equation for K, in this case indicated by K_s:

$$K_s = 7.94 + 0.142p - (3.77 + 0.236p)\frac{t_f}{1000} + (18.98 + 0.029p)\left(\frac{t_f}{1000}\right)^2 \tag{5.180}$$

where p represents pressure in bar.

Equation (5.180) is valid for $t_f = 180$–$550°C$ and for $p = 10$–100 bar.

After calculation of K_s through (5.171) it is possible to determine the value of α. With low pressure levels the value of K_s increases with temperature, as shown in Fig. 5.15, whereas it decreases with high pressure levels; at 50 bar its value is almost constant.

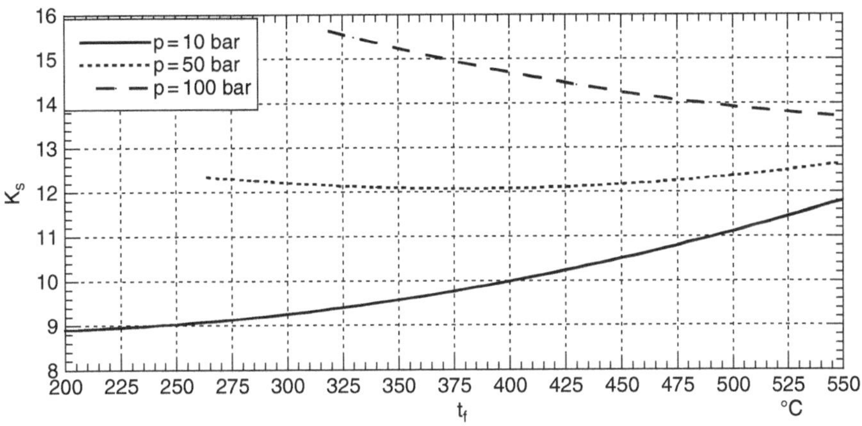

Fig. 5.15 Factor K_s for superheated stem at various pressures

Note that at high temperatures the value of K_s tends to be independent from pressure. This is due to the fact that at high temperature the values of c_p and k vary little with pressure, in contrast with what happens at lower temperatures.

5.9 Comparison Between In-Line and Staggered Arrangement

At this point, a few considerations about the two kinds of arrangement of the tubes in relation to heat transfer are overdue.

Let us consider two banks, one with in-line arrangement and one with staggered arrangement, and tubes with the same diameter and equal values of the transverse and longitudinal pitch that are hit by the same fluid.

Under these conditions, the velocity of the fluid is identical, so that both banks have the same value of Reynolds' number.

Based on (8.220) the value of α differentiates itself in both instances only as far as the different value of f_a referred to the same values of s_t/d_o, s_l/d_o and Re for the two types of arrangement.

Figures 5.12 and 5.13 clearly show that in most cases the values of f_a are higher with the staggered arrangement. This is especially true for low values of Re.

This should lead to the conclusion that staggered arrangements are definitely preferable to in-line arrangements. Nonetheless, this would be a hasty conclusion because it would neglect other aspects of the phenomenon. In fact, as we shall see in Chap. 8, the pressure drop throughout the bank depends on a similar arrangement factor f_a as a function of the same parameters.

For the two banks considered earlier, even a pressure drop differentiates itself solely by the different value of this arrangement factor. A comparison between the values in both arrangements shows that the in-line arrangement is clearly preferable because the f_a values relative to pressure drops are lower, and that translates into less pressure drops.

Thus, even though the in-line arrangement yields a lower value of α, the pressure drop is smaller. This equals a less favourable situation as far as heat transfer and the cost of the bank (greater surface with equal heat transfer), but a much more favourable situation as far as the cost of the fan and the energy absorbed by it.

Therefore, a comparison between the two arrangements must consider both aspects of the phenomenon and factor in the plant cost (bank and fan), as well as running costs (energy absorbed by the fan). In other words, it is a question of choosing the solution carrying the best cost optimization.

The criteria to help evaluate the opportunity of one arrangement over the other lead to a comparison of the values of α, given equal pressure drops.

A complete process requires setting the other parameters while keeping in mind construction and runtime constraints.

The issue is clearly complex and requires different case by case decisions. The following constitutes potential comparison criteria.

Often the width of the space available to the tube bank is set in advance. Assessing this situation and assuming that the pressure drop is the same with both arrangements leads to the following. The transverse pitch can be smaller for in-line tubes given the cross-sectional area of fluid passage can be reduced, thus increasing the speed. This way the number of tubes increases per row and the number of rows decreases, leading to a more compact bank.

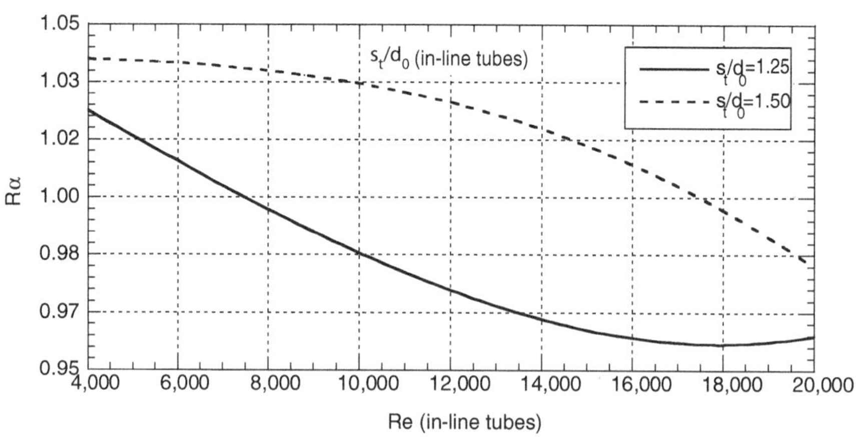

Fig. 5.16 Comparison between in-line and staggered tubes ($s_l/d_o = 1.5$)

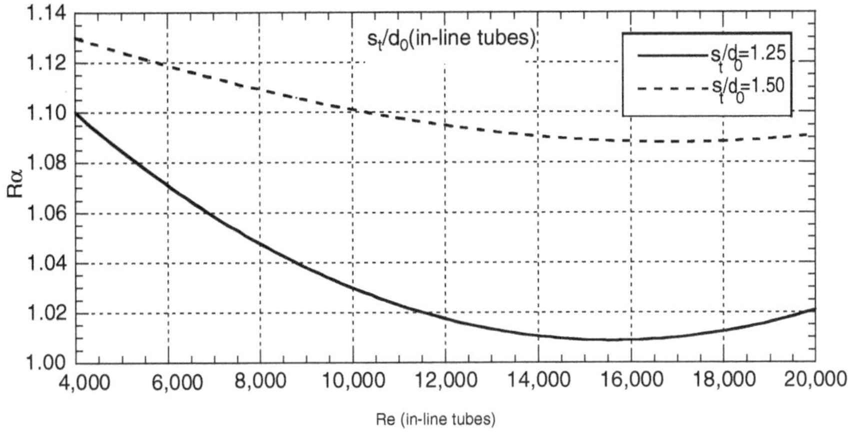

Fig. 5.17 Comparison between in-line and staggered tubes ($s_l/d_o = 2.0$)

At this point we consider the two values of the heat transfer coefficient and indicate the ratio between heat transfer coefficient with in-line and staggered tubes with $R\alpha$.

This results in the diagrams of Figs. 5.16, 5.17 and 5.18, where both Re and s_t/d_o refer to in-line tubes.

Clearly, the two arrangements are only slightly different for $s_l/d_0 = 1.5$, but the in-line arrangement is more favorable for values higher than s_l/d_0.

The analysis above is conducted in reference to one particular situation which does not rule out other possibilities, and is independent from specific construction and runtime requirements the designer may face. Nonetheless, and will all due reservations, it is possible to conclude that the in-line arrangement contrary to the simple analysis of f_a is at least to be considered very carefully to identify the best solution.

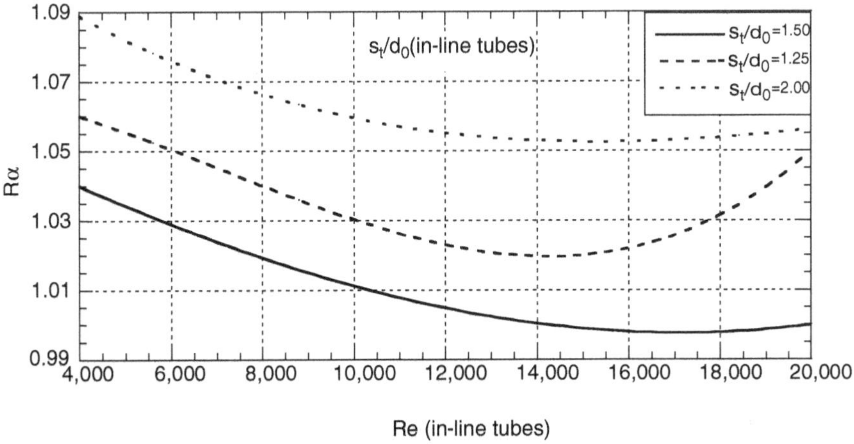

Fig. 5.18 Comparison between in-line and staggered tubes ($s_l/d_o = 3.0$)

5.10 Heat Transfer to a Single Tube

Equations about heat transfer to a single tube in horizontal position were studied by researchers who reached very similar conclusions.

Based on Reiher's experience:

$$\mathrm{Nu} = 0.35\mathrm{Re}^{0.56}. \tag{5.181}$$

Based on Hughes' experience:

$$\mathrm{Nu} = 0.326\mathrm{Re}^{0.555}. \tag{5.182}$$

Due to its cautionary approach and the way the research was conducted, we prefer the latter.

From (5.182) and bringing the exponent to 0.56:

$$\alpha = 0.326\frac{k}{d_o}\mathrm{Re}^{0.56}; \tag{5.183}$$

this can be written as follows:

$$\alpha = 0.326\frac{k}{\mu^{0.56}}\frac{G^{0.56}}{d_o^{0.44}}. \tag{5.184}$$

Considering air between 0 and 300°C, instead of (5.184) we can write that

$$\alpha = \left(3.54 + 0.433\frac{t_f}{100}\right)\frac{G^{0.56}}{d_o^{0.44}}. \tag{5.185}$$

For some types of gas between 0 and 500°C these are the corresponding equations:

Carbon dioxide (CO_2)

$$\alpha = \left[2.38 + 0.989\frac{t_f}{100} - 0.075\left(\frac{t_f}{100}\right)^2\right]\frac{G^{0.56}}{d_o^{0.44}}; \tag{5.186}$$

Carbon monoxide (CO)

$$\alpha = \left[3.425 + 0.585\frac{t_f}{100} - 0.032\left(\frac{t_f}{100}\right)^2\right]\frac{G^{0.56}}{d_o^{0.44}}; \tag{5.187}$$

Methane (CH_4)

$$\alpha = \left[6.143 + 1.934\frac{t_f}{100} - 0.0284\left(\frac{t_f}{100}\right)^2\right]\frac{G^{0.56}}{d_o^{0.44}}. \tag{5.188}$$

For flue gas between 50 and 1200°C:

$$\alpha = K_g \frac{G^{0.56}}{d_o^{0.44}} \tag{5.189}$$

with

$$K_g = 3.41 + 0.0076m + (4.426 + 0.03m)\frac{t_f}{1000} - (1.275 - 0.05m)\left(\frac{t_f}{1000}\right)^2 \tag{5.190}$$

where m stands for mass moisture percentage between 0 and 12%.

In (5.185), (5.186), (5.187), (5.188), (5.189) and (5.190) α is in W/m^{2}K and t_f is the film temperature in °C.

5.11 Heat Transfer to Finned Tubes

Finned tubes are widely deployed in industry.

Close finning with fins made of steel of constant thickness is possible only if the fluid is "clean" and there is no risk of corrosion. This is true for flue gas created by combustion of natural gas.

Cast iron finned tubes or steel tubes with cast iron finned muffs can also be used. In this case the fins have variable thickness from the bottom to the top.

Note that the use of finned tubes makes sense only if the heat transfer coefficient of the internal fluid is clearly higher than the one of the external fluid (for instance, water inside and flue gas outside). This rules out the use for air heaters because heat transfer coefficients are comparable. We are generally referring to tapered fins (Fig. 5.19) because the fins with constant thickness constitute a special case within this category.

Because of the thermal gradient along the fin and the following increase in temperature from the bottom to the top, the fin is hit by less heat flux compared to the one hitting the tube. An efficiency factor of the fin ideally reducing its surface is introduced to take this into account.

This is possible as follows.

With reference to Fig. 5.19 the generic thickness x_f of the fin is given by

$$x_f = x_{f2}\left[1 + \frac{\frac{x_{f1}}{x_{f2}} - 1}{1 - \frac{r_1}{r_2}}\left(1 - \frac{r}{r_2}\right)\right]. \tag{5.191}$$

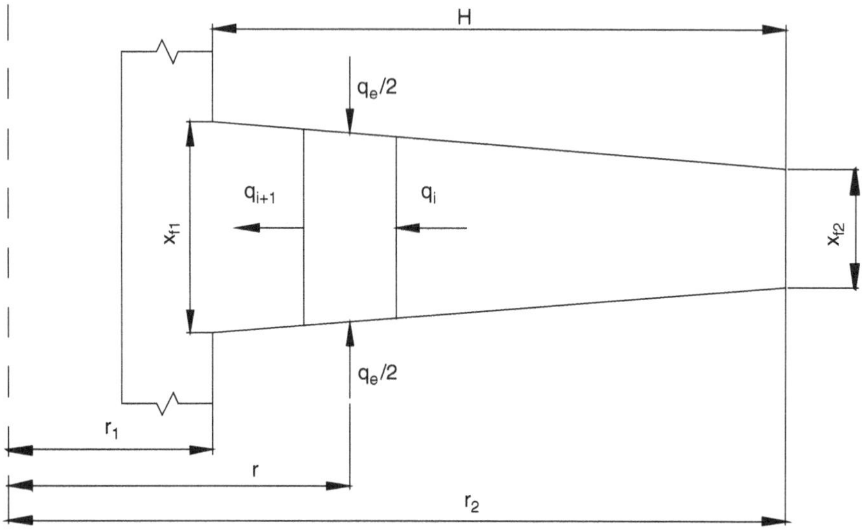

Fig. 5.19

We introduce factor ϑ given by:

$$\vartheta = \frac{\frac{x_{f1}}{x_{f2}} - 1}{1 - \frac{r_1}{r_2}}. \tag{5.192}$$

Therefore, the generic thickness of the fin is given by

$$x_f = x_{f2}\left[1 + \vartheta\left(1 - \frac{r}{r_2}\right)\right]. \tag{5.193}$$

Consider the ith section of radius r_i; the heat q_i crossing the section is given by

$$q_i = 2\pi r_i k x_{f2}\left[1 + \vartheta\left(1 - \frac{r}{r_2}\right)\right]\left(\frac{dt}{dr}\right)_i \tag{5.194}$$

where t is the generic temperature of the fin and k the thermal conductivity of the material. The temperature of the fluid transferring heat to the fin is t', while r_1 and r_2 are the radius at the bottom and the top of the fin. We assume that

$$\chi = \frac{r}{r_2} \tag{5.195}$$

$$\varphi = \frac{t' - t}{t' - t_2} \tag{5.196}$$

where t_2 as the temperature of the fin in position 2 (r=r_2).

Equation (5.194) can be written as follows.

$$q_i = -2\pi k x_{f2} \left[1 + \vartheta\left(1 - \chi_i\right)\right]\left(t' - t_2\right)\chi_i \left(\frac{d\varphi}{d\chi}\right)_i. \tag{5.197}$$

By analogy, if we consider the (i+1)-th section where its smaller radius differs from the radius r_i of Δr, we may write that

$$q_{i+1} = -2\pi k x_{f2} \left[1 + \vartheta\left(1 - \chi_{i+1}\right)\right]\left(t' - t_2\right)\chi_{i+1} \left(\frac{d\varphi}{d\chi}\right)_{i+1}. \tag{5.198}$$

On both circular crowns of width Δr that make up the sides of the fin, the external fluid transfers the heat q_e given by

$$q_e = 4\pi r_i \Delta r \alpha \left(t' - t_i\right) = 4\pi r_2^2 \alpha \left(t' - t_2\right)\chi_i \varphi_i \Delta\chi, \tag{5.199}$$

where α is the heat transfer coefficient of the external fluid.
Thermal balance requires that

$$q_{i+1} = q_i - q_e. \tag{5.200}$$

Then, through a series of steps we obtain

$$\left(\frac{d\varphi}{d\chi}\right)_{i+1} = \left\{\left[1 + \vartheta\left(1 - \chi_i\right)\right]\left(\frac{d\varphi}{d\chi}\right)_i - A\varphi_i\Delta\chi\right\}\frac{\chi_i}{\chi_{i+1}\left[1 + \vartheta\left(1 - \chi_{i+1}\right)\right]}. \tag{5.201}$$

given that

$$A = \frac{2\alpha r_2^2}{k x_{f2}}. \tag{5.202}$$

We introduce factor B given by

$$B = \sqrt{\frac{2\alpha H^2}{k x_{f2}}} \tag{5.203}$$

where H is the height of the fin.
Then

$$A = \left(\frac{B}{1 - \frac{r_1}{r_2}}\right)^2. \tag{5.204}$$

Note that at the top of the fin ($r = r_2$) we have $\chi = 1$ and $\varphi = 1$. In addition,

$$q_2 = -2\pi k x_{f2} \left(t' - t_2\right) \left(\frac{d\varphi}{d\chi}\right)_2 = 2\pi r_2 x_{f2} \alpha \left(t' - t_2\right), \tag{5.205}$$

given that q_2 is the heat transferred to the fin through the circular crown with a width equal to the thickness.

From (5.205) we obtain

$$\left(\frac{d\varphi}{d\chi}\right)_2 = -\frac{\alpha r_2}{k} = -C. \tag{5.206}$$

The sequence is as follows.

Presetting a very small value of $\Delta\chi$, and knowing the values of φ and $(d\varphi/d\chi)$ for $\chi = 1$ (top of the fin), we proceed towards its bottom by computing $(d\varphi\,/\,d\chi)_{i+1}$ through (5.201), the new value of $\chi_{i+1} = \chi_i - \Delta\chi$, the new value of $\varphi_{i+1} = \varphi_i - (d\varphi/d\chi)\Delta\chi$, and so on and so forth down to the bottom.

The heat q_1 exiting the fin that corresponds to the heat that the external fluid transfers to the fin, is equal to:

$$q_1 = -2\pi k x_{f1} \left(t' - t_2\right) \left(\frac{r_1}{r_2}\right) \left(\frac{d\varphi}{d\chi}\right)_1. \tag{5.207}$$

If the entire fin were at temperature t_1, i.e., the one at the bottom, the transferred heat q_0 would be equal to

$$q_0 = 2\pi \left(r_2^2 - r_1^2 + r_2 x_{f2}\right) \alpha \left(t' - t_1\right) \tag{5.208}$$

The efficiency factor of fin E_f, equal to the ratio q_1/q_o is therefore given by

$$E_f = \frac{-2\left(\frac{r_1}{r_2}\right)\left(\frac{d\varphi}{d\chi}\right)_1}{A\varphi_1 \frac{x_{f2}}{x_{f1}} \left[1 - \left(\frac{r_1}{r_2}\right)^2 + 2\frac{C}{A}\right]}. \tag{5.209}$$

We see that the value of E_f depends on r_2/r_1, B, C and the ratio x_{f1}/x_{f2}. As far as C, its influence is modest and we can conventionally assume that $C = 0.05$.

Tables 5.7, 5.8, 5.9 and 5.10 show the values of E_f for x_{f1}/x_{f2}=1, 1.5, 2 and 2.5, respectively.

A comparison of the values of E_f in the four Tables highlights the beneficial effect of the variability in thickness of the fins. In fact, the value of E_f increases with an increase of x_{f1}/x_{f2}.

Note, though, that the thickening of the heat flux lines in correspondence of the tube reduce the efficiency factor with respect to the above.

Table 5.7 Efficiency factor for fins (x_{f1}/x_{f2}=1)

B	r_2/r_1	E_f	B	r_2/r_1	E_f
0.20	1.20	0.979	0.80	1.20	0.813
	1.40	0.972		1.40	0.797
	1.60	0.965		1.60	0.784
	1.80	0.959		1.80	0.771
	2.00	0.954		2.00	0.760
0.40	1.20	0.940	1.00	1.20	0.741
	1.40	0.930		1.40	0.724
	1.60	0.922		1.60	0.708
	1.80	0.914		1.80	0.693
	2.00	0.907		2.00	0.680
0.60	1.20	0.881	1.20	1.20	0.672
	1.40	0.869		1.40	0.653
	1.60	0.858		1.60	0.636
	1.80	0.848		1.80	0.620
	2.00	0.839		2.00	0.606

Table 5.8 Efficiency factor for fins ($x_{f1}/x_{f2} = 1.5$)

B	r_2/r_1	E_f	B	r_2/r_1	E_f
0.20	1.20	0.984	0.80	1.20	0.855
	1.40	0.979		1.40	0.842
	1.60	0.973		1.60	0.831
	1.80	0.969		1.80	0.820
	2.00	0.964		2.00	0.811
0.40	1.20	0.955	1.00	1.20	0.795
	1.40	0.948		1.40	0.780
	1.60	0.941		1.60	0.766
	1.80	0.935		1.80	0.753
	2.00	0.930		2.00	0.742
0.60	1.20	0.910	1.20	1.20	0.734
	1.40	0.900		1.40	0.717
	1.60	0.891		1.60	0.701
	1.80	0.883		1.80	0.687
	2.00	0.876		2.00	0.675

Taking this phenomenon into account, we determine that the computation procedure shown below makes it possible to obtain the values of E_f that correspond well to reality for fins with constant thickness.

We have

$$W = B(1 + \frac{x_{f2}}{2H}) \tag{5.210}$$

where B, x_{f2} and H are previously considered quantities.

Then,

$$V = \tanh W/W; \tag{5.211}$$

$$Z = V\,(0.7 + 0.3\,V)\,. \tag{5.212}$$

Table 5.9 Efficiency factor for fins ($x_{f1}/x_{f2} = 2$)

B	r_2/r_1	E_f	B	r_2/r_1	E_f
0.20	1.20	0.987	0.80	1.20	0.881
	1.40	0.983		1.40	0.870
	1.60	0.978		1.60	0.860
	1.80	0.974		1.80	0.851
	2.00	0.971		2.00	0.843
0.40	1.20	0.964	1.00	1.20	0.829
	1.40	0.958		1.40	0.816
	1.60	0.952		1.60	0.804
	1.80	0.947		1.80	0.793
	2.00	0.943		2.00	0.783
0.60	1.20	0.927	1.20	1.20	0.775
	1.40	0.919		1.40	0.760
	1.60	0.911		1.60	0.746
	1.80	0.905		1.80	0.733
	2.00	0.898		2.00	0.722

Table 5.10 Efficiency factor for fins ($x_{f1}/x_{f2} = 2.5$)

B	r_2/r_1	E_f	B	r_2/r_1	E_f
0.20	1.20	0.989	0.80	1.20	0.899
	1.40	0.985		1.40	0.889
	1.60	0.982		1.60	0.881
	1.80	0.978		1.80	0.873
	2.00	0.975		2.00	0.866
0.40	1.20	0.970	1.00	1.20	0.853
	1.40	0.965		1.40	0.841
	1.60	0.960		1.60	0.830
	1.80	0.956		1.80	0.821
	2.00	0.952		2.00	0.812
0.60	1.20	0.938	1.20	1.20	0.805
	1.40	0.931		1.40	0.790
	1.60	0.925		1.60	0.778
	1.80	0.919		1.80	0.766
	2.00	0.914		2.00	0.756

Finally,

$$E_f = Z\left(1 + 0.02\,W^3\right)\left[0.45\,(Z-1)\log_e\left(\frac{d_0 + 2H}{d_0}\right) + 1\right]. \tag{5.213}$$

where d_0 is the external diameter of the tube without fins.

For fins with variable thickness, it is recommended to compute E_f based on (5.213) and to multiply the resulting value for the following corrective factor:

$$1 + 0.12\left(\frac{x_{f1}}{x_{f2}} - 1\right)W^{1.75}. \tag{5.214}$$

At this point, we need to determine how to compute the heat transfer coefficient of the flue gas, as well as the overall heat transfer coefficient in relation to the finned tubes.

We recommend the following.

We calculate

$$C_1 = 0.25\mathrm{Re}^{-0.35}; \tag{5.215}$$

$$C_3 = 0.2 + 0.65e^{-0.25H/b} \tag{5.216}$$

for in-line tubes, or

$$C_3 = 0.35 + 0.65e^{-0.25H/b} \tag{5.217}$$

for staggered tubes.

In (5.216) and (5.217) b is the space between the fins (equal to the pitch minus the average thickness).

Then,

$$C_5 = 1.1 - \left(0.75 - 1.5e^{-0.7N}\right) e^{-2s_l/s_t} \tag{5.218}$$

for in-line tubes, otherwise

$$C_5 = 0.7 + \left(0.70 - 0.8e^{-0.15N^2}\right) e^{-s_l/s_t} \tag{5.219}$$

for staggered tubes.

In (5.218) and (5.219) N is the number of rows run through by the gas, s_l and s_t are the longitudinal and transversal pitches. If $N \geq 6$, $N = 6$ will be used.

Then we compute

$$J = C_1 C_3 C_5 \left(\frac{d_0 + 2H}{d_0}\right)^{0.5} \left[\left(t'_m + 273\right)/(t_c + 273)\right]^{0.25}, \tag{5.220}$$

where t'_m is the average temperature of the flue gas, and t_c is given by

$$t_c = t''_m - 0.3(t'_m - t''_m); \tag{5.221}$$

where t''_m is the average temperature of the internal fluid. All temperatures are expressed in °C.

The heat transfer coefficient α' of flue gas is equal to

$$\alpha' = J\mathrm{Re}\,\mathrm{Pr}^{1/3}\,\frac{k}{d_o}. \tag{5.222}$$

Re and Pr stand for the numbers of Reynolds and Prandtl, K for the thermal conductivity of the material and d_0 for the external diameter of the tube without fins.

Recalling the significance of Re and Pr, (5.222) can be written as follows:

$$\alpha' = JG\frac{c_p^{1/3}k^{2/3}}{\mu^{2/3}}, \tag{5.223}$$

where G is the mass velocity.

Based on (5.223) we developed a few equations for air, a few types of gas and flue gas.

For air between 0 and 300°C, recalling (5.7), (5.8) and (5.9) we obtain:

$$\alpha = \left[1228.7 + 17\frac{t_b}{100} + 2.44\left(\frac{t_b}{100}\right)^2\right]JG. \tag{5.224}$$

Moreover, for $t_b = 0 - 500$°C:
Carbon dioxide (CO_2)

$$\alpha = \left[935 + 181.1\frac{t_b}{100} - 17.5\left(\frac{t_b}{100}\right)^2\right]JG; \tag{5.225}$$

Carbon monoxide (CO)

$$\alpha = \left[1226 - 152.2\frac{t_b}{100} + 37.87\left(\frac{t_b}{100}\right)^2\right]JG; \tag{5.226}$$

Methane (CH_4)

$$\alpha = \left[1848 + 998.9\frac{t_b}{100} - 89.4\left(\frac{t_b}{100}\right)^2\right]JG. \tag{5.227}$$

As far as flue gas we refer you to (5.17), (5.18) and (5.19).

Based on (5.223) we may write that

$$\alpha' = K_g JG; \tag{5.228}$$

given that

$$K_g = 1200 + 8.251m + (317.23 + 3.063m)\frac{t_b}{1000} - (76.14 - 9.687m)\left(\frac{t_b}{1000}\right)^2 \tag{5.229}$$

where m is the mass moisture percentage of gas.

In (5.224), (5.225), (5.226), (5.227),(5.228) and (5.229) α is in W/m^2K and t_b is the bulk temperature in °C.

Now, referring to a meter of tube, we indicate the surface of the fins with S_f, the external tube surface not related to the fins with S_b, the total surface of the finned tube ($S_f + S_b$) with S_0, the average surface with S_m, and the internal surface of the tube with S_i, the over-all heat transfer coefficient relative to S_0 is given by

$$U_o = \frac{1}{\left(\frac{1}{\alpha'} + f_o\right)\frac{S_0}{E_f S_f + S_b} + \frac{x}{k}\frac{S_0}{S_m} + \left(\frac{1}{\alpha''} + f_i\right)\frac{S_0}{S_i}}. \tag{5.230}$$

In this case x is the thickness of the tube, α'' the heat transfer coefficient of the internal fluid, f_o and f_i the potential fouling factors relative to both the outside and inside of the tube.

If these are steel tubes covered by finned cast iron muffs, (5.230) changes as follows.

$$U_o = \frac{1}{\left(\frac{1}{\alpha'} + f_o\right)\frac{S_0}{E_f S_f + S_b} + \frac{x_c}{k_c}\frac{S_0}{S_{mc}} + R\frac{S_0}{S_e} + \frac{x_s}{k_s}\frac{S_0}{S_{ms}} + \left(\frac{1}{\alpha''} + f_i\right)\frac{S_0}{S_i}} \tag{5.231}$$

In (5.231) x_c and x_s represent the thickness of the cast iron tube and the steel tube, k_c and k_s represent the thermal conductivity of cast iron and steel, S_{mc} and S_{ms} the average surfaces of the cast iron and steel tubes. S_i and S_e are the internal and external surface of the steel tube. R stands for thermal resistance that should be expected, assuming that the adherence between steel and cast iron tube will not be perfectly tight. Its value can be assumed to be equal to $0.7 - 1.4 \times 10^{-3}\text{m}^2\text{K/W}$.

5.12 Boiling Liquids

5.12.1 Boiling Liquids Outside the Tubes

Scholars who researched heat transfer to a liquid boiling outside the tubes focussed mostly on water, the most important one of all possible fluids with regards to boiling.

We consider evaporators in general and more specifically boiling of water in steam generators.

Studies on boiling outside the tubes were generally done by filling up a tank with water and dipping tubes or coils filled with condensing vapor that heated the surrounding water through the tube wall.

Electrical resistors dipped in water were used at times, too.

First of all, researchers considered the thermal flux given by the ratio between heat q transferred in one time unit and the exchange surface S. The flux will be considered as kW/m^2.

Generally, thermal flux has been correlated to the difference in temperature Δt between the wall of the tube and the boiling fluid.

Figure 5.20 shows a typical curve of the phenomenon obtained with water at atmospheric pressure heated with a platinum wire passed through by electricity. This allows us to identify the characteristics of the phenomenon and to derive numerical data that point to the entity of the quantities at play.

The curve has many significant areas.

In area $A - B$ the heat transfer occurs by natural convection. In fact, even though the water is saturated, the build-up of steam bubbles which should take place right

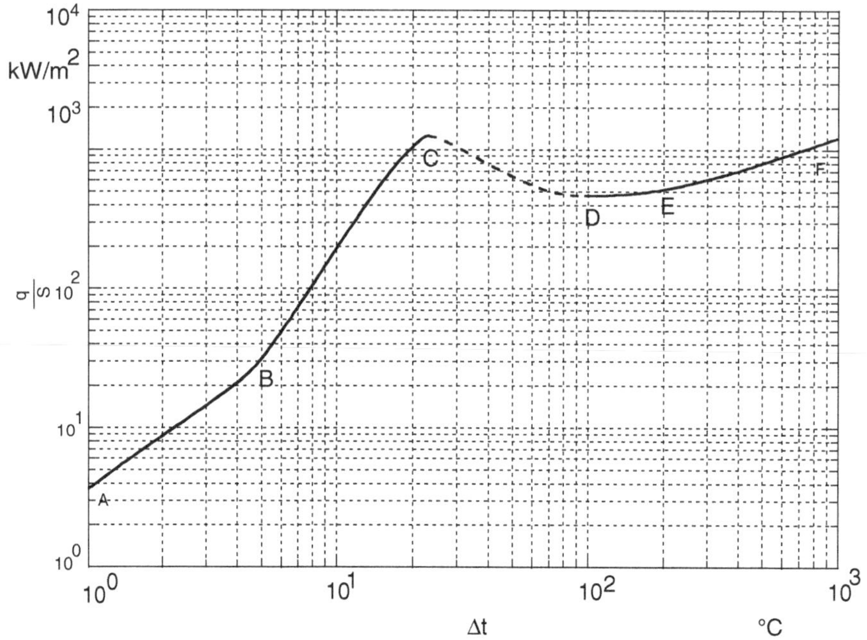

Fig. 5.20 Water boiling outside tubes (p=1 bar)

away is in fact delayed. It becomes manifest only when the temperature of the liquid exceeds the saturation temperature by a few degrees.

Since this is natural convection the heat transfer coefficient is proportional to $\Delta t^{0.25}$ (remember (5.42)) and the thermal flux is therefore proportional to $\Delta t^{1.25}$. In fact, $q/S = \alpha \Delta t$.

Area $B - C$ is where the bubble boiling takes place (Fig. 5.21); the bubbles of steam leave the wall of the tube, get bigger and are substituted by the surrounding water. The increase of thermal flux produces modest increases of Δt since the value of the heat transfer coefficient is extremely high and increasing with Δt, as we shall see.

Bubble boiling continuously increases as shown in Fig. 5.20, until peak heat flux is reached (point C) in correspondence of a difference in temperature Δt called critical Δt.

Area $C - D$ is characterized by vapor film forming on parts of the surface, and an increase in Δt means a decrease in thermal flux. The flux reaches minimum level in point D.

In area $D - E - F$ the build-up of the vapor film in contact with the surface is complete and we enter the state of film boiling.

It is interesting to obtain the values of the heat transfer coefficient from Fig. 5.20 to understand how the latter is influenced by the different situations to be faced and get a rough sense of the stakes at play.

The result is Fig. 5.22 which leads to the following comments.

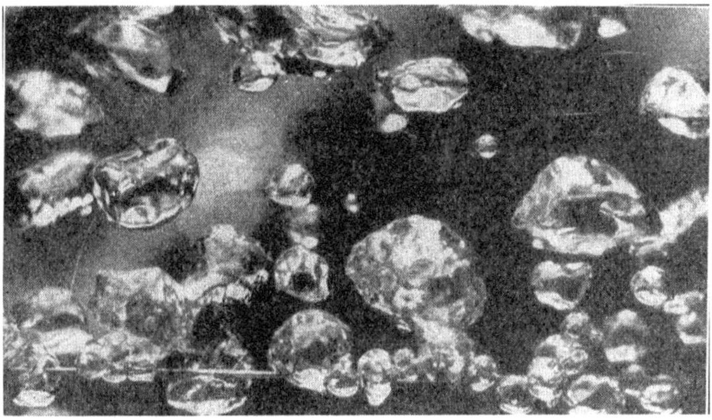

Fig. 5.21 Bubble boiling

In area $A - B$ we register a modest increase of α with Δt. In fact, we may write that by approximation with α in W/m^{2}K , $\alpha = 3590\Delta t^{0.25}$; for area $B - C$ by approximation we may write that $\alpha = 700\Delta t^{1.36}$. Finally, for area $D - E - F$ by approximation $\alpha = 73000\Delta t^{-0.6}$.

The increase of α, going from point $B$to point C, is remarkable. The heat transfer coefficient goes from about 5000 W/m^{2}K in point B to 52000 W/m^{2}K in point C. The increase in bubbles and in their dimension makes the heat transfer easier and easier, as far as variations of Δt in connection with variations of the thermal flux.

In point C the production of steam is so intense that the bubbles are unable to separate themselves, and this leads to the build-up of a film of steam.

This abruptly reduces the value of α, as shown in Fig. 5.22. The result by approximation is $\alpha = 46\times 10^6 / (q/S)^{1.5}$. It is not surprising that an increase in flux produces an increase in steam and in film thickness that the heat must cross. As a result, the

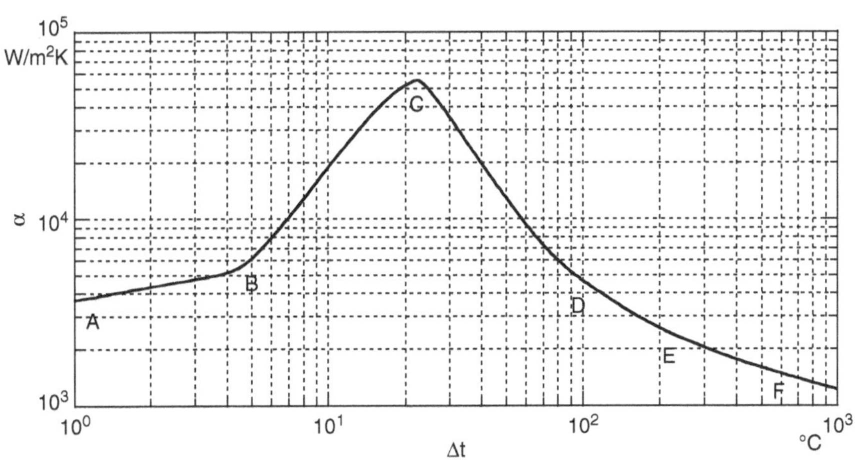

Fig. 5.22 Boiling water heat transfer coefficient (outside tubes)

value of α decreases. This reduction would seem greater than the actual increase in thickness. In fact, if the increase in thickness of the film were the only factor the value of α should only be inversely proportional to the value of q/S.

The curve in Fig. 5.20 and the conclusions that were drawn do certainly not complete the observation of the phenomenon. The curve in question is qualitatively significant, yet quantitatively it represents only one case study among many.

It suffices to think that the peak flux obtained during experiments with water generally varies between 550 and 1250 $\mathrm{kW/m^2}$. As you can see, this is a rather wide range. On the other hand, the critical Δt is always stable around 25°C. As a result, in correspondence of the peak flux α varies between 22000 and 50000 $\mathrm{W/m^2K}$. As you can see, the curve in Fig. 5.22 is located in the upper area of the range.

The following mean values of the peak flux indicated by $(q/S)_p$ for other liquids are:

Ethilic Acetate: $(q/S)_p = 190\,\mathrm{kW/m^2}$
Benzene: $(q/S)_p = 200\,\mathrm{kW/m^2}$
Heptane: $(q/S_p) = 170\,\mathrm{kW/m^2}$
Ethanol: $(q/S)_p = 300\,\mathrm{kW/m^2}$
Propanol: $(q/S)_p = 280\,\mathrm{kW/m^2}$
Methanol: $(q/S)_p = 350\,\mathrm{kW/m^2}$
Butanol: $(q/S)_p = 310\,\mathrm{kW/m^2}$

As far as the heat transfer coefficient relative to film boiling, Bromley defined a theoretical equation based on laminar motion of steam and heat transfer by conduction through the film.

If we assume that the liquid moves freely with the steam the equation in question will be:

$$\alpha = 0.724\left(\frac{grk_s^3\rho_s(\rho_l - \rho_s)}{d_o\mu_s\Delta t}\right)^{0.25}. \qquad (5.232)$$

It is very similar to (5.234) developed by Nusselt for condensing steam, as we shall see in Sect. 5.12.2.

In (5.232) g stands for gravity acceleration in $\mathrm{m/s^2}$; r is the vaporization heat in J/kg; k_s is the thermal conductivity of steam in W/m K; ρ_l e ρ_s are the densities of liquid and steam in $\mathrm{kg/m^3}$; $d_o$is the outside diameter of the tube in m, μ_s the dynamic viscosity of steam in kg/ms, and Δt the difference in temperature between tube wall and fluid in °C.

If we consider water at 1 bar with $d_o = 10\,\mathrm{mm}$ (experiments were conducted with tubes with a very small diameter), based on (5.232) we obtain

$$\alpha = \frac{813}{\Delta t^{0.25}}; \qquad (5.233)$$

and with $\Delta t = 200$°C the outcome is $\alpha = 216\,\mathrm{W/m^2K}$.

This value is much lower than those obtained from the curve in Fig. 5.22. Under the same conditions for that curve, and based on the same considerations made at the time, we would get $\alpha = 73000/200^{0.6} = 3038\,\mathrm{W/m^2K}$; for $\Delta t = 500°\mathrm{C}$ in both instances we obtain $\alpha = 172\,\mathrm{W/m^2K}$ and $\alpha = 1753\,\mathrm{W/m^2K}$. The difference has been reduced but the values are still quite different.

Equation (5.233) was supported by experimental data on heat transfer through a film of steam. Nonetheless, the results are perplexing, and the question is whether it can be applied to the phenomenon at hand.

Moreover, if α is proportional to $\Delta t^{-0.25}$, q/S is proportional to $\Delta t^{0.75}$ and $\alpha$is inversely proportional to $(q/S)^{1/3}$. It seems that the dependence of α from thermal flux is too modest, when considering that the entity of the flux determines the thickness of the film of steam, which in turn is crucial in determining the value of α.

Pressure has a considerable influence on the value of the peak flux. The phenomenon was studied by Cichelli and Bonilla who focused their attention on organic liquids and suggested a curve that would be valid for all these fluids and that highlights this influence. Further research demonstrated that this curve is valid for water, as well.

Instead of showing the original curve, we prefer to present another one because it better demonstrates the impact of pressure.

It is included in Fig. 5.23. The abscissa shows the ratio between pressure and critical pressure, the ordinate the ratio between peak flux at the examined pressure indicated by $(q/S)_p$, and peak flux at atmospheric pressure indicated by $(q/S)_{pa}$.

As you see, the ratio above reaches a maximum of about 4 in correspondence of a pressure equal to a third of the critical pressure. For water this peak is about 70 bar and equal to 2200–5000 $\mathrm{kW/m^2}$. These are extremely high values.

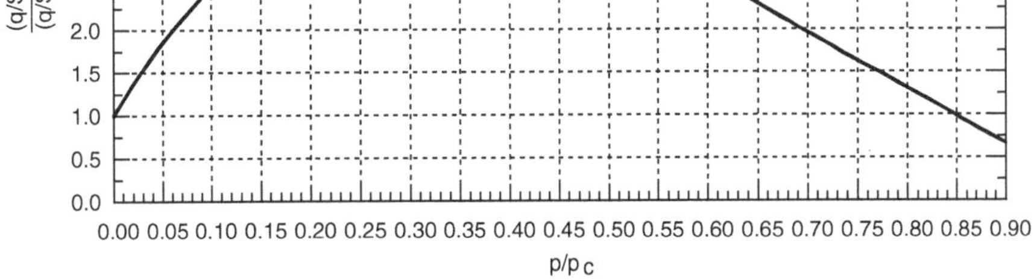

Fig. 5.23 Pressure influence on the peak flux

In conclusion, note that with bubble evaporation of different liquids at atmospheric pressure the heat transfer coefficient may vary considerably from one liquid and one type of surface to another, Δt being equal. The differences are smaller in reference to thermal flux. Finally, the diameter of the tube is basically irrelevant.

In our view the designer who must assign a value to the heat transfer coefficient of boiling water outside the tubes is not assisted by this.

The topic is complex, and full of uncertainties and incongruities. In addition, the situation may complicate things much more than this book may lead you to believe. Experimentation focuses on simple situations that may be far from reality.

For instance, it suffices to think of the tubes of smoke tube boilers with flue gas flowing inside and dipped in saturated water.

The steam bubbles forming around the tubes move upward to occupy the upper steam chamber. During this process they encounter the tubes of the upper rows. Therefore, the situation is especially complex.

Fortunately, an exact estimate of the heat transfer coefficient is not necessary for the following reasons.

What is important with regard to heat transfer is the value of the overall heat transfer coefficient. The latter includes (see (1.18) and (1.20)) both the heat transfer coefficient of the internal fluid and, in our case, the external heat transfer coefficient of the boiling water.

The heat transfer coefficient of the boiling water is always considerably greater than that of the heating fluid, be it air, any gas, flue gas and also condensing vapor; the order of magnitude of the ratio between the two heat transfer coefficients may be equal to 50–100.

Under these conditions the influence of the value of the heat transfer coefficient of boiling water on the overall heat transfer coefficient is minimal and even a macroscopic error in its estimate does not impact the value of the overall heat transfer coefficient in a meaningful way.

Our recommendation is to adopt a fixed conservative value of $\alpha = 6000\,\mathrm{W/m^2K}$; this value is only slightly greater than the value of the heat transfer coefficient in point B of the curve in Fig. 5.20.

5.12.2 Boiling Liquids Inside the Tubes

The study of boiling liquids inside the tubes is more complex than that of tubes with boiling liquid on the outside.

In fact, beyond boiling there is the impact of the movement of the fluid. The latter may be natural or forced.

It is natural if it occurs spontaneously, as a result of variations in density of the fluid when passing from the liquid phase to the formation of a liquid-steam mixture; it is forced if it occurs as a result of a thrust by a pump.

A typical example of natural movement of the fluid is with water in water tube boilers with natural circulation.

In this case tubes full of water which feed the lower headers come down from the drum located in the upper area of the boiler. This is the starting point for steam generating tubes in which part of the water turns into steam creating a water-steam mixture. The different density of the water in the tubes filled with water and of the water-steam mixture in the steam generating tubes starts natural circulation. The velocity of the mixture is not high.

If the water in the steam generating tubes comes from a pump instead, there is forced circulation. The velocity of the steam-water mixture can also be high and depends on the characteristics of the circuit and of the circulation pump.

Even boiling inside the tubes can produce film boiling phenomenon, as we saw during boiling outside the tubes. Even in this case, if a certain value of thermal flux is exceeded, bubble boiling stops, and a film of steam builds up in contact with the wall, followed by a sudden increase of the difference in temperature between wall and fluid.

Next we will examine the behavior of boiling water in the tubes, since this is of greatest interest.

The quantities that condition the occurrence of this phenomenon are the water vapor ratio (or the mass percentage of steam contained in the water-vapor mix), the pressure, the mass velocity and the thermal flux. The latter is crucial in terms of the lesser or greater danger of the phenomenon, as far as the temperature difference between the internal wall of the tube and the fluid it creates.

Figure 5.24 highlights the impact of the water-vapor ratio and the thermal flux under constant pressure and constant mass velocity.

As the flux increases, the phenomenon starts for decreasing values of the water vapor ratio and causes increasing temperature differences. In the specific case, this difference is about proportional to the cube of the flux.

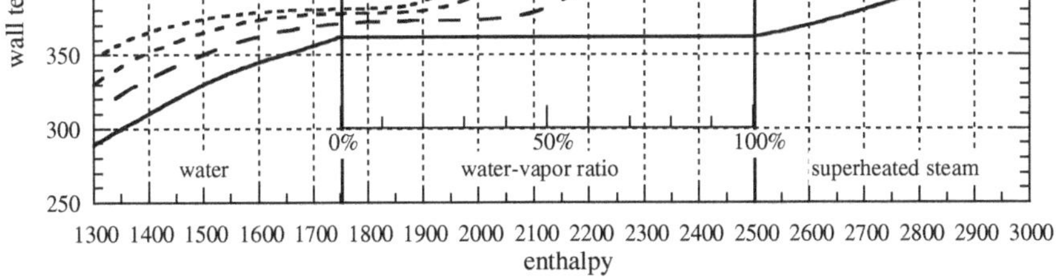

Fig. 5.24 Influence of water-vapor ratio and thermal flux on wall temperature

This strong impact of the flux on the temperature difference is easily explainable by observing that the increase in flux increases the thickness of the steam film because of the high level of boiling. Besides the greater amount of heat to transfer through the film, there is a greater thickness to penetrate. The value of Δt is therefore influenced twice by the increase in flux.

Once the maximum value is reached, Δt decreases as the water-vapor ratio increases, thus the velocity of the mix until it reaches the value corresponding to the heat transfer coefficient of the superheated steam.

The influence of pressure, water-vapor ratio and flux per constant mass velocity is represented by Fig. 5.25.

The curves indicate the conditions that start film boiling. Under equal pressure, the phenomenon starts with decreasing water-vapor ratios as the flux increases, as already shown in Fig. 5.24. Under constant flux the increase in pressure starts the film boiling with decreasing water-vapor ratios. Under high level flux or high pressure it may even occur through the mere presence of water (dotted areas of the curves). The diagram refers to a mass velocity of 830 kg/m^2s.

The influence of mass velocity, water-vapor ratio and flux can be examined in Fig. 5.26. The constant, absolute pressure is equal to 186 bar.

Note that as the mass velocity varies, the value of the water-vapor ratio responsible for the start of the phenomenon first decreases, moving from low velocity to medium velocity to increase further if the velocity reaches high values.

The purely qualitative diagram, given that it identifies only the conditions under which the phenomenon starts to occur without quantifying its entity, may lead to mistaken conclusions. In fact, the increase in mass velocity is always favourable to push back or eliminate the danger of heavy superheating of the tube as a result of film boiling.

To reduce the danger of film boiling, it is possible to use special tubes that are ribbed on the inside (see Fig. 5.27). The tendency to film boiling can be fought

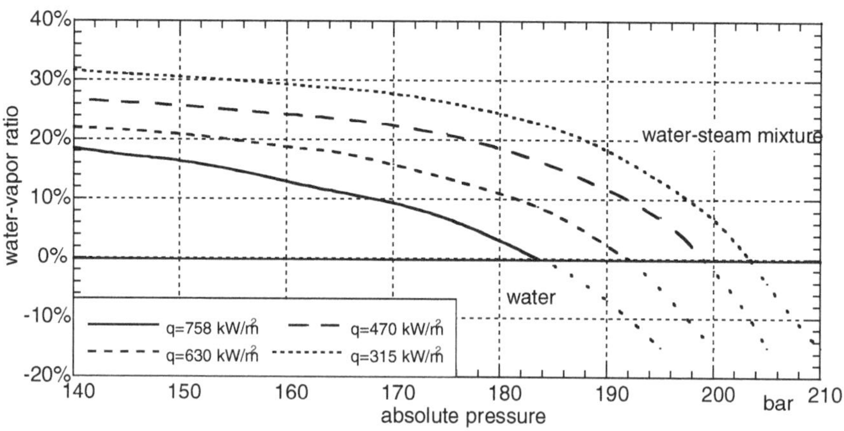

Fig. 5.25 Influence of pressure, water-vapour ratio and thermal flux on film boiling

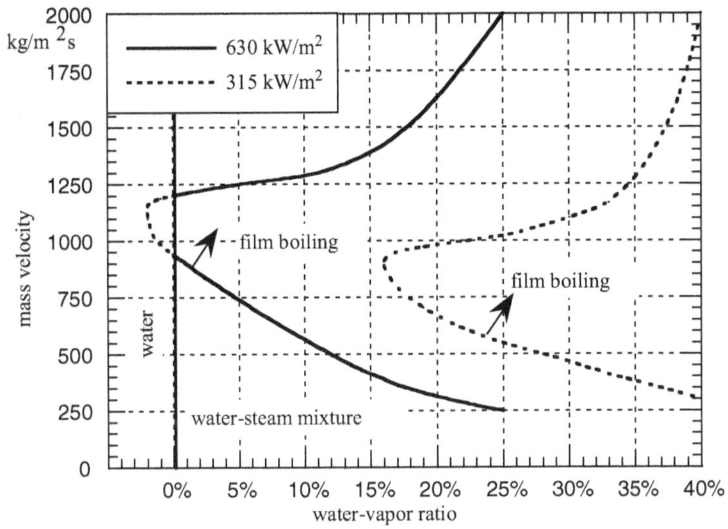

Fig. 5.26 Influence of mass velocity, water-vapor ratio and thermal flux on film boiling

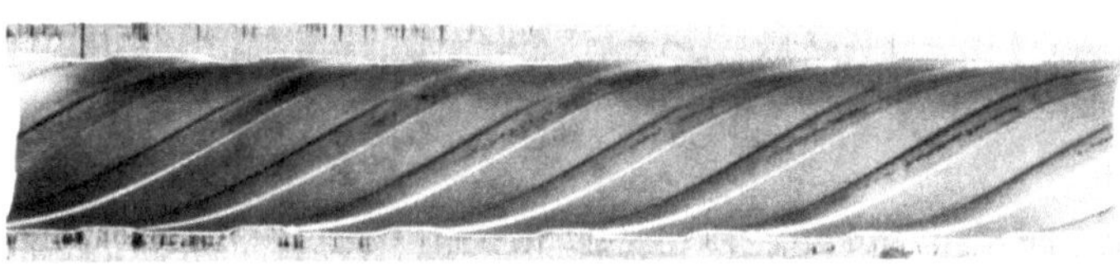

Fig. 5.27 Ribbed tube

by interrupting the continuity of the wall, and by provoking turbulent flow in the boundary layer.

In terms of the value to assume for the heat transfer coefficient of boiling water (of course, with bubbles) inside the tubes, major producers recommend to set a fixed conservative value equal to 12000 W/m^2K.

In terms of the little importance of a potential error, even a macroscopic one as far as the heat transfer coefficient of boiling water, with regard to the overall heat transfer coefficient, what was said in the section on boiling outside the tubes is even truer.

5.13 Condensing Vapors

When saturated steam enters in contact with a colder surface part of the steam condenses and forms a film of condensate in contact with the wall.

The steam continues to yield heat to the wall through the film, thus causing an increase in condensation, and an ensuing increase in thickness of the film.

The film moves with laminar motion and the heat passing through has the characteristics of heat by conduction. Therefore, the thickness of the film is crucial.

The thickness of the film increases as its motion slows down, so that there are considerable differences between horizontal and vertical tubes. Moreover, in a vertical tube the thickness of the film is greater towards the bottom because it is influenced by the liquid flowing from the top. As a result, the heat transfer coefficient on the bottom is lower compared to the upper area. During computation it will be necessary to assume a mean value of α.

Even the characteristics of the surface influence the phenomenon because if it is rough this favors the increase in thickness of the film which occupies the spaces created by the roughness of the wall without producing any motion.

The conditions for condensation have been explained and mathematically developed by Nusselt.

These were some of the points he clarified on this topic.

(a) The heat transfer is by condensation only, since the temperature of the film in contact with the steam has the same temperature. This rules out convection and radiation.
(b) The film moves with laminar motion, and the heat is transferred by conduction.
(c) The thickness of the film is conditioned by its mean velocity and by the amount of condensate moving along the wall.
(d) The velocity of the film is the result of a balance between the tangential force of friction and the weight of the film.
(e) The amount of condensate depends on the heat transfer. In turn, this depends on the thickness of the film, and on the difference in temperature between wall and steam, according to the laws of thermal conduction.

    Nusselt's equation relative to the mean value of α along a vertical wall is given by

$$\alpha = \frac{4}{3}\left(\frac{grk_f^3\rho_f^2}{4H\mu_f\Delta t}\right)^{0.25} \tag{5.234}$$

The temperature of reference is the mean temperature of the condensate which is the average between the wall and steam temperatures. The quantities are indicated with the subscript *f*.

Equation (5.234) is very similar to (5.232) which referred to boiling liquid under conditions of film boiling.

In (5.234) g is gravity acceleration in $\mathrm{m/s^2}$, r the vaporization heat in J/kg, k_f the thermal conductivity of water in W/m K, ρ_f the density of water in $\mathrm{kg/m^3}$, $H$the height of the tube or of a vertical wall in m, μ_f the dynamic viscosity in kg/ms, and Δt the difference in temperature between steam and wall in °C.

Equation (5.234) may be written as follows:

$$\alpha = 1.669\frac{r^{0.25}k^{0.75}\rho^{0.5}}{\mu^{0.25}}\left(\frac{1}{H\Delta t}\right)^{0.25}. \tag{5.235}$$

The term before the brackets consists of quantities that depend on temperature under constant pressure only. Therefore, it is possible to develop an equation to substitute (5.235), i.e.,

$$\alpha = \frac{6956 + 6388\frac{t_f}{100} - 1821\left(\frac{t_f}{100}\right)^2}{(H\Delta t)^{0.25}}, \tag{5.236}$$

with t_f =80–300°C

Conventionally assuming that t_f is equal to the saturation temperature minus 20°C , the value of α can depend on absolute pressure instead of temperature.

This leads to

$$\alpha = \frac{10814 + 89.4p - 0.959p^2}{(H\Delta t)^{0.25}} \tag{5.237}$$

with p =1–100 bar.

In the previous equations α is in $\mathrm{W/m^2K}$.

It can be interesting to establish the dependence of the heat transfer coefficient from thermal flux q/S, given that q is the heat transfer in time unit and S the surface of thermal exchange.

Based on (5.235), by indicating the heat transfer coefficient with $K/(H\Delta t)^{0.25}$ the thermal flux is equal to $q/S = (K/H)\,(H\Delta t)^{0.75}$; then

$$(H\Delta t)^{0.25} = \left(\frac{H}{K}\frac{q}{S}\right)^{1/3}; \tag{5.238}$$

finally,

$$\alpha = \left[\frac{K^4}{H\,(q/S)}\right]^{1/3}. \tag{5.239}$$

With reference to pressure, with q/S in $\mathrm{kW/m^2}$ in the end we obtain:

$$\alpha = \frac{23920 + 270p - 2.89p^2}{\left[H\,(q/S)\right]^{1/3}} \tag{5.240}$$

Here are a few examples. With $H = 1$ m, $t_f = 160$°C and Δt =60°C from (5.236) we obtain $\alpha = 4502\,\mathrm{W/m^2K}$. If the pressure is equal to 5 bar instead, the other conditions being the same, from (5.237) we obtain $\alpha = 4050\,\mathrm{W/m^2K}$. Finally, if the pressure is equal to 100 bar and the thermal flux is equal to 80 kW/m2, (5.240) leads to $\alpha = 5110\,\mathrm{W/m^2K}$.

With regard to benzol vapor at atmospheric pressure Nusselt recommends the following equation:

$$\alpha = \frac{2082}{(H\Delta t)^{0.25}}, \tag{5.241}$$

and for ethylic alcohol vapor at atmospheric pressure the following:

$$\alpha = \frac{2675}{(H\Delta t)^{0.25}} \tag{5.242}$$

The heat transfer coefficient is always expressed in $\text{W}/\text{m}^2\text{K}$.

As far as the tubes arranged horizontally, according to Nusselt, the heat transfer coefficient α_o is obtained from the following equation, where α_v is the heat transfer coefficient computed for the vertical tube.

$$\alpha_o = 0.77\left(\frac{H}{d_i}\right)^{0.25}\alpha_v. \tag{5.243}$$

As you see, for $H/d_i \geq 2.85$, as is generally the case, equation (5.243) leads to a higher heat transfer coefficient for horizontal tubes.

On various occasions researchers have pointed out that equation (5.234) generally produces values of α below the actual ones. Therefore, the resulting values are conservative values.

This happens mostly when the condense produces drops without building a uniform layer of condensate in contact with the wall. In that case the heat transfer is increased, and the value of the heat transfer coefficient goes up.

Chapter 6
Radiation

6.1 Introduction

When considering an isolated body at a certain temperature we establish that it radiates energy; this energy increases with temperature, according to a law that will be discussed later on.

The energy propagates by electromagnetic waves within a certain range of frequencies.

If there is no matter between the body in question and another body, or if the matter is *transparent*, i.e., viable to be crossed by radiation, the energy emitted by the first body, or part of it, hits the second body. In some instances, the energy passes through it, but generally the energy is partly absorbed and partly reflected by the second body.

The latter in turn transfers heat by radiation to the first one. If the energies emitted and absorbed by both bodies are different, there is an energy exchange by radiation between them. If a body absorbs all the incident energy it is called *black body.*

By analogy, the name originates from the black color because of the behavior of the body of this color versus light energy.

Luminous electromagnetic waves (in other words, those visible to our own eye) cover a range between wave length $\lambda = 0.39 \times 10^{-6}$ m and $\lambda = 0.78 \times 10^{-6}$ m, i.e., between 0.39 and 0.78 μm; if the body absorbs all the luminous energy reaching it, it appears to be black.

Yet as we shall see, luminous energy interests only a small part of all electromagnetic radiations (with wave length from 0 to ∞) related to energy transfer by radiation.

The black body, as intended here, is such if it absorbs all incident energy, instead of only the luminous one. We will establish that the highest values for the emitted energy occur for quite different wavelengths compared to luminous ones.

Then, a body of black color is such due to luminous radiation, but it can reflect energy to the non luminous area, and this means that it does not fit the definition of a black body like the one we described.

In reality, an absolutely black surface does not exist. The absorption of incident energy is never 100%; a smaller or bigger part of it is always reflected.

D. Annaratone, *Engineering Heat Transfer*, DOI 10.1007/978-3-642-03932-4_6,

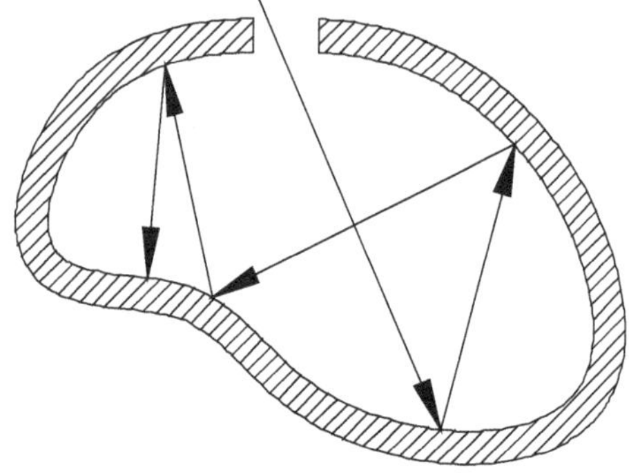

Fig. 6.1 Hollow body with hole

The only instance of a practically black surface would be the section of a hole communicating with a cavity (Fig. 6.1).

The energy entering the cavity through the hole hits its walls and, given the fact that they are basically not black, part of the energy is absorbed and part of it is reflected. The latter hits the walls to be absorbed and reflected. Clearly, in the end all the energy is absorbed by the walls of the cavity except for an irrelevant fraction exiting the hole.

6.2 The Laws of Radiation

6.2.1 Planck's Law

As we know, the energy emitted by the black body is distributed along the entire spectrum for wavelengths from 0 to ∞.

At a certain temperature of the black body a value for the intensity of the radiation corresponds to each wavelength.

This intensity of monocromatic radiation indicated by E_λ can be obtained through the following equation demonstrated by Planck:

$$E_\lambda = \frac{A\lambda^{-5}}{e^{B/\lambda T} - 1}. \tag{6.1}$$

In (6.1) $A = 3.7403 \times 10^{-16}$ Wm^2 and $B = 0.014387$ mK.

The intensity of monocromatic radiation E_λ is in W/m^3, λ is the wavelength in m and T the absolute temperature in K.

Equation (6.1) is not easy to use. If we wish to obtain E_λ in W/m^3 and the wavelength is expressed in mm instead of m, we obtain the following values for A and B: $A = 0.37403$; $B = 14.387$.

Equation (6.1) allows us to build the curves of Fig. 6.2 that shows the values of E_λ for the different wavelengths at three different temperatures.

As expected, an increase in temperature produces an increase in radiation intensity.

In addition, the curves have a peak. This means that there is a wavelength for which the monocromatic radiation intensity is at its maximum. The value of this wave length will be identified through Wien's law.

Finally, by increasing the temperature the peak moves to the left, i.e., towards shorter wavelengths. Wien's law will show this, too.

The total energy E emitted by the black body in the time unit and by surface unit (therefore expressed in $\mathrm{W/m^2}$) is obtained as follows:

$$E = \int_0^\infty E_\lambda d\lambda. \tag{6.2}$$

We digitally computed the integration in (6.2). The results for some temperatures, are shown in Table 6.1 under the label "Planck". The values obtained by (6.7) thanks to Stefan-Boltzmann are also included under the label "Stefan". This equation is about the heat radiated by the black body.

Note the optimal correspondence between the two series of values. Those derived from (6.7) are only slightly greater, and interestingly enough the values would perfectly coincide if (6.7) adopted the value 5.669355 instead of 5.672 for $\sigma \times 10^8$.

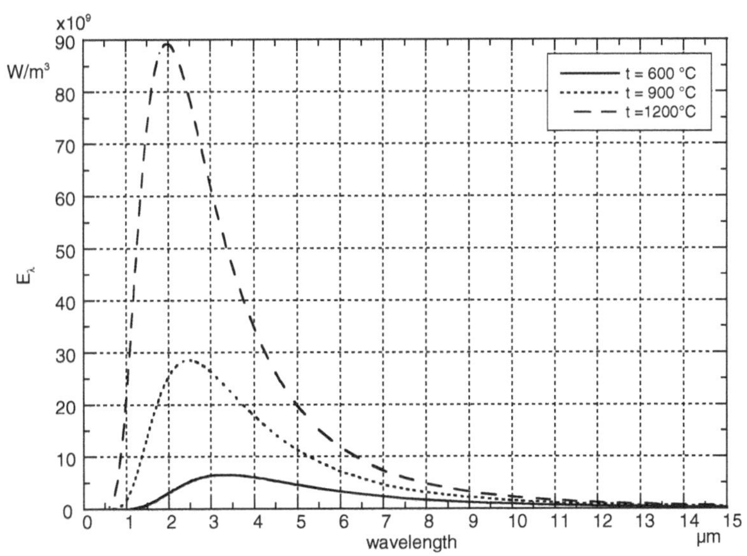

Fig. 6.2 Monocromatic radiation according to Plank

Table 6.1 Comparison between the integral of Planck's curves and Stefan-Boltzmann's equation

t(°C)	Planck (W/m^2)	Stefan (W/m^2)
400	11641	11646
500	20258	20267
600	32953	32968
700	50846	50869
800	75193	75228
900	107386	107436
1000	148954	149023
1100	201561	201655
1200	267006	267131
1300	347228	347390
1400	444297	444504
1500	560423	560684

6.2.2 Wien's Law

Through the derivative of (6.1) it is possible to obtain the wave length with the maximum intensity of monocromatic radiation.

If it is indicated by λ_{max} and we adopt

$$y = \frac{B}{\lambda_{max} T}, \tag{6.3}$$

we reach the maximum value of E_λ when

$$e^y = \frac{5}{5 - y}. \tag{6.4}$$

Equation (6.4) is satisfied for $y = 4.96512$.

Recalling the value of B we obtain

$$\lambda_{max} T = 0.002898 \text{ mK}. \tag{6.5}$$

Equation (6.5) stands for Wien's law, or the law of displacement.

As you see, λ_{max} is inversely proportional to T.

In actual instances, the values of λ_{max} are always well above the wavelengths corresponding to visible radiations. For example, even with a temperature of 2000°C (2273 K) we have $\lambda_{max} = 1.27 \times 10^{-6}$ m, or 1.27 μm which is well above the visible range. Only when we refer to the sun with a temperature of the photosphere of 6000 K we obtain $\lambda_{max} = 0.48 \times 10^{-6}$ m, or 0.48 μm which is within the visible range.

Both (6.1) and (6.5) make it possible to obtain the maximum value of the intensity of monocromatic radiations. By indicating it with $E_{\lambda\,max}$, it is given by

$$E_{\lambda\,\max} = 1.2866 \left(\frac{T}{10}\right)^5. \tag{6.6}$$

As you see, the influence of T is very strong. For example, if the temperature increases from 500°C (773 K) to 1200°C (1473 K) the value of $E_{\lambda\,\max}$ increases 25 times.

6.2.3 Stefan-Boltzmann's Law

Theoretical studies as well as experimentation led Stefan and Boltzmann to the following equation about heat radiation by the black body:

$$q = \sigma T^4 \tag{6.7}$$

where q is the radiated heat during the time unit and by surface unit in $\mathrm{W/m^2}$, while σ is the Stefan-Boltzmann constant equal to $5.672 \times 10^{-8}\,\mathrm{W/m^2\,K^4}$.

If two black bodies, at absolute temperature T_1 and T_2, respectively, face each other so that they radiate heat to one another, heat q transferred during the time unit from the warmer body to the colder one is equal to

$$q = \sigma S\left(T_1^4 - T_2^4\right); \tag{6.8}$$

where q is in W and S is either the radiating or the radiated surface in $\mathrm{m^2}$.

Equation (6.8) is valid both for emission or absorption of the radiated heat. In fact, it is irrelevant whether temperature T_2 is greater or smaller than T_1; if $T_2 > T_1$ Eq. (6.8) makes it possible to compute the heat absorbed by surface S at temperature T_1.

Preference generally goes to the following expression of q because of easier computation:

$$q = \sigma S\left[\left(\frac{T_1}{100}\right)^4 - \left(\frac{T_2}{100}\right)^4\right] \tag{6.9}$$

where $\sigma = 5.672\,\mathrm{W/m^2\,K^4}$.

As far as choosing surface S, here are the rules. If two surfaces exchange heat by radiation, and one of them is completely surrounded by the other one, the smaller surface, i.e., the surface of the internal body must be introduced in (6.8), regardless of the fact that the latter receives or gives heat. The same principle is true if a smaller surface is so close to a bigger surface that all the heat radiated by the smaller surface necessarily hits the bigger surface. Finally, if two surfaces face each other by forming an angle, the radiated surface receives only part of the total heat coming from the radiating surface. In this case we must refer to Lambert's law that will be discussed later on.

So far we only considered black bodies. Real bodies actually present an intensity of monocromatic radiations that differs from that of a black body.

Now, if $E_{\lambda 0}$ indicates the intensity of monocromatic radiation of the black body and E_λ that of the actual body, the ratio $E_\lambda/E_{\lambda 0}$ is called monocromatic emittance and indicated by ε_λ; its value varies with λ and T, as shown in Fig. 6.3. *Grey body* is the body for which value ε_λ is constant for all temperatures and all wavelengths, indicated by ε.

The grey body according to this definition does not exist but generally it is possible to study real bodies as if they were grey.

To that extent we introduce a mean value of ε in reference to the global behavior of the body versus radiation. Thus, $\varepsilon = q/q_0$, given that q stands for the heat radiated by the body in question, and q_0 for the heat radiated by the black body.

This makes the emittance independent from the wavelength. Moreover, in many instances ε may be considered a constant even with regard to T.

The emittance ranges from 0 to 1, as we shall see.

The absorption power or *black level* is the ratio $B = q_a/q_i$ between absorbed heat and incident heat on the body. Of course, it ranges from 0 to 1, and it is equal to unity for the black body.

In view of all this, if we consider any body with emittance equal to ε the heat emitted by radiation is given by

$$q = \varepsilon \sigma T^4. \tag{6.10}$$

Equation (6.10) substitutes (6.7) if the body is not black.

Equation (6.10) which is at the base of the following starts from the assumption that the body may be considered grey. This implies that the emittance is a constant of the body in question. In other words, the ratio between the energy emitted by the body and that emitted by the black body is independent from both wavelength and temperature. Often this finds no correspondence in reality.

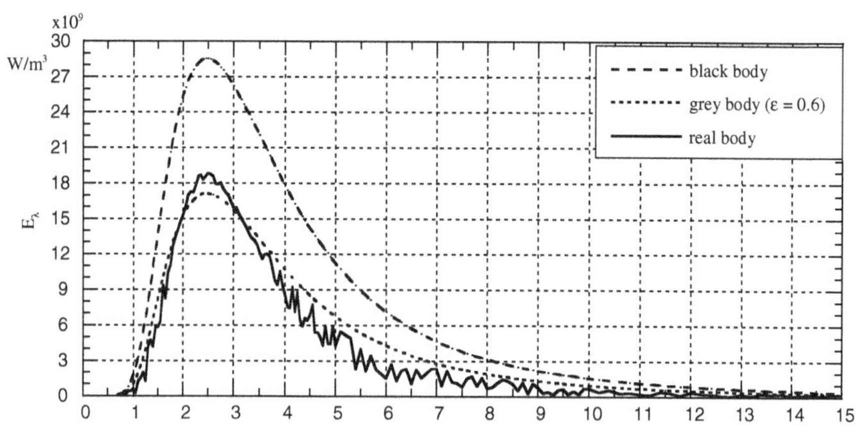

Fig. 6.3 Real and grey body

The dependence from the wavelength is not important if we only focus on the comparison between the heat transfer from the body in question and that from the black body. The phenomenon is global. All types of wavelength implying heat transfer are involved. Then it is right to express emittance as the ratio between the heat emitted by the body and the one emitted by the black body. The situation is completely different when looking at the dependence of emittance from temperature. It would be necessary to either give up the constant value of ε or to modify the exponent of T. With radiation it is customary to elevate absolute temperature to the fourth power, so it would be more logical to vary the value of ε with temperature. Unfortunately, case studies in research are limited and fragmentary and do not produce clear and safe guidelines for actual instances. Therefore, the only alternative consists of reasonable compromise. For example, when reliable data are available it is possible to adopt different values of ε for different temperature ranges.

6.2.4 Kirchhoff's Law

Figure 6.4 shows a black body with surface S_1 completely contained inside another black body with surface S_2; both bodies are at the same temperature T.

If surface S_1 has no concave parts, all the heat emitted by it reaches surface S_2.

The heat emitted by S_1 is equal to $q_1 = \sigma S_1 T^4$; the one emitted by S_2 is equal to $q_2 = \sigma S_2 T^4$. All the heat q_1 hits surface S_2 and is absorbed by it because it is a black body. Only a fraction ρ of heat q_2 reaches surface S_1 and is absorbed by it.

Given that both bodies have the same temperature, according to the second principle of thermodynamics heat transfer between them cannot take place. Therefore,

$$\rho q_2 = q_1; \tag{6.11}$$

and then $\rho = S_1/S_2$.

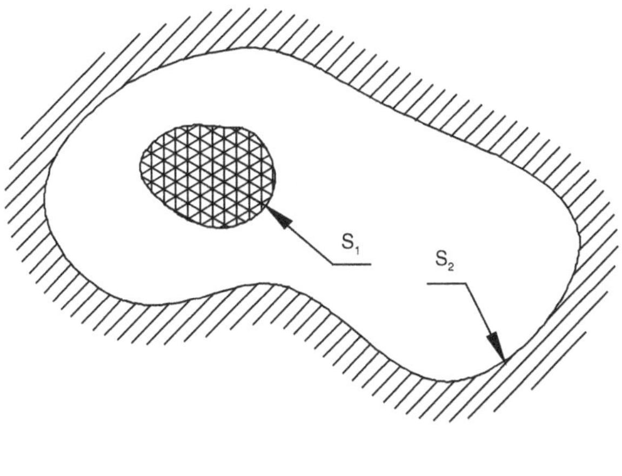

Fig. 6.4 Body fully contained in other body

At this point we substitute the inside black body with a body having the same geometry but which is not black. It is characterized by emittance ε and a black level B.

The heat emitted by S_2 reaching S_1 is again fraction ρ of the total emitted heat, i.e., $\rho\sigma S_2 T^4 = \sigma S_1 T^4$.

Considering that this is not a black body, the heat leaving surface S_1 consists of the emitted heat $\varepsilon\sigma S_1 T^4$, as well as the reflected heat $(1 - B)\sigma S_1 T^4$.

Still, according to the second principle the heat reaching the inside body must equal the leaving one. Therefore,

$$\sigma S_1 T^4 = \varepsilon\sigma S_1 T^4 + (1 - B)\,\sigma S_1 T^4; \tag{6.12}$$

and then $\varepsilon = B$.

This is Kirchhoff's law, according to which the emittance and the black level of a body are equal. Given that the value of B ranges from 0 to 1, even the emittance varies among these values.

Because of this equality, any one factor may be used to identify either the absorbed or the emitted heat. From now on we will only use the black level as the characteristic factor of the grey body.

6.2.5 Lambert' Law – Black Bodies Arranged in any Which Way

We consider an elementary black surface dS_1 emitting heat (Fig. 6.5). Then we consider the plane containing this surface and the hemisphere with the center in dS_1 and radius R, limited by the mentioned plane.

Surface dS_1 radiates onto the entire half space containing the hemisphere.

We take any radius of the sphere forming with the normal to the surface dS_1 the angle φ in addition, in correspondence of this radius we have an elementary surface dS_2 on the sphere (Fig. 6.5).

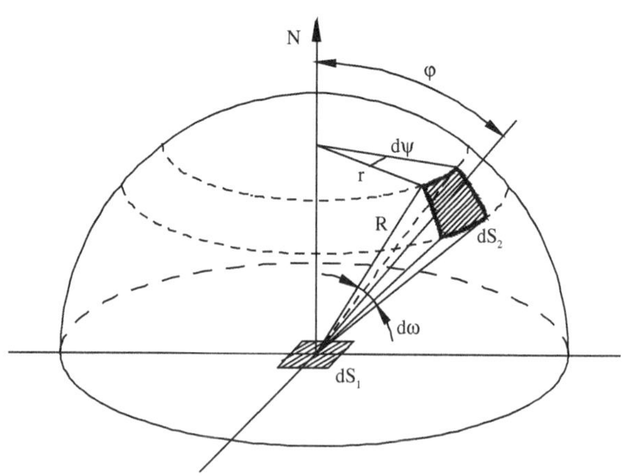

Fig. 6.5 Generally oriented surfaces

If $d\omega$ indicates the solid angle under which surface dS_2 is seen by dS_1, we have

$$dS_2 = R^2 d\omega; \tag{6.13}$$

then

$$d\omega = \frac{dS_2}{R^2}. \tag{6.14}$$

dq stands for the total heat emitted during the time unit by surface dS_1, and dq_φ for the part of this heat hitting surface dS_2.

The heat emitted during the time unit from a unitary surface by a solid unitary angle is called radiation intensity I. Heat dq_φ hitting surface dS_2 is therefore equal to

$$dq_\varphi = I dS_1 d\omega; \tag{6.15}$$

then

$$I = \frac{dq_\varphi}{dS_1 d\omega}. \tag{6.16}$$

I reaches its maximum value in the direction normal to the surface dS_1, and we indicate it with I_n

Lambert' law or the law of the cosine states that

$$I = I_n \cos\varphi. \tag{6.17}$$

Based on (6.14), (6.15) and (6.17) we have:

$$dq_\varphi = I_n \cos\varphi dS_1 d\omega = I_n \cos\varphi \frac{dS_1 dS_2}{R^2}. \tag{6.18}$$

r indicates the radius of the parallel where dS_2 is located and $d\psi$ indicates the angle which subtends one of the sides of this surface on the parallel; the angle $d\varphi$ subtends the other side. Then, surface dS_2 is equal to

$$dS_2 = R d\varphi r d\psi. \tag{6.19}$$

On the other hand, $r = R\sin\varphi$; then, based on (6.18), and recalling (6.19),

$$dq_\varphi = I_n \cos\varphi \sin\varphi d\varphi d\psi dS_1. \tag{6.20}$$

Now, if we want to achieve the total heat dq radiated by dS_1, we must integrate dq_φ and ensure that the integration is done with respect to ψ between 0 and 2π, and with respect to φ between 0 and $\pi/2$.

On the other hand,

$$\int_0^{\pi/2} \cos\varphi \sin\varphi d\varphi = \left|\frac{\sin^2\varphi}{2}\right|_0^{\pi/2} = \frac{1}{2}; \tag{6.21}$$

and

$$dq = I_n \frac{1}{2} 2\pi dS_1 = \pi I_n dS_1 \tag{6.22}$$

On the other hand, heat dq is the entire heat emitted by dS_1 in the time unit, so that if T is the absolute temperature of dS_1, given that the surface is black, we have:

$$dq = dS_1 \sigma \left(\frac{T}{100}\right)^4 \tag{6.23}$$

with $\sigma = 5.672\,\mathrm{W/m^2\,K^4}$.

Therefore, we obtain

$$I_n = \frac{\sigma \left(\frac{T}{100}\right)^4}{\pi}. \tag{6.24}$$

Equation (6.18) may be written as follows:

$$dq_\varphi = \sigma \left(\frac{T}{100}\right)^4 \cos\varphi \frac{dS_1 dS_2}{\pi R^2}. \tag{6.25}$$

Contrary to previous assumptions, note that surface dS_2 may generally not be normal to the connection of the centers of dS_1 and dS_2 (Fig. 6.6).

The heat reaching dS_2 is always conditioned by the solid angle under which the surface is seen by dS_1, even though in this case

$$d\omega = \frac{dS_2 \cos\beta}{R^2}. \tag{6.26}$$

Recalling (6.18) which includes $d\omega$ and from which the subsequent equations until (6.25) are derived, we establish that the latter must be modified as follows:

$$dq_\varphi = \sigma \left(\frac{T}{100}\right)^4 \cos\varphi \cos\beta \frac{dS_1 dS_2}{\pi R^2}. \tag{6.27}$$

If the surfaces are finite we must integrate (6.27) with respect to S_1 and S_2.

The total heat emitted by surface S_1 at temperature T_1 during the time unit is equal to

$$q_1 = \sigma S_1 \left(\frac{T_1}{100}\right)^4. \tag{6.28}$$

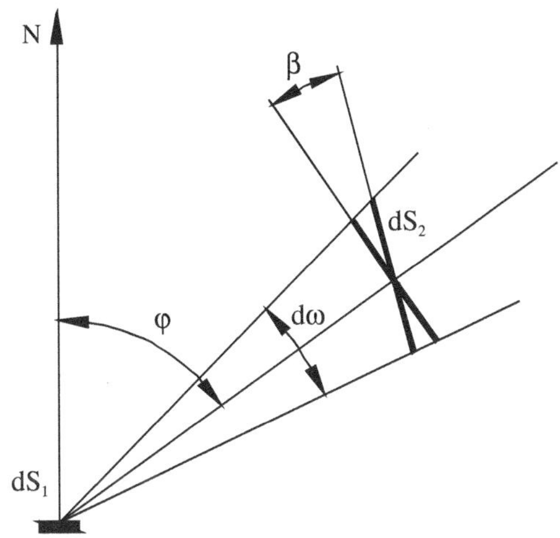

Fig. 6.6 Inclined surface with regard the axis

Part of this heat reaches surface S_2; q_{1-2} indicates this heat and F_{1-2} the factor, called form factor, by which q_1 must be multiplied to obtain q_{1-2}.

Recalling (6.27) we have

$$q_{1-2} = F_{1-2}\sigma S_1 \left(\frac{T_1}{100}\right)^4 = \sigma \left(\frac{T_1}{100}\right)^4 \int_{S_1}\int_{S_2} \frac{\cos\varphi \cos\beta dS_1 dS_2}{\pi R^2}; \qquad (6.29)$$

then

$$F_{1-2} = \frac{1}{S_1}\int_{S_1}\int_{S_2} \frac{\cos\varphi \cos\beta dS_1 dS_2}{\pi R^2}. \qquad (6.30)$$

By analogy, the heat emitted by S_2, at temperature T_2 reaching S_1 is given by

$$q_{2-1} = F_{2-1}\sigma S_2 \left(\frac{T_2}{100}\right)^4 \qquad (6.31)$$

given that

$$F_{2-1} = \frac{1}{S_2}\int_{S_1}\int_{S_2} \frac{\cos\varphi \cos\beta dS_1 dS_2}{\pi R^2}. \qquad (6.31a)$$

As you can see, $F_{1-2}S_1 = F_{2-1}S_2$.

Thus, the heat transfer from one body to the other is equal to

$$q = q_{1-2} - q_{2-1} = FS\sigma \left[\left(\frac{T_1}{100}\right)^4 - \left(\frac{T_2}{100}\right)^4\right] \qquad (6.32)$$

given that FS can be equal to either $F_{1-2}S_1$ or $F_{2-1}S_2$.

If the bodies are not black and are characterized by a black level equal to B_1 and B_2, (6.32) must include a mean black level B_m the value of which is a function of B_1 and B_2 depending on geometry.

6.3 Plane Surfaces Facing Each Other

Let us consider two plane surfaces facing each other (Fig. 6.7) at temperature T_1 and T_2, respectively.

If the surfaces are black, all the heat emitted by each surface is absorbed by the other. If $T_1 > T_2$ the heat emitted by the surface at temperature T_1 is greater than the one emitted at temperature T_2. This means that there is heat transfer by radiation from the warmer surface to the colder. This heat transfer within the time unit can be computed through (6.9) rewritten as follows:

$$q = \sigma S \left[\left(\frac{T_1}{100} \right)^4 - \left(\frac{T_2}{100} \right)^4 \right] \tag{6.33}$$

where q is in W, S (in m^2) represents each surface; they are equal, face each other and $\sigma = 5.672\,W/m^2\,K^4$.

If the surfaces are not black, the heat emitted by both surfaces is equal to, respectively,

$$q_1 = B_1 \sigma S \left(\frac{T_1}{100} \right)^4 ; \tag{6.34}$$

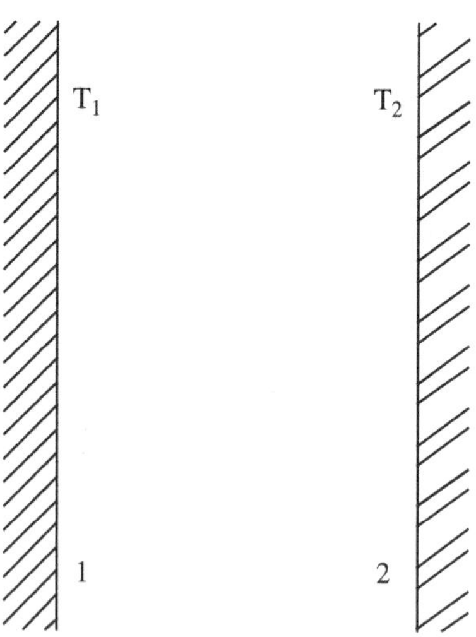

Fig. 6.7 Plane surfaces facing each other

$$q_2 = B_2 \sigma S \left(\frac{T_2}{100} \right)^4; \tag{6.35}$$

where B_1 and B_2 are the black levels of both surfaces.
q_{i1} and q_{i2} indicate the heat reaching each surface, and q_{o1} as well as q_{o2} indicate the heat leaving both surfaces (see Fig. 6.8).

The heat reaching a surface is, of course, the one coming from the other one. Then,

$$q_{o1} = q_{i2}; \quad q_{o2} = q_{i1}. \tag{6.36}$$

In both cases the fraction of reflected heat is equal to $(1 - B_1)$ and $(1 - B_2)$. Therefore, the heat leaving both surfaces is equal to, respectively,

$$q_{o1} = q_1 + (1 - B_1)\, q_{i1}; \tag{6.37}$$

$$q_{o2} = q_2 + (1 - B_2)\, q_{i2}. \tag{6.38}$$

The heat exchange is equal to the difference between the heat leaving the surface at greater temperature T_1 and the heat reaching it, i.e.,

$$q = q_{o1} - q_{i1} = q_{o1} - q_{o2}. \tag{6.39}$$

Based on (6.37) and (6.38) we obtain:

$$q_{o1} = q_1 + (1 - B_1)\, q_{o2}; \tag{6.40}$$

$$q_{o2} = q_2 + (1 - B_2)\, q_{o1}. \tag{6.41}$$

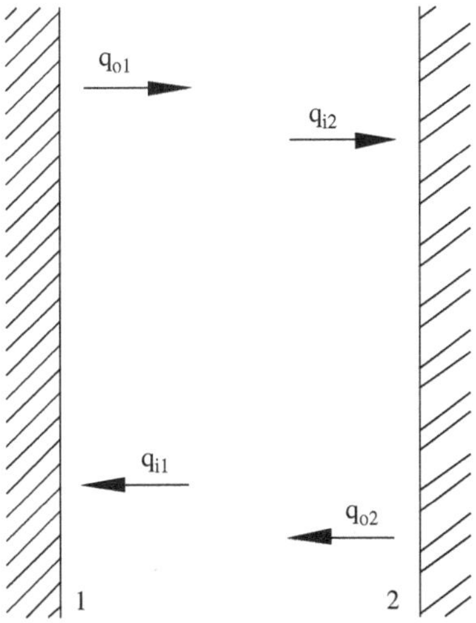

Fig. 6.8

Then

$$q_{o1} = q_1 + (1 - B_1)\, q_2 + (1 - B_1)\,(1 - B_2)\, q_{o1}; \tag{6.42}$$

$$q_{o1} = \frac{q_1 + (1 - B_1)\, q_2}{1 - (1 - B_1)\,(1 - B_2)}. \tag{6.43}$$

By analogy,

$$q_{o2} = \frac{q_2 + (1 - B_2)\, q_1}{1 - (1 - B_1)\,(1 - B_2)}. \tag{6.44}$$

Therefore, based on (6.39)

$$q = \frac{B_2 q_1 - B_1 q_2}{B_1 + B_2 - B_1 B_2}; \tag{6.45}$$

and recalling (6.34) and (6.35):

$$q = \frac{\sigma S \left[\left(\frac{T_1}{100} \right)^4 - \left(\frac{T_2}{100} \right)^4 \right]}{\frac{1}{B_1} + \frac{1}{B_2} - 1}. \tag{6.46}$$

By introducing a black level B_m which globally factors in the heat transfer between the two non black surfaces, we write that

$$q = B_m \sigma S \left[\left(\frac{T_1}{100} \right)^4 - \left(\frac{T_2}{100} \right)^4 \right]. \tag{6.47}$$

This substitutes (6.33) which is valid for black surfaces.

Thus, the global black level is given by

$$B_m = \frac{1}{\frac{1}{B_1} + \frac{1}{B_2} - 1}. \tag{6.48}$$

We establish that if one of the surfaces is black B_m corresponds to the black level of the other surface.

If there is a fluid free from convective phenomena and transparent to radiations between the two walls facing each other, the heat transfer takes place by radiation and conduction.

If x is the distance between the two walls (Fig. 6.9) the heat transfer within the time unit and by surface unit (in W/m^2) is given by

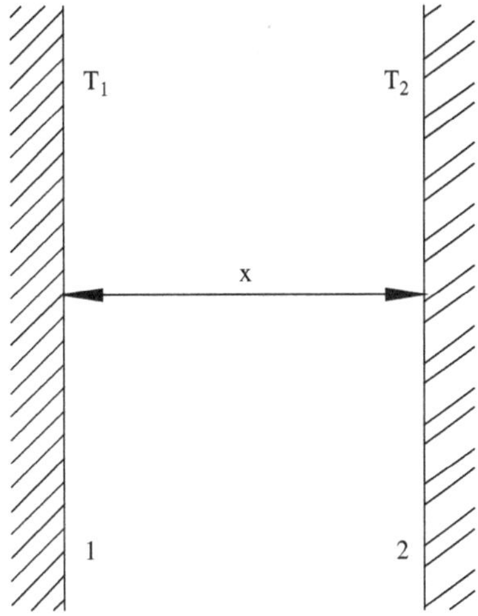

Fig. 6.9 Surfaces facing each other with interposed fluid

$$q = \frac{k}{x}(T_1 - T_2) + B_m\sigma\left[\left(\frac{T_1}{100}\right)^4 - \left(\frac{T_2}{100}\right)^4\right]. \tag{6.49}$$

It is possible to introduce an ideal radiation thermal conductivity k_r, synthetically leading to

$$q = \frac{k + k_r}{x}(T_1 - T_2). \tag{6.50}$$

Based on (6.49) we establish that

$$k_r = B_m\sigma x(T_1 + T_2)\left(T_1^2 + T_2^2\right) \times 10^{-8}. \tag{6.51}$$

By first approximation, if T_m stands for the mean known or assumed temperature between T_1 and T_2 ($T_m = (T_1 + T_2)/2$), it is possible to adopt the following value of k_r:

$$k_r = 4B_m\sigma x T_m^3 \times 10^{-8}. \tag{6.52}$$

According to a few researchers, for convective phenomena in the interspace to be ruled out, we must have

$$x \leq \sqrt[3]{\frac{30000}{t_1 - t_2}} \tag{6.53}$$

with x in mm and t_1 and t_2 in °C.

For instance, a difference in temperature of 30°C implies that the distance between the walls must never exceed 10 mm; this would apply to double-pane windows.

If convective phenomena occur in the interspace instead, the heat transfer is by radiation and convection.

If α_1 and α_2 are the heat transfer coefficients of the fluid towards both walls, the heat transfer within the time unit and by surface unit is equal to:

$$q = \frac{1}{\dfrac{1}{\alpha_1} + \dfrac{1}{\alpha_2}} (T_1 - T_2) + B_m \sigma \left[\left(\frac{T_1}{100} \right)^4 - \left(\frac{T_2}{100} \right)^4 \right]. \tag{6.54}$$

Generally, the heat transfer coefficients may be considered equal, so we assume that $\alpha = \alpha_1 = \alpha_2$. In addition, we introduce an ideal radiation transfer coefficient α_r, and we can synthesize as follows:

$$q = \left(\frac{\alpha}{2} + \alpha_r \right) (T_1 - T_2). \tag{6.55}$$

A comparison between (6.55) and (6.54) leads to

$$\alpha_r = B_m \sigma \times 10^{-8} (T_1 + T_2) \left(T_1^2 + T_2^2 \right). \tag{6.56}$$

By first approximation, by introducing the mean temperature T_m between T_1 and T_2, it is possible to adopt the following value of α_r

$$\alpha_r = 4 B_m \sigma T_m^3 \times 10^{-8}. \tag{6.57}$$

More in general, considering a wall at temperature T_1 transferring heat to an outside environment at temperature T_2 by radiation and convection, the heat transfer by time unit as well as surface unit is equal to

$$q = \alpha (T_1 - T_2) + B\sigma \left[\left(\frac{T_1}{100} \right)^4 - \left(\frac{T_2}{100} \right)^4 \right] \tag{6.58}$$

where α and B are the heat transfer coefficient and the black level relative to the wall, respectively.

6.4 Body Completely Contained in Another Body

This situation was partly discussed in the section about Kirchhoff's law (Fig. 6.4).

We saw that while all the heat emitted by the surface of the inner body S_1 hits the surface S_2 of the outer body, only part of the heat emitted by surface S_2 hits surface S_1; the part in question is equal to the ratio S_1/S_2.

Then, if the bodies are black, the heat q_{1-2} radiated by surface S_1 onto surface S_2 which completely absorbs it because it is a black body is equal to

$$q_{1-2} = \sigma S_1 \left(\frac{T_1}{100}\right)^4. \tag{6.59}$$

Instead, the heat q_{2-1} radiated by surface S_2 onto surface S_1 which completely absorbs it is equal to:

$$q_{2-1} = \sigma S_2 \left(\frac{T_2}{100}\right)^4 \frac{S_1}{S_2} = \sigma S_1 \left(\frac{T_2}{100}\right)^4. \tag{6.60}$$

Heat q which is transferred by radiation from the inside surface S_1 to the outside surface S_2 is therefore equal to:

$$q = q_{1-2} - q_{2-1} = \sigma S_1 \left[\left(\frac{T_1}{100}\right)^4 - \left(\frac{T_2}{100}\right)^4\right]. \tag{6.61}$$

Therefore, the surface of the inner body is the surface of reference.

If the surfaces are not black and B_1 and B_2 stand for the black levels of both surfaces, the heat transfer is given by

$$q = B_m \sigma S_1 \left[\left(\frac{T_1}{100}\right)^4 - \left(\frac{T_2}{100}\right)^4\right]. \tag{6.62}$$

In (6.62) B_m is the global black level; it is given by

$$B_m = \frac{1}{\frac{1}{B_1} + \frac{S_1}{S_2}\left(\frac{1}{B_2} - 1\right)}. \tag{6.63}$$

If surface S_2 is very large, the value of B_m practically coincides with B_1. As you can see, in that case only the black level of the heating body is of interest, and the heated body behaves like a black body.

This type of situation occurs, for instance, if we consider a heating body in a big room.

6.5 Solar Radiation

The temperature of the surface of the sun is about 6000 K. It emits an enormous quantity of energy, and only a fraction of it reaches the earth, i.e., about 1.8×10^{14} kW. Even though this is only a fraction of the total energy, it is equal to 10000 times the energy requirements of humanity.

This energy penetrating the atmosphere is subject to diffusion and absorption phenomena the entity of which depends on the angle of radiations with respect to the surface of the earth.

Consequently, the energy reaching the earth varies depending on latitude, season and time of day. In addition, it depends on weather conditions.

Considering the entire planet, solar radiation ranges from 0 to $1500\,\mathrm{W/m^2}$. In countries with a mild climate $1000\,\mathrm{W/m^2}$ are never reached, even under ideal seasonal and time conditions. During the winter season, as well as less favorable times of day, solar radiation may be less than one fourth of maximum levels.

Using solar energy to produce electricity or for heating through solar panels is increasingly an issue for most industrialized countries. In fact, there is quite a strong desire to differentiate the energy sources for obvious political and economic reasons to minimize the dependance on traditional sources, such as oil, natural gas, coal and nuclear energy.

Several obstacles delay the diffusion of this type of energy. First of all, the considerable cost of these plants. Its unforeseable intermittence prevents continuous and reliable availability, and this reduces its appeal in some cases, or requires expensive accumulation plants.

Moreover, the entity of solar radiation is such that they require very large surfaces to obtain significant power levels.

The future will certainly bring a constant increase in the use of this form of energy, especially if technological progress will reduce the cost of plants and also maximize the potential for their exploitation.

Yet, this form of energy will only represent a modest fraction of the global energy required for both civil and industrial purposes.

As far as the impact of solar radiation on heat transfer, we will discuss how to set up computation if it hits a wall behind which there is a room at temperature t_i (Fig. 6.10) through an example.

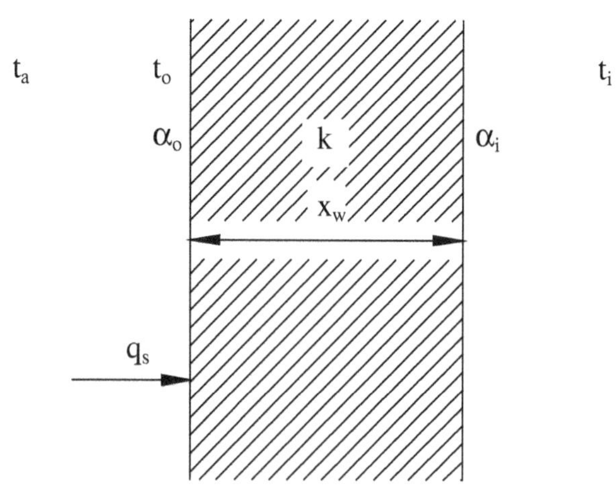

Fig. 6.10 Wall with sun radiation

q_s stands for solar radiation, t_a for the temperature of the outer environment, t_o for the outside temperature of the wall, and t_i for the temperature of the inner room. The temperatures are in reference to time unit and wall surface unit.

The heat absorbed by the wall is equal to

$$q_w = Bq_s \tag{6.64}$$

where B is the black level.

The heat transferred from the wall towards the outside is equal to

$$q_o = \alpha_o \left(t_o - t_a\right) \tag{6.65}$$

where α_o indicates the sum of the outside heat transfer coefficient and of the ideal radiation transfer coefficient to be obtained through (6.56) or (6.57). In this case, in the equations B_m must be substituted with the black level B of the wall.

The heat transfer to the inside room is equal to:

$$q_i = \frac{1}{\dfrac{1}{\alpha_i} + \dfrac{x_w}{k}} \left(t_o - t_i\right). \tag{6.66}$$

Even in this case α_i is the sum of the heat transfer coefficient and the ideal radiation transfer coefficient; x_w is the thickness of the wall.

Under balance conditions

$$q_w = q_o - q_i. \tag{6.67}$$

Therefore, based on (6.64), (6.65), (6.66) and (6.67):

$$t_o = \frac{Bq_s + \alpha_o t_a + \dfrac{1}{1/\alpha_i + x_w/k} t_i}{\alpha_o + \dfrac{1}{1/\alpha_i + x_w/k}}. \tag{6.68}$$

After determining the value of t_o, Eq. (6.66) computes heat q_i which is transferred to the inside room.

6.6 Flame Radiation

Initially, we introduce the concept of flame adiabatic temperature. It is the maximum temperature the flame could theoretically reach. In addition, and this is the most important reason for mentioning it, it will help us develop one of the computational methods that we will discuss.

If the flame developing in a closed environment did not radiate towards the surrounding walls, it would reach the temperature called *flame adiabatic temperature*, even though the naming is improper since it actually is an isobar transformation.

$H_n{}'$ indicates the heat we introduce into the furnace with the combustion of 1 kg of fuel; it is expressed in kJ/kg.

H_n indicates the net heat value of the fuel in kJ/kg; assuming that the latter is heated from room temperature t_0 to temperature t_f and that the combustion air is equally heated from t_0 to t_a, we have

$$H_n{}' = H_n + c_{pf}\left(t_f - t_0\right) - Ac_{pa}\left(t_a - t_0\right) \tag{6.69}$$

where A is the amount of air required for 1 kg of fuel in kg/kg; c_{pf} is the mean specific heat of the fuel between t_0 and t_f in kJ/kgK; c_{pa} stands for the mean isobaric specific heat of the air between t_0 and t_a in kJ/kgK; temperatures are in °C.

If t_{ad} is the adiabatic temperature of the flame, G is the amount of flue gas per kg of fuel in kg/kg, and c_{pg} is the mean isobaric specific heat of the flue gas between t_0 and t_{ad}, we have

$$H'_n = Gc_{pg}\left(t_{ad} - t_0\right). \tag{6.70}$$

t_{ad} can be computed through (6.70) combined with (6.69).

H'_n/G represents the heat introduced in the furnace for every kg of flue gas. Therefore, this ratio has the dimension of an enthalpy and we indicate it with h_g; from (6.70) we obtain the following:

$$t_{ad} - t_0 = \frac{h_g}{c_{pg}}. \tag{6.71}$$

If we assume that $t_0 = 0°\mathrm{C}$, the particular value of t_{ad} obtained from (6.71) represents the so-called theoretic combustion temperature; by indicating it with t_{tc} we obtain

$$t_{tc} = \frac{h_g}{c_{pg}} \tag{6.72}$$

In that case c_{pg} is the mean specific heat between 0°C and t_{tc}.

Based on Sect. 5.2.3 we know that specific heat is a function of t and the moisture of flue gas.

Rosin preferred to put together the diagram shown in Fig. 6.11 instead. It helps to obtain t_{tc} as a function of h_g referred to 1 Nm^3 of flue gas and the air index n, i.e., the ratio between actual combustion air and theoretical air (stoichiometric); the air index is better identifiable than moisture.

In general, the diagram can be used to compute t_{ad}. In fact, given that the mean specific heat between t_0 and t_{ad} differs from the mean specific heat between 0°C and t_{tc} in a negligible way, based on (6.71) it is possible to write that

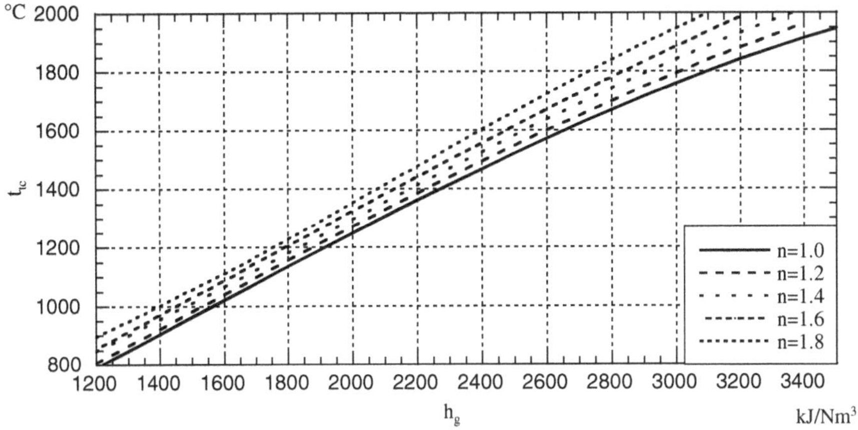

Fig. 6.11 Theoretical combustion temperature (Rosin)

$$t_{ad} = t_{tc} + t_0, \tag{6.73}$$

where t_{tc} stands for the temperature obtained from Fig. 6.11. The value of h_g to introduce is the one in (6.71), i.e., the one relative to the value of H'_n in (6.70).

In conclusion, calculation of the adiabatic temperature requires calculation of H'_n based on (6.69); by dividing this value by G_v we obtain the heat h_g per Nm^3 of flue gas to introduce in Fig. 6.11; based on the air index n, it is possible to compute t_{tc} and to determine t_{ad} through (6.73).

Regardless of its purely theoretical significance, the adiabatic temperature t_{ad} is interesting as a reference point on the more or less high temperature that the flame can reach. It will also help to compute the exit temperature of the gas from the furnace based on Konakow's computation method.

Note that above 1500°C there is considerable dissociation of the steam and the carbon dioxide in the flue gas that limits the development of heat as the temperature goes up because of the so-called dissociation heat. The impact of the dissociation cannot be ignored above 1500°C where (6.72) becomes incorrect. Note that Rosin's diagram (see Fig. 6.11) takes this phenomenon into account, albeit through some degrees of simplification. Therefore, the obtained temperatures t_{tc} are rather correct.

The temperature in the furnace is naturally lower than the adiabatic one because of heat radiation towards the walls.

Before proceeding we would like to point out the following.

If t_e represents the exit temperature of the gas from the furnace and M_f the amount of fuel burnt per time unit (in kg/s), heat q transferred to the walls of the furnace per time unit (in kW) is given by

$$q = M_f H'_n - M_f G c_{pg} \left(t_e - t_0\right) \tag{6.74}$$

where c_{pg} is the mean specific heat between t_0 and t_e.

q_s indicates the heat transferred to the furnace per radiated surface unit (in kW/m^2), and we consider heat q_i to be equal to

$$q_i = \frac{M_f H'_n}{S} \tag{6.75}$$

This is the heat introduced into the furnace in the time unit per radiated surface unit. It corresponds to the maximum heat that the flame could theoretically transfer to the walls if the flue gas left the furnace at room temperature.

Based on (6.75) and recalling the above, after a series of steps we obtain the following:

$$q_s = q_i \left[1 - \frac{c_{pg}}{h_g} (t_e - t_0) \right]. \tag{6.76}$$

As you see, there is a simple correlation between the heat transfer in the furnace per surface unit and the exit temperature of the gas from the furnace. The quantities q_i and h_g are crucial. As we shall see, they are also crucial to calculate the exit temperature of the gas from the furnace.

Note that based on (6.76), one can indifferently obtain a computational equation of q_s or t_e to identify the conditions for the heat transfer in the furnace.

From our perspective, we will introduce computational equations to determine the value of t_e. This will facilitate the comparison between the suggested computational methods, even though they are the result of very different approaches.

Many scholars tried to develop equations for the computation of heat transfer into the furnace through theoretical assumptions. Given the considerable difficulties, the results of this research are modest and unsatisfactory due to the complexity of the methodology, as well as the debatable correlation to the actual phenomenon.

Therefore, it is preferable to rely on empiric criteria because of their simplicity and acceptable closeness to reality.

At this point, we introduce two computational methods of this kind, and a method by the author derived from one of them.

They totally disregard the temperature of the radiated walls. This may seem to be in contrast with the laws of radiation where this temperature is fundamental. It is important to specify, though, that these methods are in reference to the furnace of steam generators where the walls consist of steam generating tubes.

Under these conditions, the temperature of the radiated walls which is rather low compared to the flame temperature is irrelevant. For instance, if the temperature of the walls increases from 150°C to 350°C with a hypothetical flame temperature of 1500°C, the decrease in radiated heat of about l'1,2% may be neglected.

The reader may be interested in the computation of the heat transfer even at much greater wall temperatures.

At the end of the section we will introduce a criterion to use the methods relative to furnaces of steam generators even in different situations where the values of the wall temperatures are high.

The first computational method was initially developed by Orrok and Hudson and later modified by Reid, Cohen and Corey.

Based on this modification which simplified the process, the percentage of heat transferred into the furnace compared to the introduced heat indicated by δ may be calculated through the following equation where we used the same symbolism we used earlier:

$$\delta = \frac{100}{1 + C\dfrac{\sqrt{q_i}}{h_g}} \tag{6.77}$$

where C is a factor which will be discussed later on.

Recalling (6.76):

$$\frac{\delta}{100} = \frac{1}{1 + C\dfrac{\sqrt{q_i}}{h_g}} = \frac{q_s}{q_i} = 1 - \frac{c_{pg}}{h_g}\left(t_e - t_0\right). \tag{6.78}$$

From (6.78) and after a series of steps:

$$c_{pg}\left(t_e - t_0\right) = \frac{1}{\dfrac{1}{C\sqrt{q_i}} + \dfrac{1}{h_g}}. \tag{6.79}$$

The authors of the method recommend the adoption of a mean value of C which expressing q_i in kW/m^2 and h_g in kJ/kg is equal to 155.3 kJ$^{1/2}$s$^{1/2}$mkg^{-1}.

From (6.79), also considering the entalphies of flue gases at temperatures t_e and t_0, we obtain

$$h_e - h_0 = c_{pg}\left(t_e - t_0\right) = \frac{1000}{\dfrac{6.439}{\sqrt{q_i}} + \dfrac{1000}{h_g}}. \tag{6.80}$$

Equation (6.80) already solves the problem because once the entalphy of the gases at the exit of the furnace is computed it is easy to determine their temperature.

Our preference is to make explicit t_e; to that extent we can conventionally assume 20°C for room temperature t_0, and as far as c_{pg} the value 1.164 kJ/kg K is in agreement with the actual values. From (6.80) we obtain the following:

$$t_e = 20 + \frac{1000}{\dfrac{7.495}{\sqrt{q_i}} + \dfrac{1164}{h_g}}. \tag{6.81}$$

Based on (6.81), it was possible to build the diagram shown in Fig. 6.12.

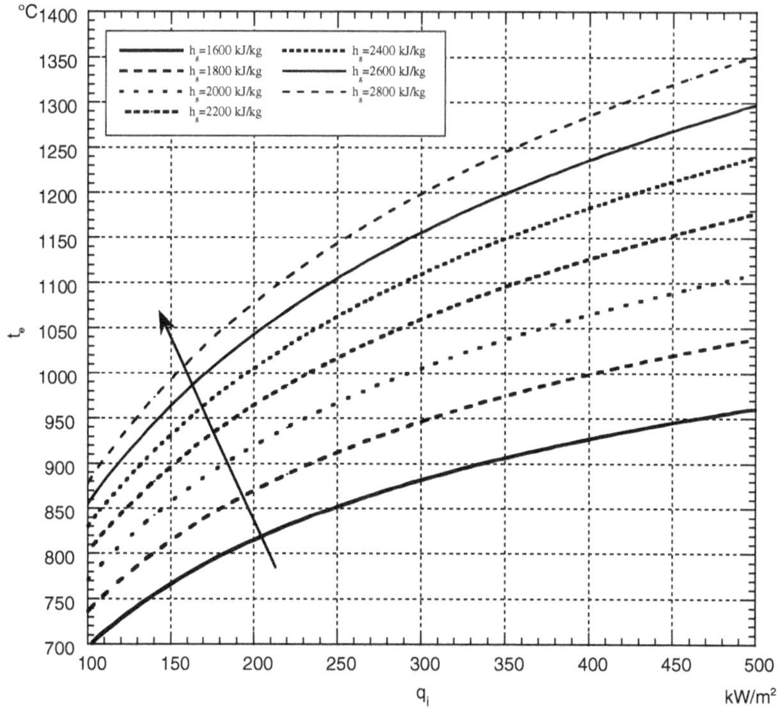

Fig. 6.12 Exit temperature from the furnace of the flue gas according to Orrok (6.81)

Equation (6.81) or the diagram, including the introduced simplifications, corresponds to Orrok's method. The inevitable errors with respect to this method do not exceed 20°C plus or minus for $t_e = 600 - 1200$°C.

Equation (6.81) is usable for combustion with fuel oil or natural gas. For coal combustion equation (6.77) is modified as follows:

$$\delta = \frac{100}{1 + f_v C \frac{\sqrt{q_i}}{h_g}} f_k. \qquad (6.82)$$

This way (6.80) changes as follows:

$$h_e = h_0 + \frac{1000 + \frac{6.439\,(1 - f_k)\,h_g}{f_v \sqrt{q_s}}}{\frac{6.439}{f_v \sqrt{q_s}} + \frac{1000}{h_g}}. \qquad (6.83)$$

The value of factor f_v depends on the percentage of surface covered by melted ashes and can be computed through Fig. 6.13 (for furnace with a hopper covered by

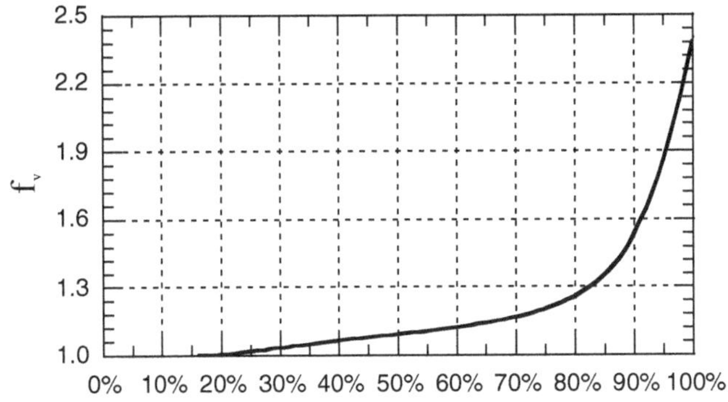

Fig. 6.13 Surface covered by melted ashes

melted ashes one can assume that $f_v = 1.08$); the factor f_k is equal to the unity for coal with a volatile matter content equal or greater than 20%, whereas if the content is under 20% it is equal to:

$$f_k = 1.34 - 0.017\,V, \tag{6.84}$$

where V is the percentage of volatile matter.

Another calculation method was developed by Konakow.

According to him,

$$\left(\frac{T_e}{T_{ad}}\right)^4 - Ko\left(1 - \frac{T_e}{T_{ad}}\right) = 0, \tag{6.85}$$

where T_{ad} is the absolute adiabatic temperature of the flame estimated regardless of the dissociation and Ko is a factor that in most cases according to Gunz can be assumed to be equal to:

$$Ko = 239\frac{q_i}{\left(\frac{T_{ad}}{100}\right)^4} \tag{6.86}$$

where $q_{\rm i}$ expressed in $\mathrm{kW/m^2}$ is given by (6.75).

Equation (6.85) facilitates the project computation. In fact, after setting temperature t_e (thus T_e) and once the adiabatic temperature t_{ad} (and therefore T_{ad}) is calculated, the value of Ko is obtained from (6.85). Based on (6.86) it is possible to compute $q_{\rm i}$ and through (6.75) one obtains the unknown surface S.

The verification, i.e., the computation of t_e knowing the surface S is not equally easy, and it must be done by trial and error.

Note that (6.86) can also be written as follows:

$$Ko = 239\frac{q_i}{\left(\frac{T_0}{100}\right)^4}\left(\frac{T_0}{T_{ad}}\right)^4. \tag{6.87}$$

Then, from (6.85)

$$\left(\frac{T_e}{T_0}\right)^4 - 239\frac{q_i}{\left(\frac{T_0}{100}\right)^4}\left(1 - \frac{T_e}{T_0}\frac{T_0}{T_{ad}}\right) = 0. \tag{6.88}$$

On the other hand, based on (6.71)

$$\frac{T_{ad} - T_0}{T_0} = \frac{T_{ad}}{T_0} - 1 = \frac{h_g}{c_{pg}T_0}. \tag{6.89}$$

Based on (6.89), from (6.88) we have:

$$\left(\frac{T_e}{T_0}\right)^4 - 239\frac{q_i}{\left(\frac{T_0}{100}\right)^4}\left(1 - \frac{T_e}{T_0 + \frac{h_g}{c_{pg}}}\right) = 0, \tag{6.90}$$

and

$$\left(\frac{T_e}{100}\right)^4 + 239q_i\left(\frac{T_e}{T_0 + \frac{h_g}{c_{pg}}} - 1\right) = 0. \tag{6.91}$$

c_{pg} represents the mean specific heat between 0°C and the theoretical combustion temperature in this case.

Note that for fuel oil without air heating $c_{pg} = 1.275 - 1.30$ kJ/kg K, whereas with air heating $c_{pg} = 1.30 - 1.32$ kJ/kg K.

For coal without air heating $c_{pg} = 1.175 - 1.20$ kJ/kg K, while with air heating we have $c_{pg} = 1.21 - 1.235$ kJ/kg K.

The values for natural gas match those for fuel oil.

With reference to fuel oil and natural gas we can assume a conventional value of c_{pg} equal to 1.3 kJ/kg K; moreover, if T_0 takes the conventional value 293 K (it corresponds to 20°C), we obtain the following from (6.91).

$$\left(\frac{t_e + 273}{100}\right)^4 + 239q_i\left(\frac{t_e + 273}{0.77h_g + 293} - 1\right) = 0. \tag{6.92}$$

Equation (6.92) is shown in Fig. 6.14.

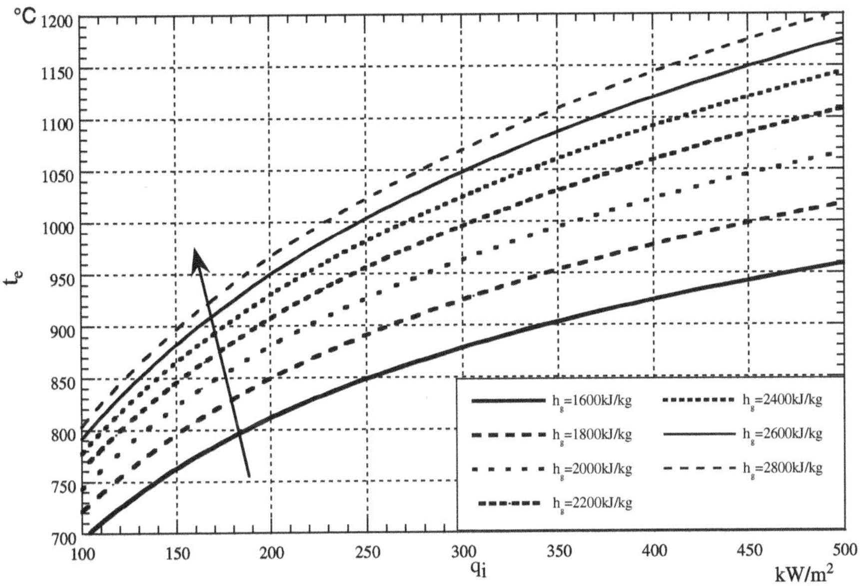

Fig. 6.14 Exit temperature from the furnace of the flue gas according to Konakow (6.92)

The diagram can be used for fuel oil, natural gas and other gaseous fuels.

Using a conventional value of c_{pg} does not introduce sensible errors, even though the actual value differs considerably from the assumed one. For instance, note that even with $c_{pg} = 1.13$ kJ/kg K instead of 1.30 kJ/kg K, and with very low values of h_g (poor fuels) that represent worst conditions in terms of made errors, Fig. 6.14 leads to a value of t_e that differs from the one computed through (6.91) by about 15°C.

Even using Konakow's method it is possible to introduce the corrective factors f_v and f_k already considered as far as Orrok's method.

The factor f_v reduces the involved surface.

Based on (6.82), note that the introduction of f_v is equal to considering a correct value of q_i given by

$$q'_i = f_v^2 q_i. \tag{6.93}$$

Then the equation of Ko must be corrected as follows:

$$Ko = 239 \frac{f_v^2 q_i}{\left(\frac{T_{ad}}{100}\right)^4}. \tag{6.94}$$

The process is not equally straightforward to factor in f_k. In fact, looking at (6.82), note how the amount of heat transferred into the furnace is proportional to f_k.

Based on Nötzlin's research the following can be written:

$$1 - \frac{T'_e}{T_{ad}} = f_k \left(1 - \frac{T_e}{T_{ad}}\right), \tag{6.95}$$

where T_e is the value of the absolute exit temperature obtained from (6.85), whereas T'_e is the actual absolute temperature that factors in the influence of f_k.

Based on Nötzlin's measurements, f_k is equal to one if the content of volatile matter is equal or greater than 20%. Otherwise

$$f_k = 1.23 - 0.012\,V, \tag{6.96}$$

where V is the percentage of volatile matter content.

A comparison of Fig. 6.12 with Fig. 6.14 shows that the resulting values of t_e differ in a more or less sensible way from one another depending on the values q_i and h_g. If the difference is equal to only 18°C for $q_i = 150\,\mathrm{kW/m^2}$ and $h_g = 1800$ kJ/kg, there is a difference of 148°C for $q_i = 500\,\mathrm{kW/m^2}$ and $h_g = 2800$ kJ/kg.

Note that low values of the two characteristic quantities occur when there is no heating of combustion air, and generally under reduced load, whereas the high values occur with a high superficial heat load in the furnace (referred to project requirements), as well as high air heating.

These differences lead to questioning which criteria are most reliable.

It is not easy to provide an answer, even though Konakow's criteria generally demonstrate more optimism about the heat transfer into the furnace, while Orrok's method is more pessimistic, especially for high values of q_i and h_g.

In view of these considerations, yet without presuming to completely solve the problem, we elaborated an equation to compute t_e in a more satisfactory way when compared Orrok's and Konakow's criteria.

Undoubtedly, the Orrok-Hudson equation is structurally better suited for a quick and simple determination of t_e, since it does not require calculation by trial and error (Konakow's method).

On the other hand, Orrok' equation in its original version is influenced by the value of factor C, usually assigned with a conventional value. The latter is the main cause for considerable, actual and not completely justifiable discrepancies between the values obtained through this equation and Konakow's equation.

So this is a spontaneous attempt to work on Orrok's equation through a value of C leading to values of t_e which are presumably closer to the reality of the phenomenon.

Therefore, we suggest the following value of C correlated to the value of h_g, i.e.,

$$C = \frac{461.9}{h_g^{0.15}}\ \mathrm{kJ^{0.65}s^{0.5}mkg^{-1.15}} \tag{6.97}$$

From (6.79) we obtain:

$$h_e = h_0 + \frac{1000}{\dfrac{2.165 h_g^{0.15}}{\sqrt{q_i}} + \dfrac{1000}{h_g}}. \tag{6.98}$$

Working as usual with the conventional value of c_{pg} which is equal to 1.164 kJ/kgK, we finally obtain the following equation:

$$t_e = 20 + \frac{1000}{\frac{2.52h_g^{0.15}}{\sqrt{q_i}} + \frac{1164}{h_g}}. \tag{6.99}$$

Equation (6.99) is shown in Fig. 6.15.

Note that the values of C obtained from (6.97) are lower than the conventional value relative to Orrok's criteria that is equal to 155.3, as we pointed out earlier.

For $h_g = 1600$ kJ/kg the result is $C = 152.7$, a slightly lower value; for $h_g = 2800$ kJ/kg the outcome is $C = 140.4$ instead; the suggested equation differs from the one obtained through Orrok's criteria in a more substantial way for high values of h_g for which the differences between the values of t_e computed according to Orrok and Konakow are greater.

Finally, note that the value of h_g for a given fuel and a potential desired air heating solely depends on the air index.

Then instead of the diagrams in Figs. 6.12, 6.14 and 6.15, it would be possible to build diagrams valid for a given fuel and a given air heating with a pencil characterized by the value of the air index or the excess of air.

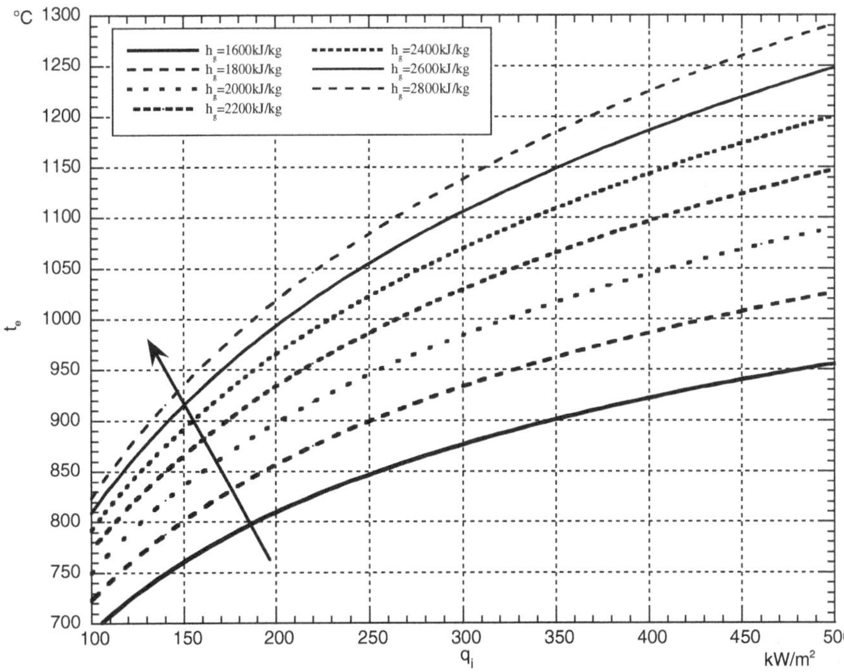

Fig. 6.15 Exit temperature from the furnace of the flue gas according to Annaratone (6.99)

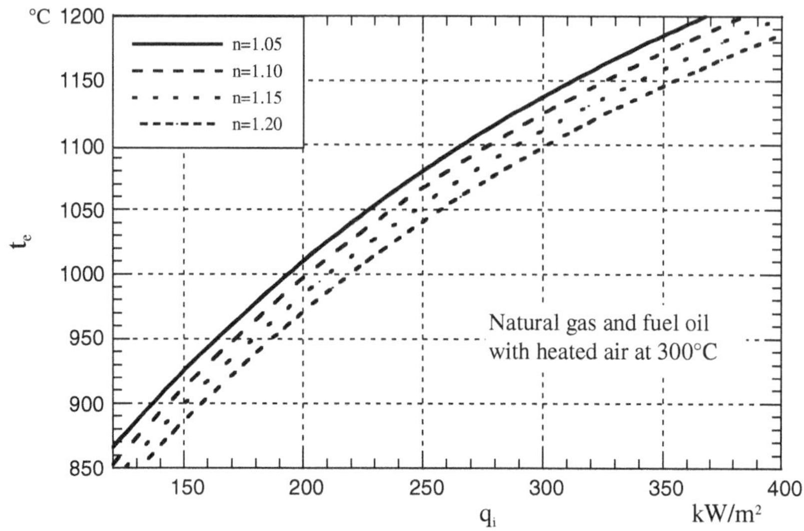

Fig. 6.16 Exit temperature from the furnace of the flue gas for power plant generators

An example of these diagrams with an experimental origin is represented by Fig. 6.16. When considering this diagram, note that with fuel oil combustion, an air index $n = 1.15$, and the fact that air is heated to 300°C, the value of h_g is about 2670 kJ/kg. Therefore, it is possible to compare this experimental diagram with those in Figs. 6.12, 6.14 and 6.15, and the outcome shows that out of these three diagrams the one closest to the experimental one (often it coincides) is the diagram in Fig. 6.15.

Naturally, the best solution, when possible, is to rely on experimental data registered on similar generators. This is what the biggest builders do by using these data to build tables or diagrams able to provide the exit temperature from the furnace or the heat transferred into it.

But the necessity to rely on calculation is always there to compare experimental data with theoretical ones, to study new constructive solutions not backed by experimental data, to do extrapolations of these data approximately knowing the influence of the different parameters (e.g., moving from the smallest already proven units to bigger units yet to be designed).

The described methods can be used to this extent, yet keeping in mind that the obtained values of t_e can in some cases differ even by $30 \div 40$°C from actual values.

At this point we need to determine how to compute the ideal radiated surface S.

Note that in the case of a tube bank that "sees" the flame, the surface S corresponds to the projected surface of the bank. Specifically, this is the surface corresponding to the cross-sectional area of the space where the bank is lodged.

The projected surface is also the reference point for the screens on the wall made with tubes side by side or with finned tubes side by side.

If the tubes are spaced instead, S is equal to the surface of the screened wall multiplied by the factor E obtained from Fig. 6.17.

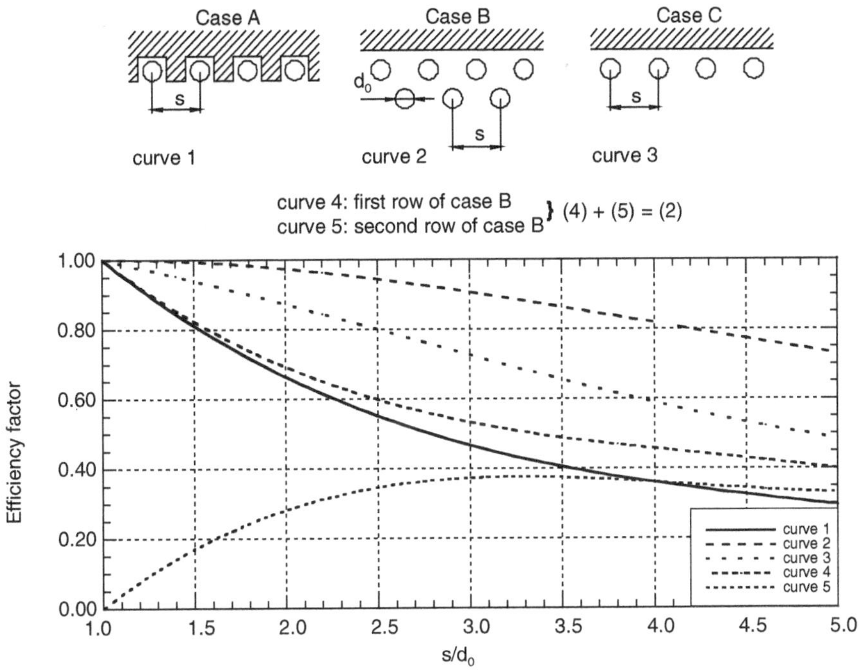

Fig. 6.17 Efficiency factor for radiated walls

In conclusion, note that based on (6.76) and (6.99) heat q_s transferred in the furnace to the walls during the time unit and by radiated surface unit is equal to:

$$q_s = q_i \frac{X}{X+1}; \tag{6.100}$$

given that

$$X = 2.165 \times 10^{-3} \frac{h_g^{1.15}}{\sqrt{q_i}}. \tag{6.101}$$

Equation (6.100) is shown in Fig. 6.18.

As we pointed out at the beginning of the section, all the suggested computational methods are in reference to the furnace of a steam generator. This includes walls coated with tubes and a wall temperature which typically never exceeds 300°C. In fact, it is often considerably lower.

The question is whether the suggested equations and whichever modifications may be used to compute the heat transfer from the flame, or the exit temperature of the flue gas from the environment in which the flame is located, in case the temperature of the walls is very high.

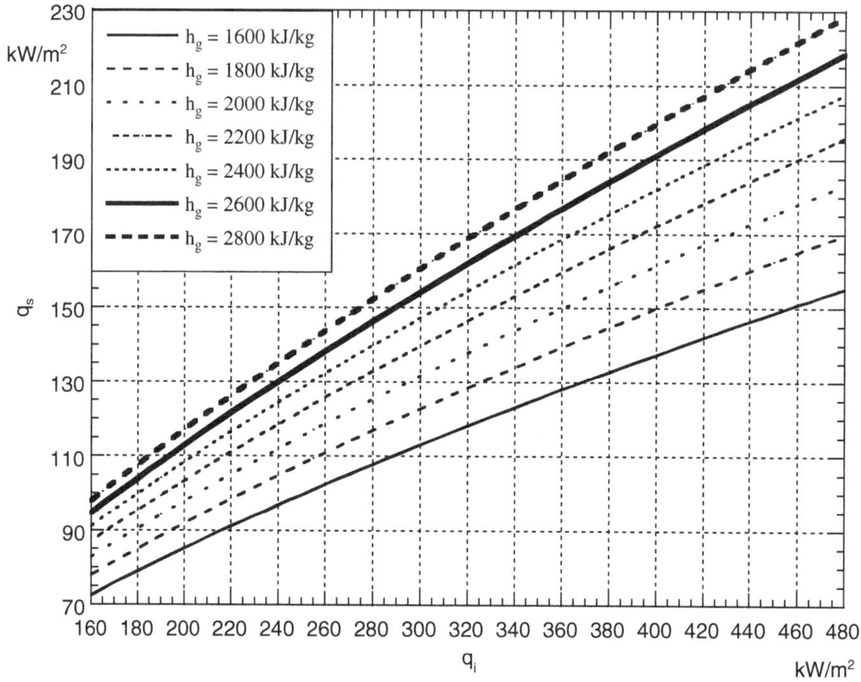

Fig. 6.18 Heat transferred in furnace

There is no easy solution to this problem. We developed an empirical criterion to be applied if the temperature of the walls exceeds 300°C. We think it will be useful to the reader, since it produces values that are fairly close to the real ones.

If t_e^* indicates the values of t_e obtained through (6.99), the exit temperature of the flue gas t_e is given by

$$t_e = 20 + \varphi\left(t_e^* - 20\right). \tag{6.102}$$

The values of φ are shown in Table 6.2.

These values were obtained based on the following empirical criterion.

First of all, we examined the situation and its known characteristics relative to heat transfer in a furnace of a steam generator.

The temperature of the radiated tubes in the furnace was assumed to be 250°C.

Since the reference temperature of the flame must necessarily range from its adiabatic temperature, and the exit temperature of the gases from the furnace (computed through (6.99)), we established that based on (6.100), exponent 4 in Stefan-Boltzmann's equation is not compatible and must be assumed to be equal to 3.8. Then, by first approximation we assumed a reference temperature of the flame equal to $(2T_e - T_{ad})/3$, given that both T_e and T_{ad} are the absolute exit temperature and the absolute adiabatic temperature, respectively. Since this temperature

Table 6.2 Factors φ

h_g	q_i	t_w	φ	φ_{80}	φ_{60}	h_g	q_i	t_w	φ	φ_{80}	φ_{60}
1600	150	400	1.023	1.059	1.110	2400	150	400	1.000	1.044	1.110
		600	1.084	1.116	1.162			600	1.058	1.100	1.161
		800	1.194	1.219	1.255			800	1.164	1.199	1.251
		1000	1.361	1.377	1.400			1000	1.325	1.352	1.391
	250	400	1.014	1.045	1.089		250	400	1.000	1.033	1.090
		600	1.051	1.080	1.120			600	1.028	1.066	1.120
		800	1.119	1.143	1.177			800	1.093	1.126	1.174
		1000	1.227	1.244	1.269			1000	1.194	1.221	1.261
	350	400	1.010	1.038	1.078		350	400	1.000	1.022	1.075
		600	1.036	1.063	1.100			600	1.011	1.046	1.096
		800	1.086	1.109	1.142			800	1.057	1.089	1.134
		1000	1.166	1.183	1.209			1000	1.131	1.158	1.197
	450	400	1.007	1.034	1.071		450	400	1.000	1.019	1.067
		600	1.029	1.053	1.088			600	1.004	1.037	1.083
		800	1.068	1.090	1.120			800	1.040	1.071	1.114
		1000	1.131	1.149	1.173			1000	1.099	1.125	1.163
2000	150	400	1.000	1.040	1.100	2800	150	400	1.023	1.073	1.144
		600	1.059	1.096	1.150			600	1.084	1.130	1.196
		800	1.164	1.195	1.240			800	1.193	1.232	1.289
		1000	1.325	1.347	1.380			1000	1.358	1.388	1.432
	250	400	1.000	1.029	1.081		250	400	1.013	1.056	1.118
		600	1.029	1.063	1.111			600	1.049	1.090	1.149
		800	1.094	1.123	1.166			800	1.115	1.151	1.204
		1000	1.197	1.220	1.254			1000	1.217	1.247	1.292
	350	400	1.000	1.020	1.067		350	400	1.009	1.048	1.105
		600	1.013	1.044	1.088			600	1.034	1.072	1.126
		800	1.060	1.087	1.127			800	1.081	1.115	1.165
		1000	1.135	1.158	1.191			1000	1.156	1.185	1.229
	450	400	1.000	1.017	1.060		450	400	1.006	1.043	1.096
		600	1.006	1.035	1.077			600	1.026	1.061	1.112
		800	1.043	1.069	1.107			800	1.063	1.095	1.143
		1000	1.103	1.125	1.158			1000	1.121	1.150	1.192

did not agree with (6.100), we introduced a corrective factor of this temperature (the factor is a function of q_i and h_g) that would take it into account. We used this new equation to compute the heat transfer from the flame with reference to a different wall temperature. This leads to a different exit temperature of the gas from the furnace. Based on these values, it was possible to obtain the values of φ. This under the assumption that the black level of the walls is comparable to the black level (roughly equal to 0.90–0.95) of the tube walls of the furnace of a steam generator.

The problem becomes even more complex if we want to consider the possibility where the black level of the walls is lower than that of the tube walls of the steam generators.

Based on Sect. 6.4 we know that if the flame develops in a rather large space the black level of the walls has no impact because they behave like a black body anyway.

But if we consider a normal furnace the outcome is different. It is impossible, though, to identify each time the impact of the black level of the walls with any kind of precision.

Even in this case we adopted an empirical criterion to correct the values of φ obtained with the computational method described above.

This way we obtained the values of φ_{80} and φ_{60} in Table 6.2. They refer to walls with a black level equal to 80%, and to 60% of the black level of tube walls of steam generators furnaces.

An additional criterion may be the following.

If R_{B} indicates the ratio between the black level of the walls and that of the generator tube walls, we adopt an ideal value S^* of the radiated surface equal to $S^* = R_B S$.

This way we consider an ideal value of q_i given by $q_i^* = q_i/R_B$. Based on q_i^*, we compute t_e^* through (6.99) and, if the wall temperature exceeds 300°C, always based on q_i^*, and also based on h_g, we obtain the value of φ in Table 6.2. As we shall see through the examples, the values of t_e obtained with this computation method are very close to those obtained by applying corrective factors, such as φ_{80} and φ_{60}. This method has two advantages. First, it is applicable to any value of R_B, secondly it is the only applicable method if the different walls of the furnace have different black level. A disadvantage is the fact that the value of q_i^* cannot be included in Table 6.2, thus preventing its use if the temperature of the walls exceeds 300°C.

Here are a few examples.

1° example:

If $h_g = 2000$ kJ/kg and $q_i = 150\,\mathrm{kW/m^2}$, from (6.99) we obtain $t_e^* = 836$°C. Assuming walls at 800°C with the same black level of the tube walls, from Table 6.2 we obtain $\varphi = 1.164$; therefore, from (6.102) we obtain that $t_e = 20 + 1.164 \times 816 = 970$°C; at this point we assume that the black level of the walls is 60% of that of the tube walls; in that case from Table 6.2 we obtain $\varphi_{60} = 1.24$; therefore, based on (6.102) we have $t_e = 20 - 1.24 \times 816 = 1032$°C.

We apply the second criterion. Given that $R_B = 0.6$, we have $q_i^* = 150/0.6 = 250\,\mathrm{kW/m^2}$; based on this value of q_i, from (6.99) we obtain $t_e^* = 946$°C, and from Table 6.2 $\varphi = 1.094$, so that from (6.102) $t_e = 20 + 1.094 \times 926 = 1033$°C. As you see, this value basically coincides with the previously obtained temperature of 1032°C.

2° example:

If $h_g = 2400$ kJ/kg and $q_i = 250\,\mathrm{kW/m^2}$, from (6.99) we obtain $t_e^* = 1023$°C; assuming that the wall temperature is equal to 1000°C and that the black level is comparable to that of the tube wall, from Table 6.2 we obtain $\varphi = 1.194$. Based on (6.102) this leads to $t_e = 20 + 1.194 \times 1003 = 1218°C$. At this point we assume that the black level of the walls is equal to 80% of that of a tube wall. From Table 6.2 we obtain $\varphi_{80} = 1.221$. Then $t_e = 20 + 1.221 \times 1003 = 1244°C$.

We apply the second criterion; given that $R_B = 0.8$, we have $q_i^* = 250/0.8 = 312.5\,kW/m^2$; based on this value of q_i, from (6.99) we obtain $t_e^* = 1080°C$; from Table 6.2 we obtain $\varphi = 1.155$, then $t_e = 20{+}1.155{\times}1060 = 1244°C$; even in this case both temperatures coincide.

3° example:

If $h_g = 2800\,kJ/kg$ and $q_i = 350\,kW/m^2$, from (6.99) we obtain $t_e^* = 1184°C$; assuming that the wall temperature is equal to 1000°C with a black level comparable to that of a tube wall, from Table 6.2 we obtain $\varphi = 1.156$; then $t_e = 20 + 1.156 \times 1164 = 1366°C$; if the black level is equal to 80% of that of the tube wall, from Table 6.2, $\varphi_{80} = 1.185$, and then $t_e = 20 + 1.185 \times 1164 = 1399°C$.

By applying the second criterion, we have $R_B = 0.8$ and $q_i^* = 350/0.8 = 437.5\,kW/m^2$; from (6.99) we obtain $t_e^* = 1252°C$; from Table 6.2 we obtain $\varphi = 1.125$, and then $t_e = 20{+}1.125 \times 1232 = 1406°C$. This value is almost identical to the previously calculated temperature of 1399°C.

6.7 Flame Radiation and Convection

If the flame develops in such a limited space that the flue gas licks the walls, heat transferred by convection is added besides heat transferred by radiation.

The typical case is the cylindrical flue of smoke tube boilers.

The heat transfer by convection represents a small percentage of the total heat transfer, but its entity cannot be ignored. Thus, it must be taken into consideration, and this is a procedure to do just that.

The computation of the exit temperature of gas from the flue requires previous calculation of maximum flame temperature t_f.

Note in this context that thermal flux in the flue varies considerably along the axis of the flue itself.

Specifically, maximum flame temperature and maximum thermal flux is reached at close distance from the burner.

This maximum value of the thermal flux is called peak flux.

The mathematical model for the calculation of temperature t_f is based on the following fundamental assumptions.

The potential heat transfer by convection in the peak flow area is ignored, since it is in any case irrelevant compared with the heat transfer by radiation from the flame.

The assumption is that the length of the area involved by the peak phenomenon is equal to 1 m per diameter of the flue which in turn is also equal to 1 m; for diameters other than 1 m it is assumed to be proportional to the square root of the diameter.

The mean thermal flux q_m due to flame radiation in the peak area is equal to:

$$q_m = \sigma\varepsilon\left[\left(t_f + 273\right)^4 - \left(t_w + 273\right)^4\right]; \tag{6.103}$$

σ stands for Stefan-Boltzmann constant, ε for the emissivity of the flame, t_f for the flame temperature and t_w for the wall temperature of the flue.

The second term in square brackets can be ignored compared to the first one. In fact, if we consider a flame temperature of 1500°C and a wall temperature of 450°C, the error made by ignoring this term is below 3%.

Taking into account the disregard for the not very significant convective phenomena, we can use the following equation:

$$q_m = \sigma\varepsilon\left(t_f + 273\right)^4. \tag{6.104}$$

The Stefan-Boltzmann constant is equal to 5.672×10^{-8} W/m^2 K^4, and so we can write that

$$q_m = 5.672\varepsilon\left(\frac{t_f + 273}{100}\right)^4 \tag{6.105}$$

with q_m expressed in W/m^2.

Based on the assumptions that were made, the heat radiated by the flame in the peak area indicated with q_r is therefore equal to:

$$q_r = q_m\pi D\sqrt{D} = 5.672\varepsilon\pi D^{3/2}\left(\frac{t_f + 273}{100}\right)^4. \tag{6.106}$$

If q is the heat introduced into the flue and M_f is, as usual, the fuel mass flow rate, H'_n is the introduced heat per mass unit of the fuel, G is the amount of flue gas per mass unit of fuel, and c_{pg} is the average specific isobaric heat of the gas between room temperature t_0 and flame temperature, we have:

$$q = M_f H'_n = M_f G c_{pg}\left(t_f - t_0\right) + q_r. \tag{6.107}$$

Assuming 20°C for temperature t_0 and recalling (6.106) we have:

$$M_f\left[H'_n - Gc_{pg}\left(t_f - 20\right)\right] = 5.672\varepsilon\pi D^{1.5}\left(\frac{t_f + 273}{100}\right)^4; \tag{6.108}$$

then

$$q\left[1 - \frac{G}{H'_n}c_{pg}\left(t_f - 20\right)\right] = 5.672\varepsilon\pi D^{1.5}\left(\frac{t_f + 273}{100}\right)^4. \tag{6.109}$$

Note that the ratio H'_n/G stands for the heat introduced into the flue per mass unit of flue gas.

As usual, indicating this with h_g, based on (6.109) we obtain:

$$q = \frac{5.672\varepsilon\pi D^{1.5}\left(\frac{t_f + 273}{100}\right)^4}{1 - \frac{c_{pg}}{h_g}\left(t_f - 20\right)} \tag{6.110}$$

As far as emissivity, experiments show that it is equal to 0.35 for flues of 3.75 ft in diameter (1.14 m).

On the other hand, if l stands for the length of the area involved in the phenomenon (which, based on the assumption of the computation is proportional to $\sqrt{D}$), based on Mac Adams' suggestion we can write that

$$\varepsilon = 1 - e^{-Cl}, \tag{6.111}$$

given that C is a constant.

It is therefore possible to write that

$$0.35 = 1 - e^{-C\sqrt{1.14}}; \tag{6.112}$$

then

$$-C\sqrt{1.14} = \log_e 0.65, \tag{6.113}$$

and finally,

$$C = 0.4. \tag{6.114}$$

Then we assume the following:

$$\varepsilon = 1 - e^{-0.4\sqrt{D}}, \tag{6.115}$$

and (6.110) is written as follows:

$$q = \frac{5.672\pi\left(1 - e^{-0.4\sqrt{D}}\right)D^{1.5}\left(\frac{t_f + 273}{100}\right)^4}{1 - \frac{c_{pg}}{h_g}\left(t_f - 20\right)} \tag{6.116}$$

The value of c_{pg} does not vary in a significant way with variations of temperature t_f for flame temperatures that are practically possible.

A fixed value of c_{pg} corresponding to a flame temperature of 1600°C for flue gas with a moisture percentage by mass of 6% is used to simplify calculation.

It is equal to 1240 J/kg K.
Thus, we can write that

$$q = \frac{5.672\pi\left(1 - e^{-0.4\sqrt{D}}\right) D^{1.5} \left(\frac{t_f + 273}{100}\right)^4}{1 - \frac{1240}{h_g}\left(t_f - 20\right)}. \tag{6.117}$$

Based on the heat introduced into the flue within the time unit (in W), the value of h_g (in J/kg), as well as the value of the diameter, it is possible to obtain temperature t_f by trial and error through (6.117).

It is also possible to develop an approximate equation to compute t_f directly.

Note that expressing q in kW, h_g in kJ/kg and introducing cross-section area A of the flue, the following equation can be obtained from (6.117):

$$\frac{q}{A} = \frac{22.69\left(1 - e^{-0.4\sqrt{D}}\right) D^{-0.5} \left(\frac{t_f + 273}{100}\right)^4}{1000 - \frac{1240}{h_g}\left(t_f - 20\right)}. \tag{6.118}$$

The analysis of (6.118) shows that the influence of the diameter is quite modest, whereas the value of h_g and the value of the ratio q/A, i.e., the heat introduced into the flue within the time unit and by cross-section area unit, are crucial.

We obtain the following approximate equation that does not include the diameter:

$$t_f = 775 + 0.18h_g + \left(0.0488h_g - 20\right)\frac{q}{1000A} - \left(0.00176h_g + 0.207\right)\left(\frac{q}{1000A}\right)^2 \tag{6.119}$$

with $q/A = 2000 - 12000\,\mathrm{kW/m^2}$, $h_g = 1800 - 2700$ kJ/kg and $D = 0.6 - 1.5$ m.

The errors made using (6.119) instead of (6.118) do not exceed 3.5% plus or minus; these errors are clearly tolerable given the quantity they refer to.

At this point, note that if we could ignore the heat transfer by convection we could write that

$$M_g c_{pg} dT = -\sigma\varepsilon\left(T^4 - T_w^4\right) dS, \tag{6.120}$$

where M_g stands for the mass flow rate of the flue gas, c_{pg} for the specific isobaric heat, T and T_w for the absolute temperatures of the flue gas and the wall of the flue, respectively.

(6.120) can be written as follows:

$$\frac{dT}{dS} = -\frac{\sigma\varepsilon\left(T^4 - T_w^4\right)}{M_g c_{pg} T^4} T^4. \tag{6.121}$$

We introduce factor K_1 given by:

$$K_1 = \frac{\sigma \varepsilon \left(T^4 - T_w^4\right)}{M_g c_{pg} T^4} \tag{6.122}$$

Therefore, (6.121) is written like this:

$$\frac{dT}{dS} = -K_1 T^4. \tag{6.123}$$

Factor K_1 varies along the flue depending on variations of T, T_w, the specific isobaric heat c_{pg} and presumably the emissivity, too. We consider it like a constant, ideally referring to mean values of the different quantities. Note that it will not be necessary to compute it, as we shall see later on.

Assuming that

$$T^{-3} = -3y, \tag{6.124}$$

we obtain

$$y = -\frac{1}{3}T^{-3}; \tag{6.125}$$

$$y' = T^{-4}\frac{dT}{dS}; \tag{6.126}$$

$$\frac{dT}{dS} = T^4 y'. \tag{6.127}$$

Then, based on (6.123):

$$y' = -K_1. \tag{6.128}$$

The integral of (6.128) is given by:

$$y = -K_1 S - C \tag{6.129}$$

where C is a constant.

From (6.129) and recalling (6.124), after a series of steps we obtain:

$$T = \sqrt[3]{\frac{1}{3K_1 S - 3C}}, \tag{6.130}$$

or

$$T = \sqrt[3]{\frac{1}{3K_1 S + C_1}} \tag{6.131}$$

where C_1 is a constant.

The surface S_0 of the flue is the one between the area where the flame forms and the end of the flue, while T_f and T_e^* represent the maximum absolute temperature of the flame and the temperature of the flue gas at the exit of the flue.

For $S = 0$, $T = T_f$, and from (6.131) we obtain:

$$C_1 = \frac{1}{T_f^3}; \tag{6.132}$$

then

$$T = \sqrt[3]{\frac{1}{3K_1 S + \frac{1}{T_f^3}}}. \tag{6.133}$$

Moreover, for $S = S_0$:

$$T_e^* = \sqrt[3]{\frac{1}{3K_1 S_0 + \frac{1}{T_f^3}}}. \tag{6.134}$$

Leading through a series of steps to:

$$K_1 = \frac{1}{3S_0}\left(\frac{1}{T_e^{*3}} - \frac{1}{T_f^3}\right). \tag{6.135}$$

If we know the temperature the flue gas would have at the exit of the flue in absence of convective phenomena and also temperature T_f, it is possible to compute the value of factor K_1 regardless of difficult considerations about the value of the emissivity.

Now, if we consider the heat transfer by convection as well, we must write that

$$M_g c_{pg} dT = -\left[\sigma\varepsilon\left(T^4 - T_w^4\right) + \alpha_g\left(T - T_w\right)\right] dS, \tag{6.136}$$

where α_g stands for the heat transfer coefficient of the gas.

Equation (6.136) can be written as follows:

$$\frac{dT}{dS} = -\frac{\sigma\varepsilon(T^4 - T_w^4)}{M_g c_{pg} T^4} T^4 - \alpha_g \frac{T - T_w}{M_g c_{pg} T} T. \tag{6.137}$$

Recalling (6.122) and introducing factor K_2 given by:

$$K_2 = \alpha_g \frac{T - T_w}{M_g c_{pg} T}, \tag{6.138}$$

Equation (6.137) is written as follows:

$$\frac{dT}{dS} = -K_1T^4 - K_2T. \tag{6.139}$$

Of course, factor K_2 is a function of S because the wall temperature T_w varies along the flue, while temperature T varies along the flue, followed in turn by variations as far as the specific isobaric heat and the heat transfer coefficient.

We can consider it constant and adopt mean values for the different quantities, therefore writing that:

$$K_2 = \alpha_{gm}\frac{T_m - T_{wm}}{M_g c_{pgm} T_m}; \tag{6.140}$$

This means that the specific isobaric heat will be the average between temperature T_f and temperature T_e of the gas at the exit of the flue, and that temperature T_m which represents the mean value among those mentioned above will be used for the computation of the heat transfer coefficient of the gas.

According to the same position in (6.124) and based on (6.126) we obtain:

$$y' = -K_1 - K_2T^{-3}. \tag{6.141}$$

Therefore, always based on (6.124):

$$y' = 3K_2y - K_1 \tag{6.142}$$

Equation (6.142) is a linear differential equation and its integral is given by:

$$y = e^{3K_2S}\left(-K_1\int e^{3K_2S}dS + C_2\right); \tag{6.143}$$

then

$$T^{-3} = e^{3K_2S}\left(\frac{K_1}{K_2}e^{3K_2S} - 3C_2\right). \tag{6.144}$$

Finally,

$$T = \frac{e^{-K_2S}}{\sqrt[3]{\frac{K_1}{K_2}e^{3K_2S} + C_3}}. \tag{6.145}$$

For $S = 0$ we have $T = T_f$, and based on (6.145):

$$C_3 = \frac{1}{T_f^3} - \frac{K_1}{K_2} \tag{6.146}$$

and

$$T = \frac{e^{-K_2 S}}{\sqrt[3]{\frac{K_1}{K_2}\left(e^{3K_2 S_0} - 1\right) - \frac{1}{T_f^3}}}. \tag{6.147}$$

Thus, the exit temperature of the gas from the flue T_e is given by:

$$T_e = \frac{e^{-K_2 S_0}}{\sqrt[3]{\frac{K_1}{K_2}\left(e^{3K_2 S_0} - 1\right) - \frac{1}{T_f^3}}}. \tag{6.148}$$

Recalling (6.135) at this point and admitting that the value of K_1 may be assumed to be approximately equal to the one relative to transfer by pure radiation, we obtain:

$$T_e = \frac{e^{-K_2 S_0}}{\sqrt[3]{\frac{e^{3K_2 S_0} - 1}{3K_2 S_0}\left(\frac{1}{T_e^{*3}} - \frac{1}{T_f^3}\right) + \frac{1}{T_f^3}}}. \tag{6.149}$$

Equation (6.149) helps to compute T_e, once the values of T_e^* and T_f are known and by computing (potentially by trial and error) the value of factor K_2 which obviously depends on T_e.

The analysis of (6.149) shows that it can be substituted in a completely satisfactory way by the following and easier equation:

$$t_e = t_e^* - 4.55\sqrt[3]{t_f}\sqrt{t_e^*}K_2 S_0. \tag{6.150}$$

All temperatures are in °C; the exit temperature t_e of the flue gas from the flue is computed based on temperature t_e^* (this is the exit temperature of the gas computed based on the criteria in Sect. 6.6), the maximum temperature of the flame t_f (for which the calculation criteria were explained earlier), the value of factor K_2 (which depends on the entity of the convection phenomena), and surface S_0.

The latter must be computed as follows:

$$S_0 = \pi D\left(L - \sqrt{D}\right), \tag{6.151}$$

where L is the length of the flue; L and D are expressed in m.

Note that the heat transfer coefficient a_{gm} in the equation of K_2 must be computed based on the criteria in Sect. 5.4.

6.8 Radiation of CO_2 and Steam

A comparison between the actual heat transferred to the tubes of a bank, with the one computed based on the heat transfer coefficient of gas (Sects. 5.4 and 5.8) shows increasingly stronger differences matching the increasing temperature of the flue gas.

This finding in 1923, lead Schack to consider the possibility that the increase in transferred heat were caused by the radiation in the infrared of the steam and the carbon dioxide in the gas. In reality, both steam and CO_2 possess considerable absorption bands in the infrared field, i.e., for wave lengths greater than 0.8μ.

Calculation of this radiation performed with appropriate equations showed that it can be such that it must be included in the computation about transferred heat.

Subsequent research was done by Thomas, Möller, Schmick, Schmidt, Hottel, Mangeldorf and Eckert.

Specifically, based on studies by Hottel and Egbert it was possible to create the diagrams in Figs. 6.19, 6.20, 6.21 and 6.22.

The first two are in reference to carbon dioxide, the other two to steam.

All diagrams on the x-axis include the reference temperature in °C; the parameter of the different curves is represented by *ps* (in m× atm) where *p* stands for partial pressure of CO_2 or steam (in atm), whereas *s* indicates the mean beam length.

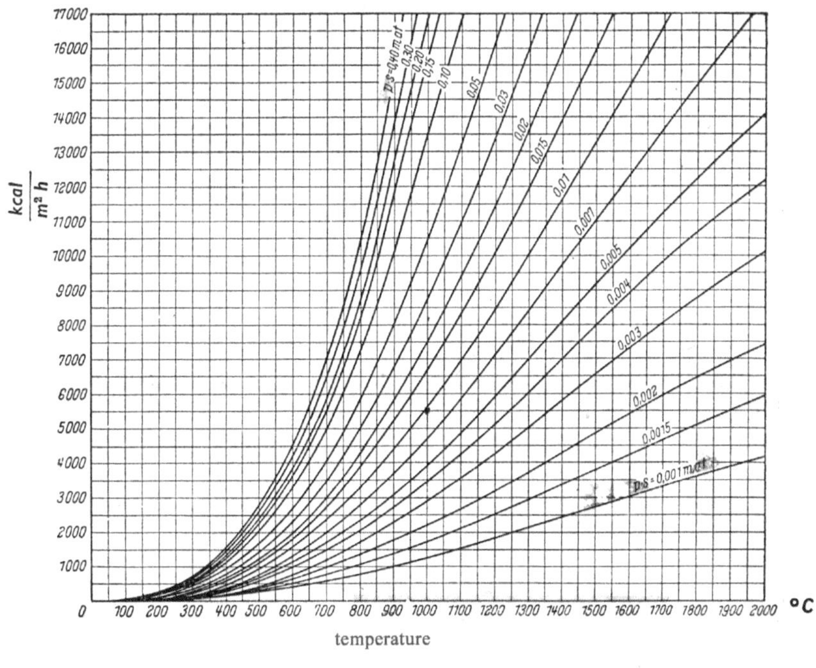

Fig. 6.19 Radiation of carbon dioxide

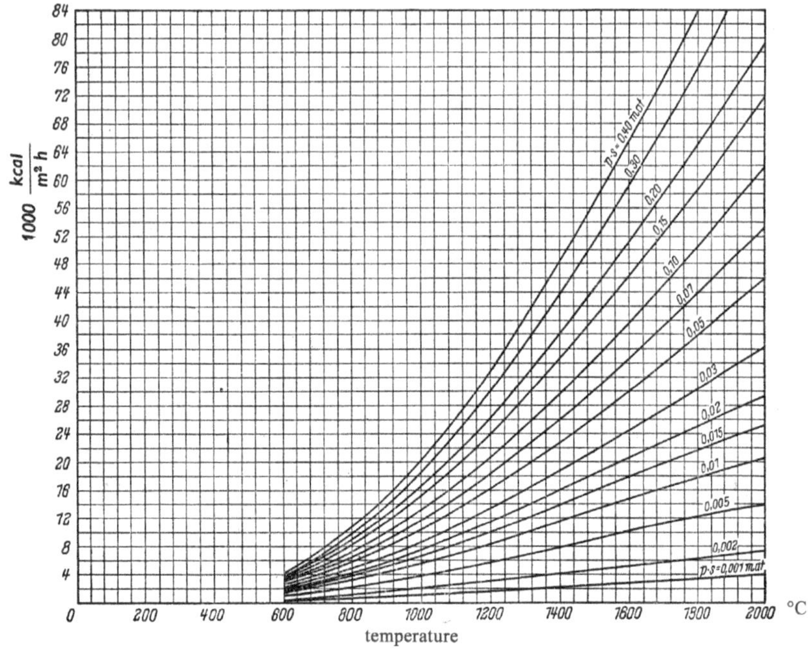

Fig. 6.20 Radiation of carbon dioxide

Partial pressure is the pressure that the gas or the steam would have if they were the only occupants of the volume occupied by the flue gas. It corresponds to the volumetric percentage of the gas or steam in question with respect to the totality of the components of flue gas. In other words, if, for instance, the volumetric percentage of CO_2 in flue gas is equal to 10%, the partial pressure is equal to 0.1 atm.

The mean beam length (in m) is a characteristic quantity of the phenomenon which depends on the geometry; the values will be specified as follows.

The y-axis indicates the heat by the time and surface unit radiated by the gas or by the steam at the temperature indicated on the x-axis towards a black wall at a temperature of 0 K.

Using the diagrams requires the following.

As far as CO_2, the value of its radiation can be read on the y-axis of Fig. 6.19 or Fig. 6.20; this value indicated by q'_{CO_2} corresponds to the temperature of the CO_2 itself. Then we read the value of the radiation corresponding to the temperature of the radiated wall indicated by q''_{CO_2}. If the black level is equal to B, heat q_{CO_2} radiated to the wall is equal to:

$$q_{CO_2} = B\left[q'_{CO_2} - \left(\frac{T'}{T_w}\right)^{0.65} q''_{CO_2}\right] \tag{6.152}$$

where T' and T_w are the absolute temperatures of CO_2 and wall, respectively.

By analogy, the same process is valid using the diagrams in Figs. 6.21 and 6.22.

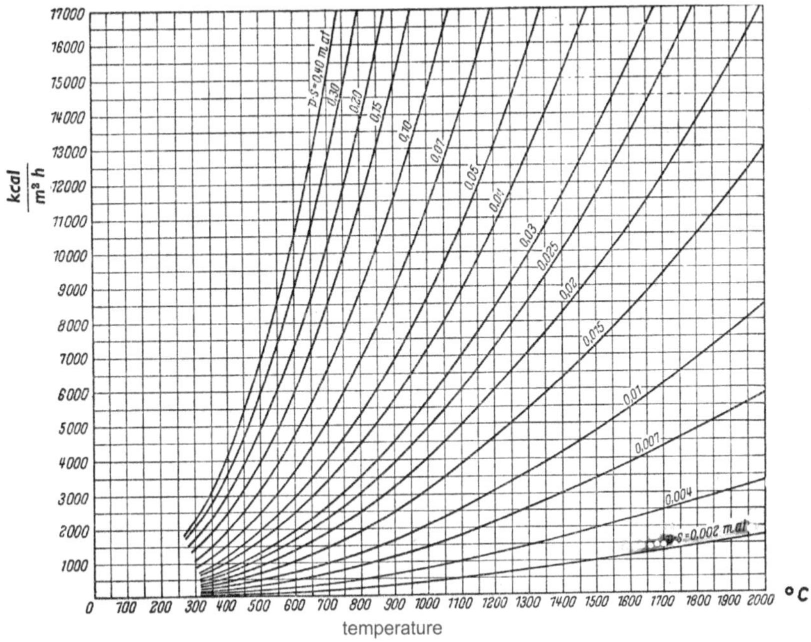

Fig. 6.21 Radiation of steam

By indicating the values taken from the ordinate with respect to the steam temperature as well as the wall temperature with q'_{H_2O} and q''_{H_2O}, the heat radiated to the wall q_{H_2O} is given by

$$q_{H_2O} = B(q'_{H_2O} - q''_{H_2O}). \tag{6.153}$$

Note that the radiation values in Figs. 6.19, 6.20, 6.21 and 6.22 are provided in kcal/m^{2}h; in order to compute them in W/m^2 they must be multiplied by 1.163.

Schack developed equations to compute the heat radiated by the CO_2 and by the steam in substitution of Hottel's diagrams discussed earlier.

If T' and T_w stand for the absolute temperatures of the gas and the radiated wall, the heat radiated by carbon dioxide can be computed through the following equation:

$$q_{CO_2} = 10.349B\left(p_{CO_2}x_r\right)^{0.4}\left[\left(\frac{T'}{100}\right)^{3.2} - \left(\frac{T_w}{100}\right)^{3.2}\left(\frac{T'}{T_w}\right)^{0.65}\right]; \tag{6.154}$$

q_{CO_2} is the heat expressed in W/m^2, B is the black level of the surface, p_{CO_2} is the partial pressure of CO_2 in atm and x_r, the so-called mean beam length expressed in m, the significance and values of which will be specified later.

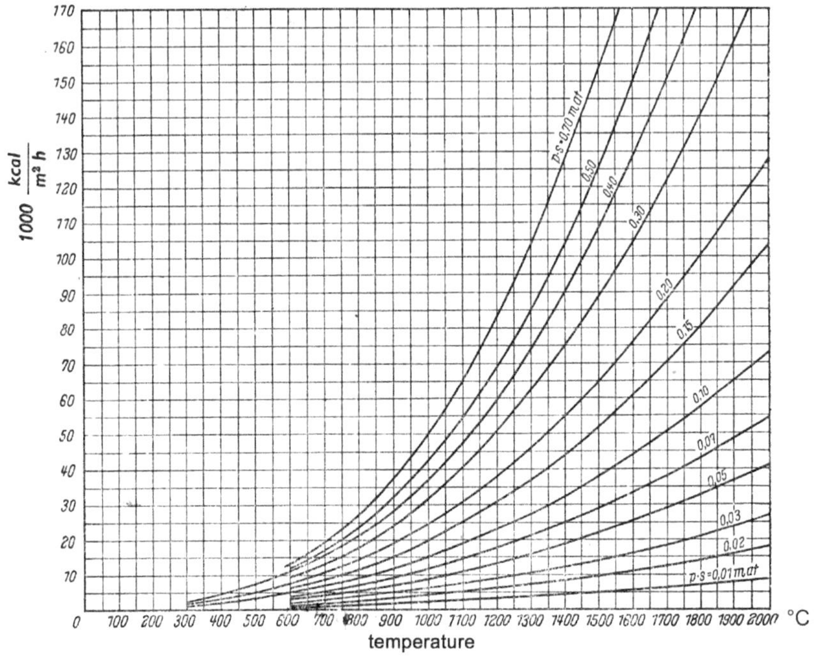

Fig. 6.22 Radiation of steam

(6.154) is valid for $p_{CO_2}x_r = 0.003 - 0.4\,\text{atm} \times \text{m}$, and $T' = 773 - 2073$ K ($t' = 500 - 1800°\text{C}$).

As far as the black level corresponding to the ratio between the heat radiated by the surface at a certain temperature, and the heat radiated by the blackbody at the same temperature, we observe that, for instance, $B = 0.745$ is the result of experiments on firebricks. Most interestingly, in the case of radiation inside a tube bank, it is possible to assume that $B = 0.90 - 0.95$.

By analogy, as far as radiation from steam, we can write the following equation:

$$q_{H_2O} = B\left(46.51 - 84.89\,p_{H_2O}x_r\right)\left(p_{H_2O}x_r\right)^{0.6}\left[\left(\frac{T'}{100}\right)^{\gamma} - \left(\frac{T_w}{100}\right)^{\gamma}\right], \qquad (6.155)$$

given that

$$\gamma = 2.32 + 1.37\sqrt[3]{p_{H_2O}x_r}. \qquad (6.156)$$

In (6.155) q_{H_20} stands for the heat radiated by H_2O expressed in W/m^2, p_{H_2O} for the partial pressure of H_2O, while the other symbols have the already known meaning.

Equation (6.155) is valid for $p_{H_2O}x_r = 0 - 0.36\,\text{atm} \times \text{m}$ and for $T' = 673 - 2173$ K ($t' = 400 - 1900$°C). In the rare instances when the equations above are invalid (e.g., the large channels of certain waste-heat generators), it is possible to use Hottel's diagrams discussed before.

The heat q_r radiated by the flue gas within the time unit and per surface unit is therefore equal to:

$$q_r = q_{CO_2} + q_{H_2O}. \tag{6.157}$$

Thus, we can compute q_r through (6.154) and (6.155).

Note that if the CO_2, and the steam are present simultaneously their radiations influence each other and reduce their global impact. The maximum deviation occurs when the ratio CO_2/H_2O is about 0.6.

This reduction is significant only for high values of the mean beam length; for $px_r = 0.03$ (having indicated the sum of both partial pressures with p) the reduction amounts to 2%; for $px_r = 0.61$ the reduction is equal to 7.8%; in most cases it may be ignored, and we will do that by validating (6.157).

A second observation is about the potential presence of SO_2 in the flue gas (fuels containing sulphur, such as fuel oils); even SO_2 transfers heat by radiation. The values are similar to those of carbon dioxide, and it may be compared to the latter during computation.

If we introduce an ideal heat transfer coefficient α'_r that identifies the radiated heat, we may write that

$$q_r = q_{CO_2} - q_{H_2O} = \alpha'_r \left(t' - t_w\right), \tag{6.158}$$

where t' and t_w are temperatures (in °C) corresponding to T' and T_w.

From (6.158) we have:

$$\alpha'_r = \frac{q_{CO_2} + q_{H_2O}}{t' - t_w}. \tag{6.159}$$

The availability of α'_r is very useful as far as computation requirements. In fact, the heat transfer by radiation is given by (6.158), whereas the one by convection is equal to:

$$q_c = \alpha'_c \left(t' - t_w\right), \tag{6.160}$$

where α'_c stands for the heat transfer coefficient computed based on Chap. 5.

The total heat transferred by the gas to the tubes is therefore equal to:

$$q = q_c - q_r = \left(\alpha'_c - \alpha'_r\right)\left(t' - t_w\right). \tag{6.161}$$

At this point, if we introduce the total ideal heat transfer coefficient equal to:

$$\alpha' = \alpha'_c + \alpha'_r; \tag{6.162}$$

the transferred heat is given by:

$$q = \alpha' \left(t' - t_w\right). \tag{6.163}$$

Thus, α' stands for the heat transfer coefficient to be introduced into the equations, such as (1.18) or (1.20), to compute the overall heat transfer coefficient U.

This explains why it is so practical for a computer to have the ideal heat transfer coefficient by radiation. This is, of course, a quantity called heat transfer coefficient only because of its dimension, but in fact it is not really so, because this is about the radiated heat and not the one transferred by convection.

Let us consider the following range of the different quantities:

$$\begin{aligned} px_r &= 0.01\text{–}0.36\,\text{atm} \times \text{m} \\ \beta = p_{H_2O}/p_{CO_2} &= 0.3-2 \\ t_w &= 200-600^\circ\text{C} \\ \Delta t_w &= 200-1000^\circ\text{C} \\ B &= 0.95 \end{aligned}$$

where p is the sum of the partial pressures of carbon dioxide and steam, and Δt_w is the difference in temperature between the flue gas and the radiated wall.

Within this range it is possible to rapidly determine the value of α'_r by assuming that

$$\alpha'_r = K_r \overline{\alpha}'_r \tag{6.164}$$

The values of K_r can be taken from Table 6.3, those of $\overline{\alpha}'_r$ from Fig. 6.23. The potential errors range from –7% to +5%. These values are acceptable considering that the errors are typically small. The greater ones occur at the edges of the range.

With regard to partial pressures, if the fuel is a mix of hydrocarbons we have:

$$p_{CO_2} = \frac{0.01866\,(100 - H)}{8.882n + (0.17573n + 0.05558)\,H}; \tag{6.165}$$

$$p_{H_2O} = \frac{0.11117H}{8.882n + (0.17573n + 0.05558)\,H}. \tag{6.166}$$

given that H is the mass hydrogen percentage in the fuel and n the air index.

Table 6.4 was created based on (6.165) and (6.166) and may be used even if the fuel is not simply a mix of hydrocarbons provided the percentages of the components other than carbon and hydrogen are modest.

Table 6.3 K_r factor

px_r (atm × m)	$\beta = p_{H_2O}/p_{CO_2}$							
	0.3	0.4	0.5	0.7	1.0	1.3	1.6	2.0
0.005	0.108	0.107	0.106	0.104	0.102	0.099	0.098	0.096
0.010	0.145	0.145	0.144	0.143	0.141	0.139	0.137	0.135
0.015	0.175	0.175	0.174	0.174	0.172	0.171	0.169	0.167
0.020	0.200	0.200	0.200	0.200	0.200	0.198	0.197	0.196
0.025	0.221	0.222	0.223	0.224	0.225	0.224	0.223	0.222
0.030	0.241	0.243	0.244	0.246	0.248	0.248	0.248	0.247
0.035	0.259	0.262	0.264	0.267	0.270	0.271	0.272	0.271
0.040	0.276	0.280	0.283	0.287	0.291	0.293	0.294	0.295
0.045	0.292	0.297	0.301	0.306	0.311	0.315	0.316	0.318
0.050	0.308	0.313	0.318	0.325	0.331	0.335	0.338	0.340
0.060	0.327	0.344	0.350	0.360	0.369	0.375	0.379	0.383
0.070	0.365	0.374	0.381	0.394	0.406	0.414	0.420	0.425
0.080	0.390	0.402	0.411	0.426	0.441	0.452	0.459	0.466
0.090	0.415	0.428	0.439	0.457	0.476	0.488	0.497	0.506
0.100	0.439	0.454	0.467	0.488	0.509	0.524	0.535	0.545
0.120	0.484	0.504	0.520	0.547	0.575	0.594	0.608	0.621
0.140	0.527	0.551	0.571	0.603	0.638	0.662	0.679	0.696
0.160	0.568	0.596	0.620	0.659	0.700	0.728	0.748	0.768
0.180	0.608	0.640	0.668	0.712	0.759	0.792	0.815	0.837
0.200	0.646	0.683	0.714	0.765	0.818	0.854	0.880	0.904
0.240	0.720	0.766	0.804	0.866	0.930	0.972	1.003	1.031
0.280	0.791	0.845	0.891	0.962	1.035	1.083	1.115	1.143
0.320	0.859	0.921	0.973	1.054	1.134	1.184	1.216	1.242
0.360	0.925	0.995	1.053	1.141	1.226	1.275	1.305	1.325
0.400	0.989	1.067	1.130	1.224	1.309	1.356	1.385	1.410

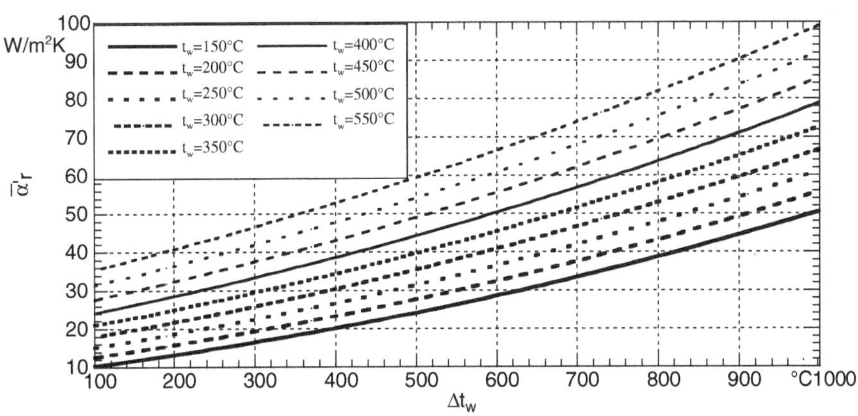

Fig. 6.23 Ideal heat transfer coefficient for the radiation in a cavity

Table 6.4 Partial pressure of CO_2 and H_2O in flue gas

$H\%$	n			
	1.0	1.1	1.2	1.3
p_{CO_2} (atm)				
0	0.2101	0.1910	0.1751	0.1616
5	0.1766	0.1609	0.1478	0.1367
10	0.1500	0.1370	0.1261	0.1167
15	0.1284	0.1175	0.1082	0.1003
20	0.1105	0.1012	0.0934	0.0867
25	0.0954	0.0875	0.0808	0.0751
p_{H_2O} (atm)				
5	0.0554	0.0505	0.0464	0.0429
10	0.0993	0.0907	0.0834	0.0773
15	0.1350	0.1235	0.1138	0.1055
20	0.1646	0.1508	0.1391	0.1291
25	0.1895	0.1738	0.1605	0.1490

The values of p and β relative to flue gas for the most frequently used fuels are roughly as follows:

Pulverized coal	$p = 0.17 - 0.20\,\text{atm}$	$\beta \approx 0.4$
Fuel oil	$p = 0.20 - 0.22\,\text{atm}$	$\beta \approx 0.8$
Natural gas	$p = 0.24 - 0.26\,\text{atm}$	$\beta \approx 2$

As far as the mean beam length x_r, if we refer to a tube bank we can use Figs. 6.24 and 6.25 where s_t and s_1 are the transversal and the longitudinal pitch of the tubes, respectively, and d_0 is the outside diameter of the tubes.

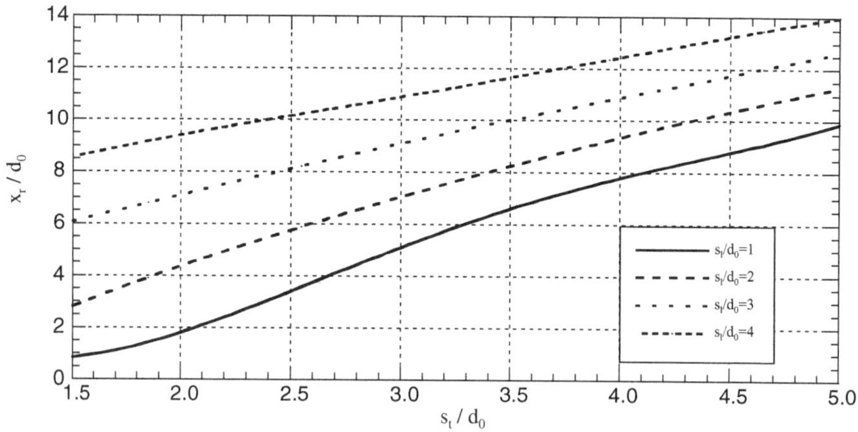

Fig. 6.24 Ratio x_r/d_0 for in-line tubes

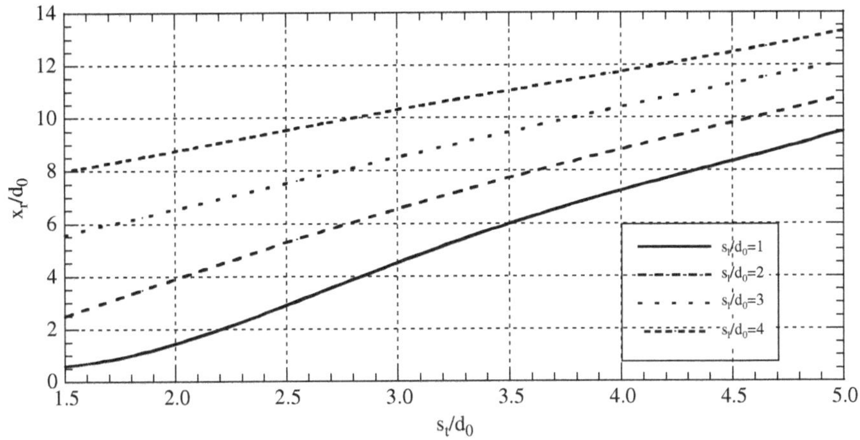

Fig. 6.25 Ratio x_r/d_0 for staggered tubes

If the gases are in a cavity, the mean beam length may be computed through the following equation:

$$x_r = 3.6\frac{V_c}{S_c}, \tag{6.167}$$

where V_c is the volume of the cavity and S_c its surface.

In Table 6.5 the values obtained from (6.167) are compared with those computed directly by different researchers.

We establish the perfect coincidence or, for one among the examined instances, the considerable approximation of the values computed through (6.167).

Specifically, as far as the inside of the tubes we have:

$$x_r = 0.9d_i; \tag{6.168}$$

where d_i stands for the inside diameter.

Table 6.5 Values of γ ($x_r = \gamma a$)

	Characteristic dimension a	γ	
		Equation 6.167	Direct computing
Sphere	Diameter	0.6	0.6
Cylinder of infinite length	Diameter	0.9	0.9
Cylinder with h=d	Diameter	0.6	0.6
Space between parallel planes	Distance between planes	1.8	1.8
Cube	Corner length	0.6	0.6
Parallelepiped with sides in ratio 1:2:6	Shorter corner length	1.08	1.06

Chapter 7
Heat Exchangers and Tube Banks

7.1 Introduction

In Chap. 1 when we indicated the law regulating heat transfer from a heating fluid to a heated fluid through a separation wall (see Sect. 1.13) we referred to two fixed temperatures t' and t'' of both fluids.

But if the heating fluid yields heat to the other fluid, it cools down and the heated fluid receiving heat becomes warmer.

In other words, the heat transfer is followed by variability of the temperatures of the two fluids. This is what happens in heat exchangers and tube banks.

This fact leads to a series of consequences to be discussed in the following sections.

Here are some preliminary considerations.

The variability of the temperatures of the two fluids implies the necessity to identify a mean difference in temperature to allow the correct calculation of the heat transfer.

The specific heat of the fluids which is crucial for the amount of cooling of the heating fluid and for the heating of the heated fluid, varies with temperature. It will be necessary to introduce a mean specific heat, and this requires familiarity with the enthalpy of fluids.

The overall heat transfer coefficient, assumed to be constant in Chap. 1, actually varies with temperature, since the heat transfer coefficients of both fluids vary with it. Therefore, it will be necessary to decide to which temperatures to refer the value of the heat transfer coefficients or the overall heat transfer coefficient for a correct computation of the heat transfer.

The way in which the two fluids interact with each other is crucial. There are two classic types of interaction, one with the fluids in parallel flow and one with the fluids in counter flow. In the first case the heated fluid enters the heat exchanger in the same location of the heating fluid, whereas in the second case the heated fluid enters the heat exchanger where the heating fluid is exiting it.

These situations that simplify the computation of the mean temperature difference will be discussed in Sect. 7.2.

D. Annaratone, *Engineering Heat Transfer*, DOI 10.1007/978-3-642-03932-4_7,

This situation is rare. The path of the two fluids may cross the other one, or it may be a compromise between a path with cross flow and motion in parallel flow or counter flow. Therefore, in all these cases it will be necessary to factor in the actual modality of the heat exchange in ways that will be discussed later on.

We will also point out the possibility for fluids not moving in pure parallel flow or counter flow, but where the heat transfer is such that they can conventionally be considered to be in parallel flow or counter flow.

Finally, there are two types of computation for heat exchangers and tube banks. The first one is the design calculation, consisting of the identification of the exchange surface required to obtain certain results. The second one makes it possible to compute the outlet temperatures of the fluids and the transferred heat, once the exchange surface has been set. This is a verification calculation, and we will discuss both.

7.2 Mean Logarithmic Temperature Difference

If we examine two fluids in parallel flow or in counter flow, the pattern of the temperatures t' and t'' is shown in both Figs. 7.1 and 7.2 where the temperatures (and generally all the quantities) referring to the heating fluid are indicated by superscript ($'$), and those referring to the heated fluid by superscript ($''$). In addition, the temperatures at the inlet are indicated with subscript 1 and those at the outlet with subscript 2.

M' and M'' are the mass flow rates of both fluids, and c'_{pm} and c''_{pm} refer to the mean specific isobaric heat. The overall heat transfer coefficient U is assumed to be constant.

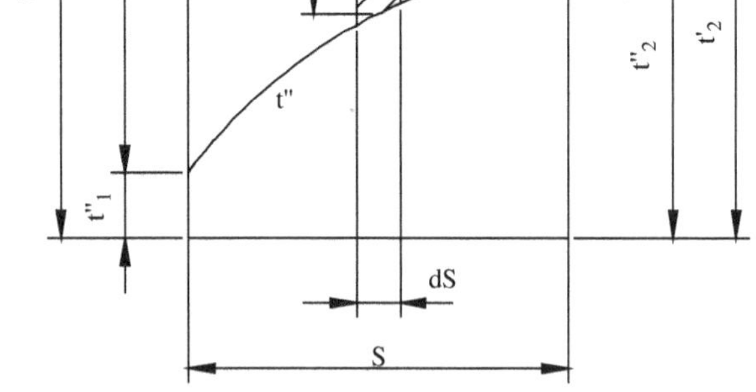

Fig. 7.1 Parallel flow

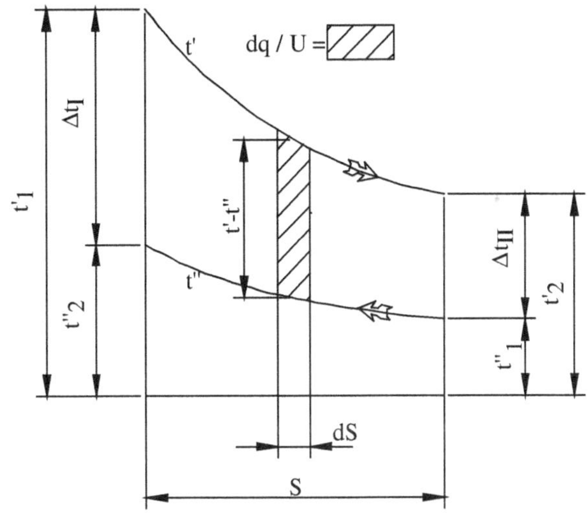

Fig. 7.2 Counter flow

The heat transferred through the elementary surface dS is given by:

$$dq = UdS(t' - t''). \tag{7.1}$$

On the other hand, given that t' decreases with the increase surface and by introducing the exchange efficiency η_e , the same value dq is equal to

$$dq = -\eta_e M' c'_{pm}\, dt'. \tag{7.2}$$

If the exchange occurs with parallel flow, given that t'' increases with S, from Fig. 7.1 we see that

$$dq = M'' c''_{pm}\, dt''. \tag{7.3}$$

Vice versa, Fig. 7.2 relative to heat transfer during counter flow shows that

$$dq = -M'' c''_{pm}\, dt''. \tag{7.4}$$

Therefore,

$$d\left(t' - t''\right) = -dq\left(\frac{1}{\eta_e M' c'_{pm}} \pm \frac{1}{M'' c''_{pm}}\right); \tag{7.5}$$

and recalling (7.1)

$$d\left(t' - t''\right) = -UdS\left(t' - t''\right)\left(\frac{1}{\eta_e M' c'_{pm}} \pm \frac{1}{M'' c''_{pm}}\right). \tag{7.6}$$

Here the plus sign indicates parallel flow and the minus sign indicates counterflow.

On the other hand

$$q = M''c''_{pm}\left(t''_2 - t''_1\right) = \eta_e M'c'_{pm}\left(t'_1 - t'_2\right). \tag{7.7}$$

Thus, with parallel flow

$$\frac{1}{\eta_e M'c'_{pm}} + \frac{1}{M''c''_{pm}} = \frac{1}{q}\left(t'_1 - t''_1 - t'_2 + t''_2\right), \tag{7.8}$$

and with counter flow

$$\frac{1}{\eta_e M'c'_{pm}} - \frac{1}{M''c''_{pm}} = \frac{1}{q}\left(t'_1 - t''_2 - t'_2 + t''_1\right), \tag{7.9}$$

The term on the right of the equal sign of both (7.8) and (7.9) (Figs. 7.1 and 7.2) is equal to:

$$\frac{\Delta t_I - \Delta t_{II}}{q}. \tag{7.10}$$

(7.6) can therefore be written as follows:

$$\frac{d\left(t' - t''\right)}{t' - t''} = -\frac{UdS}{q}\left(\Delta t_I - \Delta t_{II}\right); \tag{7.11}$$

and through integration we obtain:

$$\left|-\log_e\left(t' - t''\right)\right|_I^{II} = \frac{US}{q}\left(\Delta t_I - \Delta t_{II}\right); \tag{7.12}$$

then

$$\log_e\frac{\Delta t_I}{\Delta t_{II}} = \frac{US}{q}\left(\Delta t_I - \Delta t_{II}\right). \tag{7.13}$$

Finally,

$$q = US\frac{\Delta t_I - \Delta t_{II}}{\log_e\dfrac{\Delta t_I}{\Delta t_{II}}}. \tag{7.14}$$

The following quantity is the mean logarithmic temperature difference Δt_{ml}:

$$\Delta t_{ml} = \frac{\Delta t_I - \Delta t_{II}}{\log_e\dfrac{\Delta t_I}{\Delta t_{II}}}; \tag{7.15}$$

then

$$q = US\Delta t_{ml}. \tag{7.16}$$

The resulting equation is quite similar to (1.13) where instead of the constant difference in temperature between the heating fluid and the heated one, we have the mean logarithmic temperature difference given by (7.15) (of course, U represents U_o and U_i, respectively, depending on whether S is the outside or inside surface of the tubes (see (1.18) and (1.20)).

Another way to proceed is suggested by the fact that, if the ratio $\Delta t_I/\Delta t_{II}$ is not too high, Δt_{ml} does not considerably differ from the mean arithmetic temperature difference equal to:

$$\Delta t = \frac{\Delta t_I + \Delta t_{II}}{2}. \tag{7.17}$$

Therefore, we can write that

$$\Delta t_{ml} = \chi \frac{\Delta t_I + \Delta t_{II}}{2}. \tag{7.18}$$

Based on (7.15) and (7.18), the corrective factor χ is given by

$$\chi = \frac{2\,(\Delta t_I - \Delta t_{II})}{(\Delta t_I + \Delta t_{II}) \log_e \dfrac{\Delta t_I}{\Delta t_{II}}}. \tag{7.19}$$

The value for χ obtained from Fig. 7.3 clearly shows the influence of $\Delta t_I/\Delta t_{II}$ on the reduction of Δt_m with respect to the mean arithmetic temperature difference.

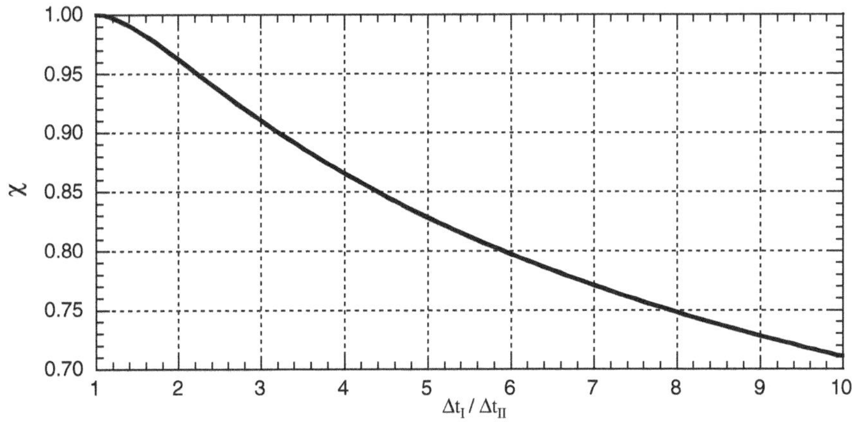

Fig. 7.3

Note that the use of this diagram combined with (7.18) leads to the exact computation of Δt_{ml}.

In the case of fluids in parallel flow, the value of the ratio $\Delta t_I / \Delta t_{II}$ is higher than with fluids in counter flow, thus the value of both χ and Δt_{ml} is smaller. Based on (7.16), it follows that a greater surface with equal transferred heat is needed.

The assumption so far was that the value of U is constant.

In fact, the heat transfer coefficients of both fluids vary with temperature, and so does the value of U. Therefore, it is a question of defining which value of U must be introduced in (7.16).

It is customary to consider the values of the heat transfer coefficients of both fluids corresponding to the average between the inlet and the outlet temperature, and to compute the overall heat transfer coefficient U based on these values of α.

This is the only recommendable (conservative) criterion for heated fluid, even though the behavior of the temperature is not linear. As far as the heating fluid, given the behavior of temperature, it is generally advisable to adopt the logarithmic average between the inlet and outlet temperatures as reference temperature. Naturally, if the film temperature must be adopted for the computation of the heat transfer coefficient, the temperature of reference must be the average between the temperature mentioned earlier and the wall temperature.

The mean logarithmic temperature of the heating fluid is given by

$$t'_{ml} = \frac{t'_1 - t'_2}{\log_e \dfrac{t'_1}{t'_2}} \tag{7.20}$$

We will come back to this topic when discussing the verification calculation.

7.3 Mean Specific Heat

Both the design and the verification calculation of the exchanger require knowledge of the mean isobaric specific heat of both fluids.

It is given by

$$c_{pm} = \frac{\int_{t_2}^{t_1} c_p dt}{t_1 - t_2}. \tag{7.21}$$

The integral in (7.21) is none other than the difference between enthalpy h_1 corresponding to temperature t_1 and enthalpy h_2 corresponding to temperature t_2. Considering that the specific heat is usually expressed in J/kg K, whereas the enthalpy is typically expressed in kJ/kg, we have

$$c_{pm} = \frac{h_1 - h_2}{t_1 - t_2} 1000. \tag{7.22}$$

To obtain the required values of c_{pm} it is thus necessary to know the enthalpies of the fluids.

Note that if it is possible to express c_p with an equation like

$$c_p = X + Yt + Zt^2 \tag{7.23}$$

after integrating c_p and recalling the usual units of measure for c_p and h (J/kg K and kJ/kg), we obtain

$$h = X\frac{t}{1000} + \frac{Y}{2}\frac{t^2}{1000} + \frac{Z}{3}\frac{t^3}{1000} \tag{7.24}$$

Equation (7.24) easily leads to the equation of the enthalpy if there is an equation of the specific heat within the type indicated by (7.23).

Now we indicate a few equations to be used for the computation of the enthalpy, always expressed in kJ/kg; the temperatures are in °C.

7.3.1 Water and Superheated Steam

The enthalpies for water and superheated steam can be taken exactly from the publication "Properties of Water and Steam in SI-Units–Springer Verlag" or from similar publications.

Yet, for the approximated computation of the enthalpy of water we can refer to (5.3) and obtain

$$h = 421.96\frac{t}{100} - 9.36\left(\frac{t}{100}\right)^2 + 5.74\left(\frac{t}{100}\right)^3 \tag{7.25}$$

valid for temperatures between 20 and 250°C.

7.3.2 Air and Other Gases

For the enthalpy of air we can refer to (5.7), thus obtaining for the enthalpy that

$$h = 1003.79\frac{t}{1000} + 37.76\left(\frac{t}{1000}\right)^2 + 72\left(\frac{t}{1000}\right)^3 \tag{7.26}$$

valid for $t = 0 - 300$°C.

The following approximated equations are valid, except for flue gas, for temperatures between 0°C and 500°C.

Oxygen (O_2)

$$h = 914.2\frac{t}{1000} + 117.7\left(\frac{t}{1000}\right)^2 + 22.8\left(\frac{t}{1000}\right)^3 \tag{7.27}$$

Nitrogen (N_2)

$$h = 1038\frac{t}{1000} + 18.4\left(\frac{t}{1000}\right)^2 + 78.13\left(\frac{t}{1000}\right)^3 \tag{7.28}$$

Carbon dioxide (CO_2)

$$h = 813.3\frac{t}{1000} + 502.3\left(\frac{t}{1000}\right)^2 - 209.5\left(\frac{t}{1000}\right)^3 \tag{7.29}$$

Carbon monoxide (CO)

$$h = 1038.4\frac{t}{1000} + 35.14\left(\frac{t}{1000}\right)^2 + 78.18\left(\frac{t}{1000}\right)^3 \tag{7.30}$$

Methane (CH_4)

$$h = 2149\frac{t}{1000} + 1550.4\left(\frac{t}{1000}\right)^2 + 136.3\left(\frac{t}{1000}\right)^3 \tag{7.31}$$

Flue gas

The enthalpy of flue gas is obtained through (5.17). It is given by

$$\begin{aligned} h = {} & (971.7 + 10.49m)\frac{t}{1000} + (162.76 - 2.49m)\left(\frac{t}{1000}\right)^2 \\ & - (25.53 - 2.02m)\left(\frac{t}{1000}\right)^3 \end{aligned} \tag{7.32}$$

In (7.32) m is the mass moisture percentage of the gas; (7.32) is valid for $t = 50 - 1200$°C and for $m = 0 - 12\%$.

7.4 Design Calculation

The design calculation consists of determining the surface S of the exchanger to obtain a certain result.

To that extent, note that for thermal balance we can write that

$$q = M''c''_{pm}\left(t''_2 - t''_1\right) = \eta_e M' c'_{pm}\left(t'_1 - t'_2\right) \tag{7.33}$$

In (7.33) q is the heat transferred to the heated fluid in the time unit in W, M' and M'' are the mass flow rates of the heating fluid and the heated fluid, respectively, in kg/s, t'_1 and t'_2 are the inlet and outlet temperatures of the heating fluid, t''_1 and t''_2 are the inlet and outlet temperatures of the heated fluid in °C, c'_{pm} and c''_{pm} are the mean isobaric specific heat of both the heating and the heated fluid in J/kg K, and η_e is the actual or assumed efficiency of the heat exchange.

In addition, we know that

$$q = US\Delta t_m. \tag{7.34}$$

For the design calculation, once M', M'',t'_1,t''_1,η_e are known, we may wish to obtain the exchange of a certain heat q; from (7.33) we obtain the temperatures t'_2 and t''_2, given that the two mean specific heat depend on the four temperatures in question. It is possible instead to impose temperature t'_2 or temperature t''_2; (7.33) still leads to the other unknown temperature and to heat q.

In any case, in the end we have the value of q and the four temperatures.

At this point, if the fluids are in parallel flow or in counter flow we compute the mean logarithmic temperature difference. If this not the case, we compute the actual mean temperature difference by multiplying the logarithmic one by a corrective factor; in any case we obtain the value of Δt_m.

Once the overall heat transfer coefficient U is computed, we obtain the necessary surface S through (7.34).

As far as the calculation of U we already pointed out in Sect. 7.2 that for the computation of the heat transfer coefficient of the heated fluid it is best to refer to the arithmetic average of both inlet and outlet temperatures, whereas for the computation of the heat transfer coefficient of the heating fluid, it is generally best to refer to the logarithmic average of the two temperatures above, the necessity to refer to film temperature when it is required for the computation of α, notwithstanding.

7.5 The Mean Difference in Temperature in Reality

In real instances the behavior of the fluids, with the exception of fluids with cross flow which are a case in itself, is usually close to the behavior of fluids in parallel flow or counter flow. In general, the most logical methodology to obtain the actual value of Δt_m is to refer to the mean logarithmic difference in temperature in parallel flow or counter flow, and to introduce a corrective factor by which to multiply this difference to obtain Δt_m.

To that extent we introduce three dimensionless factors, the same we will use for the verification calculation.

They are:

$$\psi = \frac{t'_2 - t''_1}{t'_1 - t''_1}; \tag{7.35}$$

$$\beta = \frac{\eta_e M' c'_{pm}}{M'' c''_{pm}}; \tag{7.36}$$

$$\gamma = \frac{US}{\eta_e M' c'_{pm}}. \tag{7.37}$$

Since this is a design calculation, the inlet and outlet temperatures of both fluids are known, and as a result so are the value of ψ and β.

If we consider the fluids in parallel flow, there is precise connection between the three indicated factors. In fact, based on (7.61) factor γ which is indicated by γ_p, is given by

$$\gamma_p = \frac{1}{1+\beta} \log_e \frac{1}{(1+\beta)\psi - \beta}. \tag{7.38}$$

If we consider the fluids in counter flow instead, and if $\beta \neq 1$, based on (7.70) factor γ indicated with γ_c is given by

$$\gamma_c = \frac{1}{1-\beta} \log_e \left(\frac{1-\beta}{\psi} + \beta \right). \tag{7.39}$$

If $\beta = 1$ instead, from (7.74) we obtain

$$\gamma_c = \frac{1}{\psi} - 1. \tag{7.40}$$

In real instances the value of γ meant to satisfy the imposed value of ψ, is close plus or minus from the value of γ_p or γ_c.

Based on Sects. 7.2 and 7.4, the transferred heat is equal to

$$q = \eta_e M' c'_{pm} \left(t'_1 - t'_2 \right) = US\Delta t_m = \eta_e M' c'_{pm}\, \gamma\, \Delta t_m; \tag{7.41}$$

then

$$\Delta t_m = \frac{t'_1 - t'_2}{\gamma}. \tag{7.42}$$

Given that t'_1 and t'_2 are fixed values, we establish that Δt_m is inversely proportional to γ.

If we consider the fluids in parallel flow, instead of (7.42) we must write that

$$\Delta t_{ml(p)} = \frac{t'_1 - t'_2}{\gamma_p} \tag{7.43}$$

where $\Delta t_{ml(p)}$ is the mean logarithmic temperature difference referred to fluids in parallel flow, and γ_p is obtained through (7.38).

Therefore,

$$\chi_p = \frac{\Delta t_m}{\Delta t_{ml(p)}} = \frac{\gamma_p}{\gamma}. \tag{7.44}$$

In other words, if the reference is to fluids in parallel flow, after computation of γ_p with (7.38) based on imposed values of ψ and β, the case in question is examined and the value of γ required to obtain the requested value of ψ is calculated; this way the value of χ_p is computed through (7.44) which makes it possible to compute the actual mean temperature difference for the case in question.

The procedure is similar in reference to fluids in counter flow. In that case

$$\chi_c = \frac{\Delta t_m}{\Delta t_{ml(c)}} = \frac{\gamma_c}{\gamma} \tag{7.45}$$

where $\Delta t_{ml(c)}$ is the mean logarithmic temperature difference referred to fluids in counter flow, and γ_c is obtained through (7.39) or (7.40).

Note that with reference to fluids in parallel flow, for the situation to actually be possible we must have

$$\psi > \frac{\beta}{1+\beta}. \tag{7.46}$$

If the reference is to fluids in counter flow instead, and $\beta > 1$, for the situation to actually be possible we must have

$$\psi > \frac{\beta - 1}{\beta}. \tag{7.47}$$

The described process allowed us to build a series of Tables which are included in Appendix B. We refer the reader to this section to make the comparisons discussed in the text.

We did not consider the instances where $\gamma > 6$ since they are unlikely and not advisable. In addition, we neglected those cases where the difference plus or minus between the actual mean temperature difference and the logarithmic one is under 1%, thus to be considered rather insignificant.

In the Tables of Appendix B the missing values to the left of those included correspond to impossible cases or to those where $\gamma > 6$. The missing values to the right of those included correspond to cases where the difference between Δt_m and $\Delta t_{ml(p)}$ or $\Delta t_{ml(c)}$ is less than $\pm 1\%$; for those we can assume the mean logarithmic temperature difference for Δt_m.

7.5.1 Fluids with Cross Flow

The behavior of fluids in cross flow (Fig. 7.4) is closer to that of fluids in counter flow compared to fluids in parallel flow.

So we computed the values of χ_c to include them in the Tables B.1 and B.2.

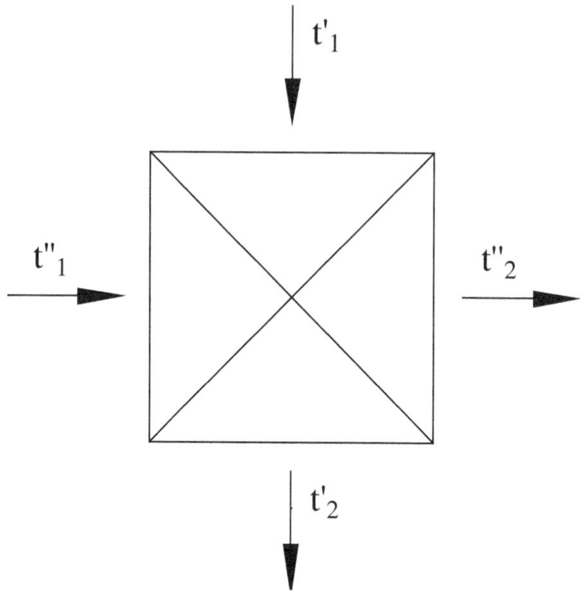

Fig. 7.4 Cross flow

7.5.2 Heat Exchangers

We consider heat exchangers with two passages of the fluid inside the tubes shown in Fig. 7.5.

As you see, there are four possible combinations indicated by the letters A, B, C and D.

Types A and B which are apparently different from one another, have in fact the same behavior and have the same value of χ.

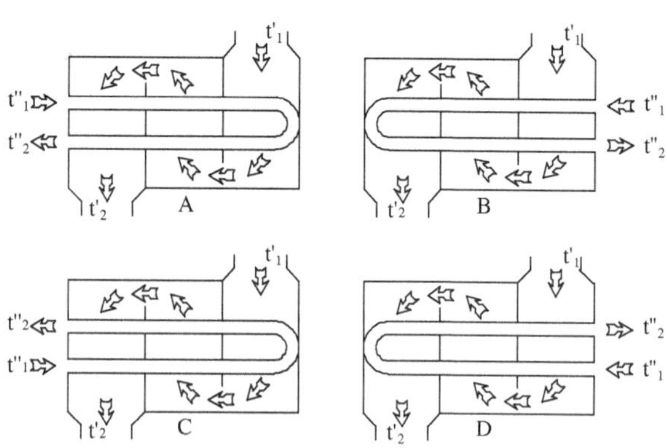

Fig. 7.5 Heat exchangers with two passages of internal fluid

This depends on the fact that each has one of the two peculiar characteristics of fluids in parallel flow. In fact, in type A the internal fluid exits the tubes in the same location in which the external fluid exits the exchanger; in type B the internal fluid enters the tubes in the same location in which the external fluid enters the exchanger; this makes their behavior absolutely identical and similar to that of fluids in parallel flow.

Similar considerations are true for types C and D.

Each one has one of the peculiar characteristics of fluid in counter flow. In fact, in type C the internal fluid enters the tubes in the same location in which the external fluid exits the exchanger. In type D the internal fluid exits the tubes in the same location in which the external fluid enters the exchanger. This makes their behavior absolutely identical and similar to that of fluids in counter flow.

Therefore, for types A and B it would be logical to calculate the value of the corrective factor χ_p , thus referring the requested mean difference in temperature Δt_m to the mean logarithmic difference relative to fluids in parallel flow. Nonetheless, to be able to compare them with types C and D we preferred to compute χ_c; for types C and D the logical solution is undoubtedly that to compute the corrective factor χ_c , thus referring Δt_m to the mean logarithmic difference in temperature relative to fluids in counter flow.

The computation of the values of χ_c is based on a few schemata and assumptions. First of all, the position of the baffles must be such that the exchange surface is divided in equal sections for the various passages of the fluid outside the tubes. Moreover, we assume that the differences in temperature of the different threads of the external fluid annul each other, due to the mixture of the threads occurring with the reversal of the direction of the flow. Thus, the temperature of the external fluid is uniform at the entrance of the new passage.

The considered range is as follows: $\beta = 0.2 - 2.0$ and $\psi = 0.05 - 0.9$. The values of χ_c for types A and B are shown in Tables B.3, B.4, B.5 and B.6. They are in reference to exchangers with 2, 3, 4 and 5 passages of the external fluid, respectively.

A single passage of the fluid outside the tubes is not considered because in that case the exchanger is reduced to a coil with two sections; its behavior is implied by the section on coils to follow shortly.

Tables B.7, B.8, B.9 and B.10 show the values of χ_c relative to types C and D for 2, 3, 4 and 5 passages of the fluid outside the tubes, respectively.

Analysis of the Tables leads us to interesting considerations.

First of all, it is not surprising that, β and ψ being equal, the value of Δt_m for types A and B is always lower than that for types C and D.

In addition, the increase in the number of passages of the fluid outside the tubes in types A and B is matched by an increase of Δt_m, whereas in types C and D it decreases. The difference in behavior between types A and B and types C and D which is rather noticeable with 2 passages of the external fluid diminishes with the increase in passages.

In view of this, if the number of passages is greater than 5 due to caution it is possible to adopt for all types the values of χ_c included in Table B.6.

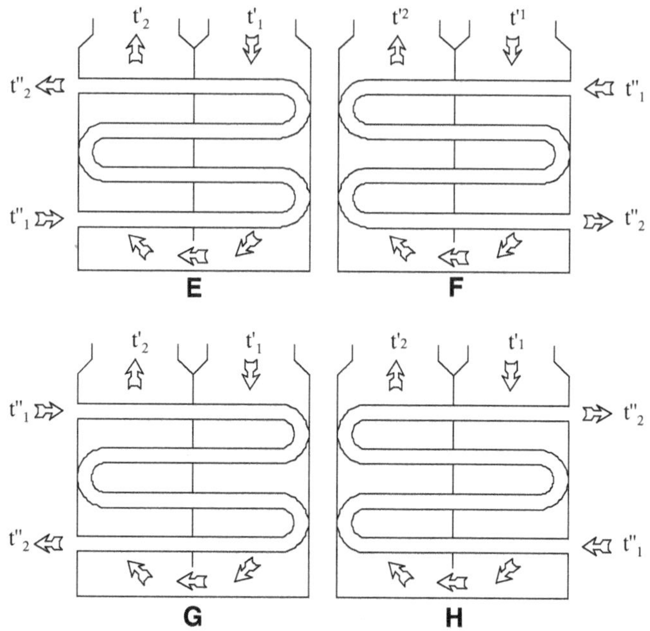

Fig. 7.6 Heat exchangers with four passages of internal fluid

Here is a fundamental word of caution. The use of corrective factors χ_C of the various Tables depends on the type of exchanger, specifically on the fact that the exchanger has one of the two typical characteristics of fluids in parallel flow or in counter flow. Then, if the passages of the external fluid are even, the outlet location of the external fluid changes with respect to Fig. 7.5. If the inlet location of the fluid inside the tubes remains unchanged, type A becomes type C and viceversa. Thus, the reference to the different Tables changes, as well.

Now we consider exchangers with 4 passages of the fluid inside the tubes (Fig. 7.6).

If there is just one passage of the external fluid, the exchanger is reduced to a coil with 4 sections. We refer you to the section on coils.

If the number of passages of the fluid outside the tubes is ≥ 3 , the behavior of these exchangers is quite similar to that of exchangers with 2 passages of the fluid inside the tubes. Thus, it is possible to use Tables B.4, B.5 and B.6 for types E and F, and Tables B.8, B.9 and B.10 for types G and H.

The situation is entirely different if the exchanger has 2 passages of the external fluid, as shown in Fig. 7.6.

The behavior of an exchanger with 4 passages of the fluid inside the tubes is considerably different from that of an exchanger with 2 passages. Tables B.3 and B.7 cannot be used. For the value of χ_C one must refer to Table B.11 for types E and F in Fig. 7.6, whereas Table B.12 will be the reference for types G and H of the same figure.

7.5.3 Coils

In the case of coils in Figs. 7.7 and 7.8, it would not be possible to speak of fluids in parallel flow or counter flow. In fact, each section of the coil is hit by the fluid outside the tubes in such a way to be considered cross flow. Therefore, the coil is the sum of elements in which the fluxes are in cross flow.

Usually, though, if the internal fluid enters the coils in correspondence of the inlet in the coil of the external fluid (Fig. 7.7), it is customary to speak of fluids in parallel flow. If the inside fluid enters the coils in correspondence of the outlet of the external fluid (Fig. 7.8), it is customary to speak of fluids in counter flow.

At this point we would like to analyze the topic in-depth by starting with the coils in Fig. 7.7.

With respect to fluids in real parallel flow they show differences in heat transfer that we would like to highlight. Based on the premises, the corrective factor χ_p is logically calculated.

The considered range is as follows: $\beta = 0.2 - 2.0$ and $\psi = 0.03 - 0.90$.

Tables B.13 and B.14 show the values of χ_p, relative to coils with 2 and 3 sections, respectively.

We establish that the value of χ_p is always greater than one. This means that the heat transfer occurs with more favourable characteristics compared to those relative to fluids in parallel flow, given that the value of Δt_m is greater than $\Delta t_{ml(p)}$.

As the number of section increases, the value of χ_p gets close to unity. If there are 4 sections there are few instances where $\chi_p > 1.02$; if the number of sections is ≥ 4, giving up the little advantage represented by $\chi_p > 1$, we recommend to adopt the mean logarithmic difference in temperature referred to fluids in parallel flow as value of Δt_m.

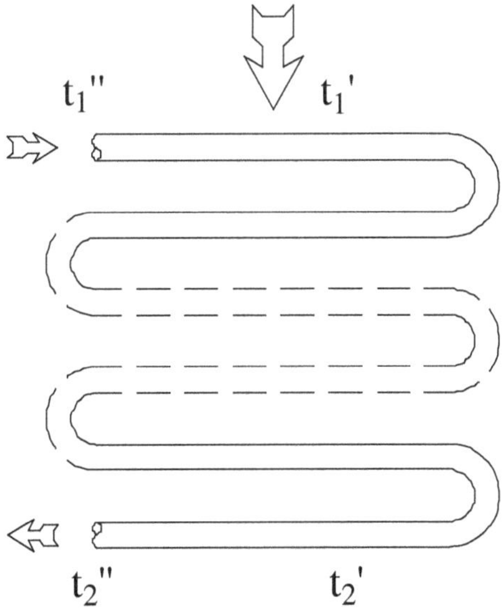

Fig. 7.7 Coils-parallel flow

Fig. 7.8 Coils-counter flow

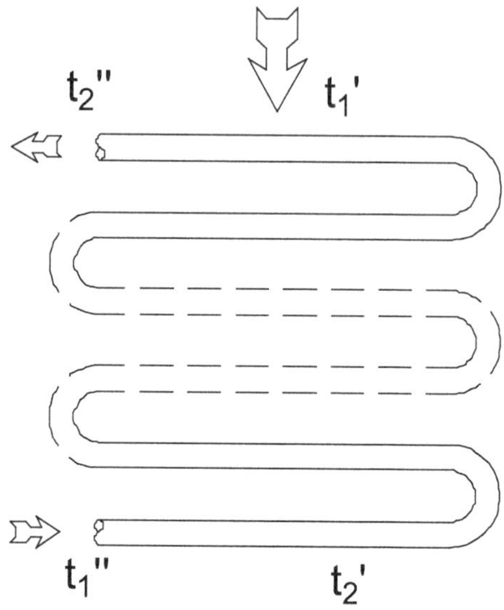

Now we consider the coils in Fig. 7.8

Naturally, in this case we calculated the values of χ_c.

The analyzed range is as follows: $\beta = 0.2 - 2.0$ and $\psi = 0.04 - 0.88$.

The values of χ_c for a number of sections equal to 2, 3, 4, 6 and 8 are shown in Tables B.15, B.16, B.17, B.18 and B.19.

As expected, we establish that the values of χ_c are all below unity. This means that the heat transfer is less favorable in comparison with fluids in counter flow, given that Δt_m is smaller than $\Delta t_{ml(c)}$.

The phenomenon is particularly noticeable when the number of sections is small, while it decreases when their number is high.

If the number of sections is ≥ 10, the situations where $\chi_c < 0.98$ are rare and unlikely. Therefore, it is possible to conclude that in reality if the number of sections is ≥ 10, the coil may be treated as if the fluids were in fact in counter flow by adopting for Δt_m the value of $\Delta t_{ml(c)}$.

7.5.4 Tube Bank with Various Passages of the External Fluid

We consider a tube bank consisting of a series of straight tubes; a fluid flows inside the tubes, while another fluid hits the bank outside with a series of passages created through dividing baffles. If there is only one passage of the fluid outside the tubes, these are fluids in cross flow, and we refer the reader to the appropriate section.

The classic device of this type is the recuperative air heater at the end of a steam generator. From now on we will refer to this device but keeping in mind that this type of exchanger can be used even with other fluids, generally gaseous ones.

In air heaters the flue gas is generally located inside the tubes while the air hits the bank outside, but nothing stands in the way of the opposite solution.

The external fluid can enter the heater in correspondence of the inlet to the tubes of the internal fluid, or viceversa with the external fluid entering the heater in correspondence of the exit from the tubes of the internal fluid. Figure 7.9 represents an air heater of the first kind with three passages of the external fluid. Figure 7.10 represents an air heater of the second kind instead.

Clearly, with the first kind the behavior of the fluid through the heater recalls the typical behavior of fluids in parallel flow, whereas the second type is similar to that of fluids in counter flow.

In fact, with these devices it is customary to speak of fluids in parallel flow or in counter flow even this is not exactly true. This topic requires in-depth analysis.

Therefore, we will refer to χ_p for the first type and to χ_c for the second one.

We could consider using the values of χ_p and χ_c already obtained for the coils. In fact, if the fluid flowing through the tubes of the heater were compared to the fluid hitting the coil, and the fluid hitting the tubes of the heater with the fluid flowing through the coil, the analogy is evident. Still, we must consider that while the temperature of the internal fluid is unique in any position along the coil, the temperature of the fluid hitting the tube bank varies not only depending on the direction of the flux, but also transversally to it. Then the values of the factors cited in relation with the coils are only approximated values.

Another very simple procedure could be as follows. If we assume that the heater is represented by a series of sections where the motion of the fluids is in cross flow, the values of χ_c relative to cross flow can be used for every section, and in the end a global value of χ_p or χ_c is reached to solve the problem. Even this method, though, contains an error. The values of χ_c in Tables B.1 and B.2 are based on uniform temperatures at the inlet of both fluids, while those at the outlet are the average temperatures of the various threads at the exit. In our case we can hypothesize that the temperature of the external fluid is uniform at the inlet of every passage, given the mixture of the threads occurring with the reversal of the flux, but

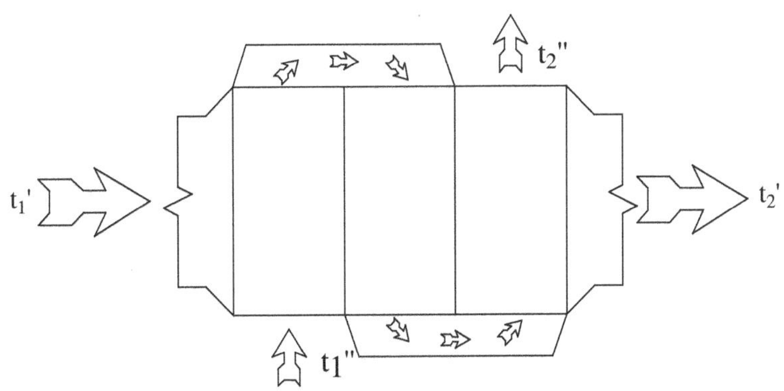

Fig. 7.9 Tube bank with several passages of the external fluid-parallel flow

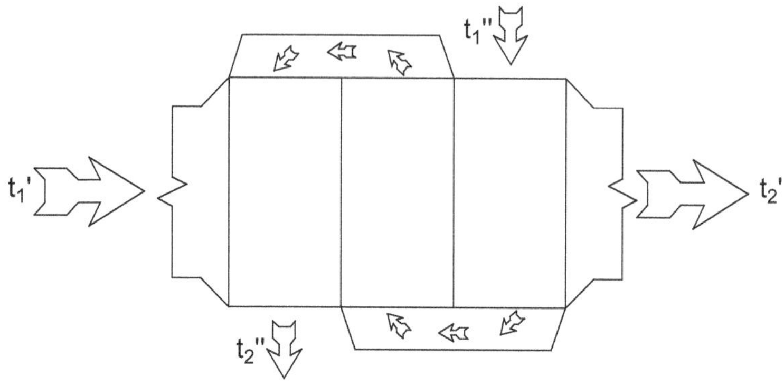

Fig. 7.10 Tube bank with several passages of the external fluid-counter flow

this is certainly not true for the fluid flowing in the tubes. For the latter the division in sections is purely formal because every tube is in one piece where the fluid takes on its own temperature condition which changes from tube to tube. In view of this, even this method leads to values of χ_p or χ_c yielding only approximated computation.

To obtain more realistic values of χ it is therefore necessary to do a more in-depth analysis to factor in these facts. This is what was done leading to the values in Tables B.20, B.21, B.22, B.23 and B.24.

We considered the following range: $\beta = 0.2 - 2.0$ and $\psi = 0.05 - 0.90$.

For the air heater in parallel flow in Fig. 7.9 the Tables B.20 and B.21 list the values of χ_p for 2 and 3 passages of the external fluid. Of course, they are greater than unity, given that the heat transfer is more favourable than in the case of fluids in parallel flow. We establish that the values of χ_p with 2 passages are greater compared to those with 3 passages. Therefore, the solution with 3 passages is less favourable. Finally, if the passages are $\geq$ 4 our advice is to give up the modest advantage represented by the fact that in some cases $\chi_p > 1$ by adopting the mean logarithmic temperature difference relative to fluids in parallel flow for Δt_m.

Tables B.22, B.23 and B.24 show the values of χ_c relative to the air heater in counter flow in Fig. 7.10, respectively, with 2, 3 and 4 passages of the external fluid. The examined range include: $\beta = 0.2 - 2.0$ and $\psi = 0.05 - 0.90$.

Of course, the factors χ_c are below unity since the heat transfer is less favourable compared to the one with fluids in counter flow.

We see that even with 4 passages the difference between Δt_m and the mean logarithmic temperature difference may sometimes reach 5%, and it is advisable to take this fact into account.

We did not pursue the investigation any further by examining even solutions with a number of passages greater than 4 given that they are unlikely. In case solutions of this type were adopted, we recommend to conservatively refer to the values of χ_c listed in Table B.24.

7.6 Verification Calculation

7.6.1 General Considerations

The verification calculation is in reference to an exchanger, or a tube bank, that were already sized either permanently or temporarily, so that the exchange surface S is known. The verification calculation computes the unknown outlet temperatures of both fluids and the transferred heat.

If the exchanger is sized permanently the verification calculation is done to verify the performance of the exchanger under conditions other than those it was designed for.

The verification calculation can be used even in substitution of the design calculation. This procedure is actually fairly widespread.

In that case the exchange surface is temporary, even though it is possible to compute the heat transfer coefficients of the fluids and the overall heat transfer coefficient. The verification calculation makes it possible to evaluate the performance of the exchanger and to modify its surface if it does not satisfy requirements until the desired result is reached.

The following section will focus on the calculation relative to fluids in parallel flow and counter flow. These two conditions are also the foundation of the behavior of fluids in real cases, provided corrective factors are introduced, as already pointed out in reference to the computation of the actual mean temperature difference between the fluids (Sect. 7.5).

With reference to real cases, even for the verification calculation it is necessary to introduce corrective factors compared to the outcome of the verification calculation for the fluids in parallel flow and counter flow.

This will be the focus of Sect. 7.6.3.

7.6.2 Fluids in Parallel Flow or in Counter flow

The symbolism from Sect. 7.2 will be used.

Based on (7.15) and (7.16) and considering fluids in parallel flow, we may write that

$$q = US\frac{(t'_1 - t''_1) - (t'_2 - t''_2)}{\log_e \dfrac{t'_1 - t''_1}{t'_2 - t''_2}}. \tag{7.48}$$

Note that

$$\frac{t'_1 - t''_1}{t'_2 - t''_2} = \frac{1 + \dfrac{t'_2 - t''_1}{t'_1 - t'_2}}{\dfrac{t'_2 - t''_1}{t'_1 - t'_2} - \dfrac{t''_2 - t''_1}{t'_1 - t'_2}}. \tag{7.49}$$

Moreover, as in Sect. 7.5 we introduce

$$\psi = \frac{t'_2 - t''_1}{t'_1 - t''_1}; \tag{7.50}$$

$$\beta = \frac{\eta_e M' c'_{pm}}{M'' c''_{pm}}; \tag{7.51}$$

$$\gamma = \frac{US}{\eta_e M' c'_{pm}}; \tag{7.52}$$

On the other hand, the transferred heat is given by:

$$q = M'' c''_{pm} \left(t''_2 - t''_1\right) = \eta_e M' c'_{pm} \left(t'_1 - t'_2\right). \tag{7.53}$$

Then, from (7.51):

$$\beta = \frac{t''_2 - t''_1}{t'_1 - t'_2}. \tag{7.54}$$

We introduce factor ε given by

$$\varepsilon = \frac{t'_2 - t''_1}{t'_1 - t'_2}. \tag{7.55}$$

From (7.50) and (7.54) we obtain:

$$\frac{t'_1 - t''_1}{t'_2 - t''_2} = \frac{1 + \varepsilon}{\varepsilon - \beta}. \tag{7.56}$$

Comparing (7.48) with (7.53) and with reference to (7.52), we have:

$$\gamma = \frac{t'_1 - t'_2}{(t'_1 - t''_1) - (t'_2 - t''_2)} \log_e \frac{t'_1 - t''_1}{t'_2 - t''_2}. \tag{7.57}$$

Recalling (7.56) and (7.51), (7.57) leads to the following:

$$\gamma = \frac{1}{1 + \beta} \log_e \frac{1 + \varepsilon}{\varepsilon - \beta}. \tag{7.58}$$

Then, from (7.58) we obtain:

$$\varepsilon = \frac{1 + \beta e^{(1+\beta)\gamma}}{e^{(1+\beta)\gamma} - 1}. \tag{7.59}$$

Based on (7.50) and (7.55):

$$\psi = \frac{\varepsilon}{\varepsilon - 1} \tag{7.60}$$

And finally, if ψ_p indicates the value of ψ for fluids in parallel flow,

$$\psi_p = \frac{e^{-(1+\beta)\gamma} + \beta}{1+\beta}. \tag{7.61}$$

If the fluids are in counter flow, instead of (7.48) we have:

$$q = US\frac{(t'_1 - t''_2) - (t'_2 - t''_1)}{\log_e \dfrac{t'_1 - t''_2}{t'_2 - t''_1}}. \tag{7.62}$$

Note that

$$\frac{t'_1 - t''_2}{t'_2 - t''_1} = \frac{\dfrac{t'_1 - t''_1}{t'_1 - t'_2} - \dfrac{t''_2 - t''_1}{t'_1 - t'_2}}{\dfrac{t'_1 - t''_1}{t'_1 - t'_2} - 1} \tag{7.63}$$

Assuming that

$$\eta = \frac{t'_1 - t''_1}{t'_1 - t'_2}, \tag{7.64}$$

and recalling (7.54) we have:

$$\frac{t'_1 - t''_2}{t'_2 - t''_1} = \frac{\eta - \beta}{\eta - 1}. \tag{7.65}$$

By analogy with (7.57) we also have:

$$\gamma = \frac{t'_1 - t'_2}{(t'_1 - t''_2) - (t'_2 - t''_1)} \log_e \frac{t'_1 - t''_2}{t'_2 - t''_1} \tag{7.66}$$

And from that, recalling (7.51) as well as (7.54):

$$\gamma = \frac{1}{1-\beta} \log_e \frac{\eta - \beta}{\eta - 1} \tag{7.67}$$

(7.67) leads to the following:

$$\eta = \frac{\beta - e^{(1-\beta)\gamma}}{1 - e^{(1-\beta)\gamma}} \tag{7.68}$$

Observing that

$$\psi = 1 - \frac{1}{\eta} \tag{7.69}$$

if ψ_c indicates the value of ψ for fluids in counter flow, we have

$$\psi_c = \frac{1-\beta}{e^{(1-\beta)\gamma} - \beta}. \tag{7.70}$$

Note that the value of ψ is undetermined if $\beta = 1$; in that case, though, we have:

$$\Delta t_{ml} = t'_2 - t''_1 = t'_1 - t''_2 \tag{7.71}$$

Therefore, based on (7.53) and (7.62)

$$\eta_e M' c'_{pm} \left(t'_1 - t'_2\right) = US \left(t'_2 - t''_1\right); \tag{7.72}$$

then

$$\frac{t'_1 - t''_1}{t'_2 - t''_1} - 1 = \gamma. \tag{7.73}$$

Thus, recalling (7.50):

$$\frac{1}{\psi} - 1 = \gamma. \tag{7.74}$$

Finally,

$$\psi_c = \frac{1}{\gamma + 1}. \tag{7.75}$$

Figure 7.11 was built based on (7.61) and Fig. 7.12 was built based on (7.70) and (7.75).

Once the value of ψ is known, recalling (7.50) temperature t'_2 is given by

$$t'_2 = t''_1 + \psi \left(t'_1 - t''_1\right). \tag{7.76}$$

If t'_2 is known, based on (7.54) we have

$$t''_2 = t''_1 + \beta \left(t'_1 - t'_2\right) \tag{7.77}$$

As far as the exchange efficiency η_e, there are no values carrying general validity because it is influenced by numerous factors.

With exchangers its value depends on the heat loss on the outside. If the exchanger is well insulated it is possible to have very high values of η_e that are

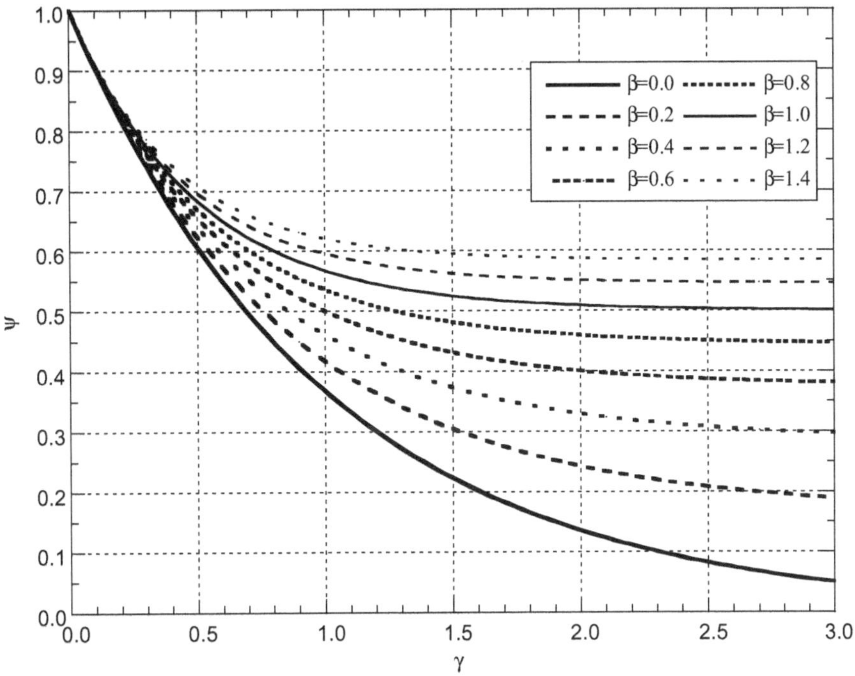

Fig. 7.11 Factor ψ for parallel flow

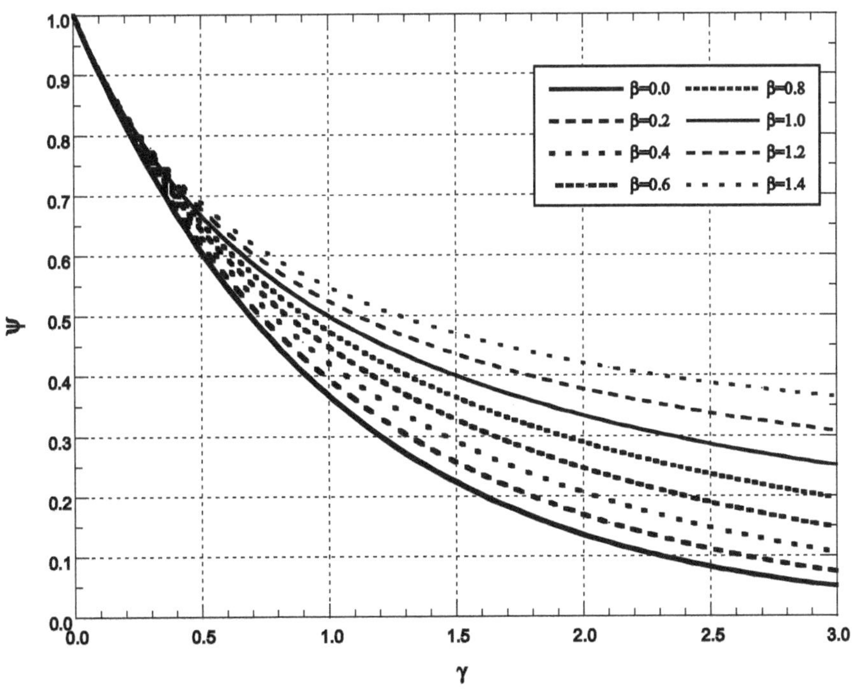

Fig. 7.12 Factor ψ for counter flow

equal to 0.98–0.99; the same criterion is true for tube banks if they are contained in a casing. The tube bank can be located in a space where its walls consist of tubes filled with another fluid; in that case the flue gas transfers part of the heat to these tubes, and the exchange efficiency referred to the bank can therefore be considerably lower than unity. In that case, though, there is no heat loss.

Sometimes, if the inlet temperature t'_1 of the heating fluid is known, the outlet temperature t''_2 of the heated fluid is imposed, while the inlet temperature t''_1 is unknown; in that case the latter temperature is calculated through the following equation

$$t''_1 = \frac{t''_2 - \beta t'_1 (1 - \psi)}{1 - \beta + \beta\psi}, \tag{7.78}$$

where temperature t'_2 is always calculated through (7.76).

It is certainly interesting to compare fluids in parallel flow and counter flow in relation to heat transfer.

Based on (7.52) and (7.53) we have:

$$q = \frac{US}{\gamma}\left(t'_1 - t'_2\right). \tag{7.79}$$

Based on (7.76) and after a series of steps we have:

$$q = \frac{US}{\gamma}\left(t'_1 - t''_1\right)(1 - \psi). \tag{7.80}$$

Based on (7.80) we establish that heat q is proportional to $(1 - \psi)$; thus, if two exchangers or tube banks are compared with one another with the same values of $U, S. \gamma, t'_1, t''_1$, one with the fluids in parallel flow and the other with the fluids in counter flow, the ratio between the transferred heats is given by (Table 7.1):

$$\frac{q_{counter}}{q_{parallel}} = \frac{(1+\beta)\left(e^{(1-\beta)\gamma} - 1\right)}{\left(e^{(1-\beta)\gamma} - \beta\right)\left(1 - e^{-(1+\beta)\gamma}\right)}. \tag{7.81}$$

Table 7.1 Ratio between transferred heat with fluids in counter flow and fluids in parallel flow

β	γ				
	0.2	0.4	0.6	0.8	1.0
0.2	1.0024	1.0085	1.0172	1.0276	1.0390
0.4	1.0047	1.0165	1.0331	1.0527	1.0742
0.6	1.0069	1.0240	1.0477	1.0755	1.1056
0.8	1.0090	1.0311	1.0611	1.0958	1.1330
1.0	1.0111	1.0377	1.0733	1.1137	1.1565
1.2	1.0131	1.0439	1.0842	1.1294	1.1763

The choice among fluids in parallel flow or counter flow is irrelevant until the difference in transferred heat in both instances amounts to a few percentage points, and the ratio is consequently roughly below 1.02 – 1.03.

We establish that these values of the ratio are matched by decreasing values of γ with increases of β.

For tube banks of a steam generator this means that in that respect the values of γ decrease passing from an economizer to a superheater, and then to an air heater. But for these devices the value of k decreases, so that we can basically conclude based on (7.81) that the ratio S/M' is crucial for deciding the type of flow.

In other words, if the surface is modest with respect to the mass flow rate of the heating fluid, it is possible to opt for the solution in parallel flow without sensible drawback with respect to the heat transfer.

In conclusion, if the heated fluid is a boiling liquid, the assumption must be $c''_{pm} = \infty$. Based on (7.61) and (7.70) (in this case they coincide since we cannot speak of fluids in parallel or counter flow) we obtain

$$\psi = e^{-\gamma} \tag{7.82}$$

The value of U, included in γ, can be calculated with considerable satisfaction by referring for α' to the mean logarithmic temperature; this corresponds to what was pointed out earlier.

It is possible to obtain even better results through the following procedure.

Note that

$$dq = U_o dS\left(t' - t_s\right), \tag{7.83}$$

where t' stands for the generic temperature of the heating fluid, t_s for the temperature of the boiling liquid, and S for the generic surface.

We may also write that

$$dq = -M' c'_p dt', \tag{7.84}$$

given that M' is the mass flow rate of the heating fluid..

Assuming that

$$y = t' - t_s, \tag{7.85}$$

we obtain

$$\frac{dy}{dS} = -\frac{U_o}{M' c'_p} y. \tag{7.86}$$

In order to factor in variations of U_0 and $c_p{}'$ with temperature variations, at this point we assume that

$$\frac{U_0}{M' c'_p} = A + B y^{\delta}, \tag{7.87}$$

where A, B and δ are constant.

Therefore,

$$\frac{dy}{dS} = -Ay - By^{\delta+1}. \tag{7.88}$$

The differential equation (7.88) is a Bernoulli equation of the following type:

$$y' = \varphi y + \psi y^n. \tag{7.89}$$

Resolving the equation we obtain

$$y^{-\delta} = -\delta e^{-\delta \int -AdS} \left(\int -Be^{\delta \int -AdS} dS + C \right) \tag{7.90}$$

where C is a constant; from (7.90) we obtain

$$y^{\delta} = -\frac{e^{-\delta AS}}{\frac{B}{A} e^{-\delta AS} + C\delta} \tag{7.91}$$

As far as constant C, note that for $S = 0$ is $t' = t_1'$; therefore, $y = t_1' - t_s$, as a result. From (7.91) we obtain

$$\left(t_1' - t_s\right)^{\delta} = -\frac{1}{\frac{B}{A} + C\delta}; \tag{7.92}$$

then

$$C = -\frac{1}{\delta}\left[\frac{B}{A} + \left(t_1' - t_s\right)^{-\delta}\right]. \tag{7.93}$$

From (7.91) and factoring in (7.93) we obtain

$$y = \frac{\left(t_1' - t_s\right) e^{-AS}}{\left[1 + \frac{B}{A}\left(1 - e^{-\delta AS}\right)\left(t_1' - t_s\right)^{\delta}\right]^{1/\delta}}. \tag{7.94}$$

Thus, given that S_0 is the heat exchanger surface, the exit temperature t_2' of the heating fluid from the bank is given by:

$$t_2' = t_s + \left(t_1' - t_s\right) \frac{e^{-AS_0}}{\left[1 + \frac{B}{A}\left(1 - e^{-\delta AS_0}\right)\left(t_1' - t_s\right)^{\delta}\right]^{1/\delta}}. \tag{7.95}$$

Note that if in (7.88) $B = 0$, thus $\gamma = AS_0$, (7.95) would be reduced to the following equation, one that is well-known in relation to our previous discussion:

$$t_2' = t_s + \left(t_1' - t_s\right) e^{-\gamma}. \tag{7.96}$$

Standard procedures are as follows.

First of all, consider the following temperature t_0', intermediate between t_1' and t_2':

$$t_0' = t_s + \sqrt{(t_1' - t_s)(t_2' - t_s)}. \tag{7.97}$$

Also, compute the values γ_1,γ_0 and γ_2 provided by (7.98):

$$\begin{aligned}\gamma_1 &= \frac{U_o S_0}{M' c_p'} \quad \text{for } t' = t_1' \\ \gamma_0 &= \frac{U_o S_0}{M' c_p'} \quad \text{for } t' = t_0' \\ \gamma_2 &= \frac{U_o S_0}{M' c_p'} \quad \text{for } t' = t_2'\end{aligned} \tag{7.98}$$

Based on (7.97) we establish that

$$A = \frac{1}{S_0} \frac{\gamma_1 \gamma_2 - \gamma_0^2}{\gamma_1 + \gamma_2 - 2\gamma_0}; \tag{7.99}$$

$$\delta = \frac{\log_e \dfrac{\gamma_1 - AS_0}{\gamma_0 - AS_0}}{\log_e \sqrt{\dfrac{t_1' - t_s}{t_2' - t_s}}}; \tag{7.100}$$

$$B = \frac{\gamma_1 - AS_0}{(t_1' - t_s)^\delta}. \tag{7.101}$$

All elements required to compute t_2' through (7.95) are available this way.

Note that the calculation procedure assumes advance knowledge of t_2' which is the final result of the calculation. Therefore, it is necessary to proceed by trial and error up to agreement between the assumed value of t_2' and the one resulting from the computation.

The values of the exit temperature of the heating fluid from the heat exchanger obtained through this method are practically exact.

With heat exchangers it is customary to consider their efficiency. It is given by the ratio between the actual heat exchange and the maximum value of the heat that the exchanger could theoretically exchange. The latter corresponds to the infinite surface, so that from (7.45) we have $\gamma = \infty$.

Note that heat q transferred into the exchanger is equal to

$$q = \eta_e M' c'_{pm} \left(t'_1 - t''_1\right)(1 - \psi) \tag{7.102}$$

as can easily be verified.

In the case of fluids in parallel flow, with $\gamma = \infty$, based on (7.61) we obtain

$$1 - \psi_p = \frac{1}{1+\beta}; \tag{7.103}$$

this is the maximum transferred heat indicated by q_∞ and equal to

$$q_\infty = \eta_e M' c'_{pm} \left(t'_1 - t''_1\right) \frac{1}{1+\beta}. \tag{7.104}$$

Under these conditions, of course we have $t'_2 = t''_2$

Efficiency E of the exchanger is therefore equal to

$$E = \frac{q}{q_\infty} = (1+\beta)\left(1 - \psi_p\right). \tag{7.105}$$

If the fluids are in counter flow it is important to distinguish the instances where $\beta \leq 1$ from those where $\beta > 1$.

If $\beta < 1$ based on (7.70), with $\gamma = \infty$, we obtain $\psi = 0$; similarly, if $\beta = 1$ based on (7.75) we obtain $\psi = 0$. Then

$$E = \frac{q}{q_\infty} = 1 - \psi_c. \tag{7.106}$$

If $\psi = 0$ based on (7.76) we determine that $t'_2 = t''_1$; if $\beta = 1$ the temperatures of the two fluids are equal in every position.

If instead $\beta > 1$ based on (7.70), with $\gamma = \infty$, we obtain $(1 - \psi) = 1/\beta$; therefore,

$$E = \frac{q}{q_\infty} = \beta\left(1 - \psi_c\right). \tag{7.107}$$

With $\gamma = \infty$ based on (7.76) and (7.77) through a series of steps we obtain $t''_2 = t'_1$.

Since ψ_p and ψ_c are a function of β and γ, it is possible to build Table 7.2 showing the values of E for various values of these two parameters for fluids in parallel flow, as well as counter flow.

7.6.3 Factor Ψ in Real Cases

Except for cross flow, in this section we will constantly refer to one of two classic ways to generate heat transfer that we illustrated earlier, i.e., the one with fluids in parallel flow and that with fluids in counter flow.

Specifically, we will refer to either one or the other type of heat exchange, when examining real cases. Therefore, we will consider the relative values of ψ (ψ_p for fluids in parallel flow and ψ_c for fluids in counter flow) by introducing corrective factors to obtain realistic values of ψ for the examined cases.

Table 7.2 – Heat exchanger efficiency

Parallel flow

β	γ 0.5	1.0	1.5	2.0	2.5	3.0
0.5	0.528	0.777	0.895	0.950	0.976	0.989
1.0	0.632	0.865	0.950	0.982	0.993	0.998
1.5	0.713	0.918	0.976	0.993	0.998	0.999
2.0	0.777	0.950	0.989	0.998	0.999	0.999

Counter flow

β	γ 0.5	1.0	1.5	2.0	2.5	3.0
0.5	0.362	0.565	0.691	0.775	0.833	0.874
1.0	0.333	0.500	0.600	0.667	0.714	0.750
1.5	0.460	0.661	0.770	0.838	0.882	0.913
2.0	0.565	0.775	0.874	0.927	0.957	0.974

In reference to fluids in parallel flow we will set $\psi = \phi_p \psi_p$ listing in the Tables the values of corrective factor ϕ_p. In reference to fluids in counter flow instead we will set $\psi = \phi_c \psi_c$ and listing in the Tables corrective factor ϕ_c.

ψ_p is obtained through (7.61), whereas ψ_c is calculated through (7.70) and (7.75). Factors β and γ included in these equations are computed through (7.51) and (7.52). As we shall see, β and γ are crucial to obtain the corrective factors, as well.

7.6.3.1 Fluids with Cross Flow

For fluids cross flow (Fig. 7.4) we directly calculated factor ψ; its values are shown in Tables C.1 and C.2 in Appendix C.

Factor ψ for cross flow can also be computed with excellent approximation through the following equation:

$$\psi = (1 - Z)\,\psi_c + Z\psi_p \tag{7.108}$$

with

$$Z = 0.5 - 0.136\,(1 + 0.24\beta)\,\sqrt[3]{\gamma} \tag{7.109}$$

As you see, the value of ψ is intermediate between ψ_p and ψ_c. The behavior of fluids with cross flow is therefore intermediate between the two classic ways, and slightly closer to the condition of fluids in counter flow.

7.6.3.2 Heat Exchanger

This is in reference to heat exchangers with two passages of the fluid inside the tubes schematized in Fig. 7.5.

As with the calculation of the actual value of Δt_m (Sect. 7.5) one must distinguish cases A and B from cases C and D.

This is the examined range: $\beta = 0.2 - 2.0$ e $\gamma = 0.2 - 3.0$.

Contrary to what we did in Sect. 7.5, for types A and B, and taking into account that their behavior is similar to that of fluids in parallel flow, we computed the values of corrective factor ϕ_p. Tables C3, C4 and C5 of Appendix C show the values of ϕ_p for heat exchangers with 2, 3 and 4 passages of the fluid outside the tubes.

Considering that the exchanged heat is proportional to $(1 - \psi)$, we ignored the cases where the corrective factor would lead to a difference of the same heat with respect to that corresponding to fluids in parallel flow under 1%. In these cases one can treat the exchanger as if the fluids were really in parallel flow. These are the missing values to the left of those listed.

We see that the values of ϕ_p are below unity; this means that the exit temperature of the heating fluid from the exchanger is lower than the one with fluids in parallel flow, and as a result the exchanged heat is greater.

If the number of passages of the external fluid is greater than 4, it is possible to adopt the values of ϕ_p in Table C.5 given that with respect to 4 passages the variations are very small.

As far as types C and D in Fig. 7.5, and taking into account that their behavior is similar to those with fluids in counter flow, we computed the values of ϕ_c and included them in Tables C.6, C.7 and C.8, respectively for 2, 3 and 4 passages of the external fluid. We refer the reader to what was said earlier about the missing values on the left of those listed.

We see that the values of ϕ_c are greater than unity; this means that the exit temperature of the heating fluid is higher than the one of fluids in counter flow, and so there is less exchanged heat.

For a number of passages greater than 4 it is possible to adopt the values of ϕ_c in Table C.8 because with respect to 4 passages the variations are very small.

We refer to the recommendation in Sect. 7.5. If the passages of the external fluid are even, the exit location of the external fluid changes with respect to Fig. 7.5.

If the entrance location of the fluid inside the tubes remains the same, type A becomes type C and viceversa. This, in turn, changes the reference to the different Tables.

Let us now consider exchangers with 4 passages of the fluid inside the tubes (Fig. 7.6).

If the number of passages of the external fluid is ≥ 3, it is possible to use the values of ϕ_p shown in Tables C.4 and C.5 for types E and F, or the values of ϕ_c in Tables C.7 and C.8 for types G and H. In fact, the differences between the corresponding values of ϕ are irrelevant or so modest that they can be ignored.

The situation is completely different if the exchanger has 2 passages of the external fluid, as shown in Fig. 7.6.

The behavior of an exchanger with 4 passages of the fluid inside the tubes is considerably different from that of an exchanger with 2 passages. Tables C.3 and C.6 cannot be used. For types E and F in Fig. 7.6 one must refer to Table C.9; for types G and H in the same figure one must refer to Table C.10.

7.6.3.3 Coils

Coils shown in Fig. 7.7 are usually considered, as already pointed out in Sect. 7.5, coils with fluids in parallel flow. By analogy, those in Fig. 7.8 are considered coils with fluids in counter flow.

Therefore, as for the computation of Δt_m, we refer the reader to these two reference situations; for coils in Fig. 7.7 we computed the values of ϕ_p whereas for those in Fig. 7.8 we computed the values of ϕ_c.

Recalling that heat transfer is proportional to $1 - \psi$, we ignored the cases where the difference between the actual heat transfer and the one relative to fluids in parallel flow or counter flow (depending on the type of coil) is below $\pm 1\%$, given that this difference can be considered insignificant.

We consider the following range: $\beta = 0.2 - 2.0$ e $\gamma = 0.2 - 3.0$.

Let us consider the coil in Fig. 7.7; if it consists of 2 sections the corrective factor ϕ_p can be obtained through Table C.11.

As you see, in some cases (the most likely ones) the value of the corrective factor is below unity which means that the heat transfer is greater than that with pure parallel flow. In other instances the opposite is true.

Now, if we consider coils with 3 or more sections, we see that the difference in transferred heat in the coil with respect to fluids in parallel flow is slightly greater than $\pm 1\%$ for few cases with high values (unlikely) of β and γ. Considering that a coil with only two sections is exceptional and unlikely, it is possible to state that coils in parallel flow behave in fact as such with regard to the value of ψ in verification calculation.

Let us now examine coils in counter flow in Fig. 7.8.

The use of corrective factors ϕ_c referred to ψ_c in relation to fluids in counter flow is shown in Tables C.12, C.13, C.14 and C.15 for coils with 2, 3, 4 and 5 sections, respectively.

As expected, we establish that the corrective factor is always greater than unity, and this means that the transferred heat is less than that corresponding to fluids in actual counter flow.

We see that there are less and less cases to consider when going from 2 to 5 sections. For a number of sections equal to 6 or more the difference in heat transferred in the coil and in fluids in counter flow slightly exceeds 1% only in few and rather exceptional instances with high values of β and γ.

In conclusion, for a number of sections equal or higher than 6 (as is usually the case) the verification calculation conducted for fluids in counter flow is correct, whereas for a smaller number of sections, if necessary, one can introduce the indicated corrective factor.

7.6.3.4 Tube Bank with Various Passages of the External Fluid

We refer to the devices in Figs. 7.9 and 7.10. As already pointed out in Sect. 7.5, the classic device of this kind is the heat recuperator located at the end of a steam generator, provided that this type of exchanger can be used also with other fluids, typically gaseous fluids.

The external fluid can enter the heater in correspondence of the entrance into the tubes of the internal fluid, or the opposite takes place with the external fluid entering the heater in correspondence of the exit from the tubes of the internal fluid. Fig. 7.9 represents an air heater of the first kind with three passages of the external fluid. Fig. 7.10 represents an air heater of the second kind instead.

Again, in the first case we speak of an air heater with fluids in parallel flow, and in the second case we speak of a heater with fluids in counter flow.

As usual, we will refer to ψ_p for the first type, and to ψ_c for the second one by introducing the usual corrective factors ϕ_p and ϕ_c.

This is the considered range: $\beta = 0.2 - 2.0$ e $\gamma = 0.2 - 3.0$

As far as the heater in parallel flow in Fig. 7.9, we establish that the corrective factors ϕ_p are independent from the number of passages of the external fluid. They are shown in Table C.16. Note that for low values of β and γ they are below unity. This means that the heat transfer is greater than that corresponding to fluids in parallel flow. Instead, for high values of β and γ the corrective factor is greater than unity. For these values the behavior of the heater is worse compared to that of fluids in parallel flow.

Tables C.17, C.18 and C.19 are about heaters in counter flow shown in Fig. 7.10, respectively with 2, 3 and 4 passages of the external fluid.

We did not consider a higher number of passages because it is unlikely. In conclusion, note that even though $\beta \leq 2$ and $\gamma \leq 3$ the values of ϕ are not within those considered in the Tables, the heater may be computed as if the fluids were, in fact, in parallel flow or in counter flow. If they are within the values of the Tables instead, it is advisable to evaluate through the values of ϕ_p and ϕ_c, if the errors made during computation of the heat transfer are to be considered acceptable when ignoring them. Note that the heater in counter flow with 3 passages of the external fluid is not to be recommended.

Finally, another possible solution for the air heater is shown in Figs. 7.13 and 7.14. The internal fluid performs 2 passages while the external fluid performs 2 or 3 passages. These solutions lead to design of a very compact heater.

These heaters cannot be defines in terms of parallel flow or counter flow. Nonetheless, given that the internal fluid enters the tubes in correspondence of the entrance into the heater of the external fluid, we take fluids in parallel flow as reference.

The issue is complex, and the adopted criterion is to consider the heater as a group of sections where fluids move with cross flow, even though this is not completely correct.

The range under consideration is as follows: $\beta = 0.9 - 1.4$ and $\gamma = 0.6 - 2.6$.

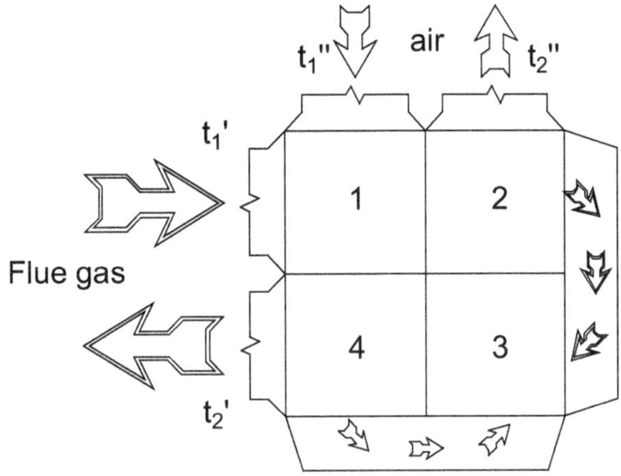

Fig. 7.13 Air heater-cross flow 4 sections

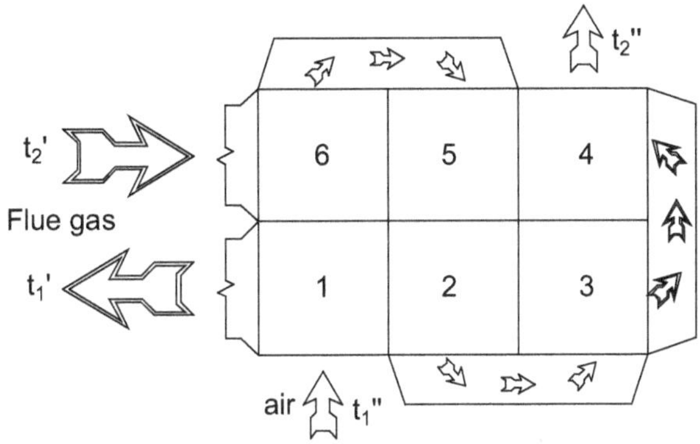

Fig. 7.14 Air heater-cross flow 6 sections

If we refer to the heater in Fig. 7.13 the corrective factor ϕ_p is obtained through Table C.20. If we refer to the heater in Fig. 7.14 instead, the values of ϕ_p are taken from Table C.21.

Chapter 8
Pressure Drops

8.1 Introduction

A fluid in a tube or a duct is subject to a pressure drop, i.e., a decrease in pressure, the decrease will grow with its velocity.

Pressure drops in a straight tube originate from the friction between fluid and wall. Beyond the velocity of the fluid, they are also influenced by density and viscosity, by the diameter of the tube (or the real or hydraulic diameter of the duct), and by the roughness of the wall. These drops will be called distributed pressure drops.

Besides these, there are pressure drops located in certain positions that depend on changes of cross-section area or direction along the path of the fluid. They are located at the inlet or at the outlet of a piping, in curves, elbows, valves, and so on; they are basically not influenced by the viscosity of the fluid and the roughness of the wall but only by velocity, density and the geometric characteristics of the element disturbing the flux. These drops will be called concentrated pressure drops.

Inside the tube the motion can be laminar or turbulent. Since normally it is turbulent, we will discuss this case first and more extensively reserving only a few considerations to laminar motion. The unlikely occurrence of laminar motion was already discussed in the section on heat transfer by convection.

When a fluid hits a tube bank there are also pressure drops in the fluid. They depend on numerous factors. First of all, the velocity of the fluid but also the density, the number of rows crossed by the fluid itself, the number of Reynolds, and the ratios between the transversal and the longitudinal pitch and the diameter of the tubes. These aspects will be discussed in Sect. 8.4.

Finally, in Sect. 8.5 we will examine pressure drops outside finned tubes.

8.2 Distributed Pressure Drops

8.2.1 Turbulent Motion

The pressure drop Δp along a straight pipe is computed through the following equation:

D. Annaratone, *Engineering Heat Transfer*, DOI 10.1007/978-3-642-03932-4_8,

$$\Delta p = \lambda \frac{L}{d_i} \rho \frac{V^2}{2}; \tag{8.1}$$

L stands for the length of the tube, d_i for the inside diameter, ρ for the density of the fluid, V for its velocity and λ for a factor that will be specified later on.

If L and d_i are in m, ρ in kg/m^3 and V in m/s, the pressure drop Δp is in Pa. Generally, it is more convenient to refer to mass velocity G.

Recalling that

$$G = V\rho, \tag{8.2}$$

(8.1) can be written as follows:

$$\Delta p = \lambda \frac{L}{d_i} \frac{G^2}{2\rho} \tag{8.3}$$

with G in kg/m^{2}s.

As far as gas, it is very convenient to refer to density under normal conditions as ρ_0.

Note that

$$\rho = \rho_0 \frac{269.5p}{T} \tag{8.4}$$

where p stands for pressure in bar and T for the absolute temperature in K.

From (8.3) based on (8.4) we obtain

$$\Delta p = 1.855 \frac{L}{d_i} \frac{G^2}{p\rho_0} \frac{T}{1000} \tag{8.5}$$

Often gases are at atmospheric pressure, and if pressure drops only amount to a few thousand Pa in the worst case, the pressure is very close to the atmospheric one and can therefore be considered equal to atmospheric pressure, by assuming that $p = 1.013$ bar. Therefore, (8.5) changes into

$$\Delta p = 1.83\lambda \frac{L}{d_i} \frac{G^2}{\rho_0} \frac{T}{1000}. \tag{8.6}$$

If the calculation of Δp involves a duct with a non circular cross-sectional area, it is required to introduce the hydraulic diameter.

By indicating the cross-sectional area of the duct with A and the wet perimeter with P (in this case it coincides with the geometric perimeter of the cross-sectional area), the hydraulic diameter is given by:

$$d_i = \frac{4A}{P}. \tag{8.7}$$

Thus, by indicating the sides of the rectangular cross-section with a and b, we have

$$d_i = \frac{2ab}{a+b}. \tag{8.8}$$

Factor λ, called friction factor, has been the object of numerous research projects. In this book we will limit our analysis to the most significant and well-known equations.

The following is the famous equation by Blasius:

$$\lambda = 0.316\,\mathrm{Re}^{-0.25} \tag{8.9}$$

where Re is the number of Reynolds; it is valid for $\mathrm{Re} \leq 10^5$.

The following equation by Nikuradse considers the range of the high numbers of Reynolds instead:

$$\lambda = 0.032 - 0.221\mathrm{Re}^{-0.237}. \tag{8.10}$$

It is valid for $\mathrm{Re} = 10^6 - 10^8$.

Based on these very simple equations, it is evident that λ depends only on the number of Reynolds.

Further equations included later on are more complete and closer to reality because they factor in the relative roughness of the walls, as well.

Relative roughness stands for the ratio between the expected higher roughness of the surface in contact with the fluid and the actual or the hydraulic diameter. By indicating it with ε Fig. 8.1 makes it possible to compute it as a function of the diameter and of the type of surface.

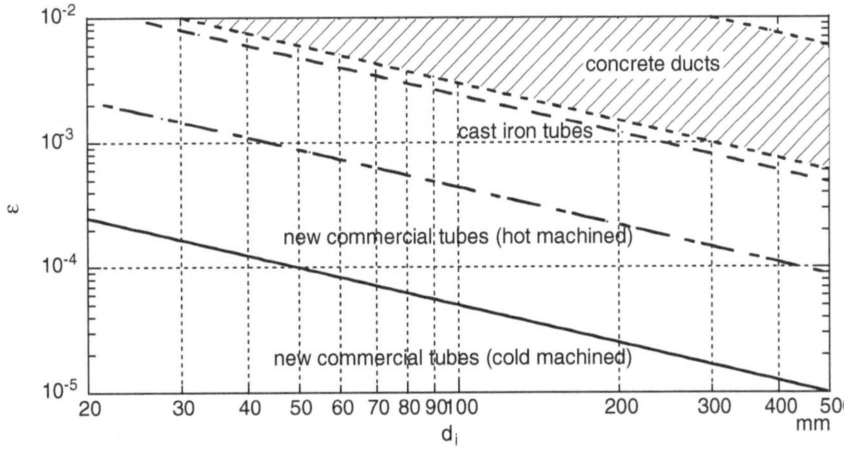

Fig. 8.1 Relative roughness ε

Karmann and Nikuradse distinguish between two borderline conditions of the flow in the pipe. The first refers to turbulent flow in a practically smooth pipe. This condition occurs when

$$\mathrm{Re}\,\varepsilon \leq 70 - 150. \tag{8.11}$$

In that case they suggest the following equation:

$$\frac{1}{\lambda} = 2\log_e\left(\mathrm{Re}\sqrt{\lambda}\right) - 0.8 \tag{8.12}$$

where λ is, of course, independent from ε.

The second boundary condition occurs with a perfectly rough pipe. This means that

$$\mathrm{Re}\,\varepsilon \geq 1000 - 2500. \tag{8.13}$$

In that case

$$\frac{1}{\sqrt{\lambda}} = 2\log_e\frac{1}{2\varepsilon} + 1.735. \tag{8.14}$$

For values of the product $\mathrm{Re}\,\varepsilon$ within $70 - 150$ and $1000 - 2500$, λ follows a not very well defined rule. Karmann and Nikuradse suggested curves that are tangential to the only curve relative to the smooth pipe and to the various curves (as a function of ε) relative to the rough pipe.

Putting the equations suggested by Karmann and Nikuradse on a diagram, and following the criteria above as far as the conditions for smooth or rough pipes, we obtain a well-known diagram in the literature that we chose not to introduce in favor of Moody's diagram which is equally famous.

Moody's diagram is shown in Fig. 8.2.

In the case of turbulent flow the values of λ obtained from Moody's diagram can also be computed through the following equation authored by Colebrook:

$$\frac{1}{\sqrt{\lambda}} = -2\log_e\left(\frac{\varepsilon}{3.7} + \frac{2.51}{\mathrm{Re}\sqrt{\lambda}}\right). \tag{8.15}$$

The calculation of λ through (8.15) is not very straightforward because it must be done by trial and error.

Still, a certain process quickly leads to a practically exact value of λ.

Note that (8.15) can be written as follows:

$$\lambda = \frac{1}{4\log_e^2\left(\dfrac{\varepsilon}{3.7} + \dfrac{2.51}{\mathrm{Re}\sqrt{\lambda}}\right)}. \tag{8.16}$$

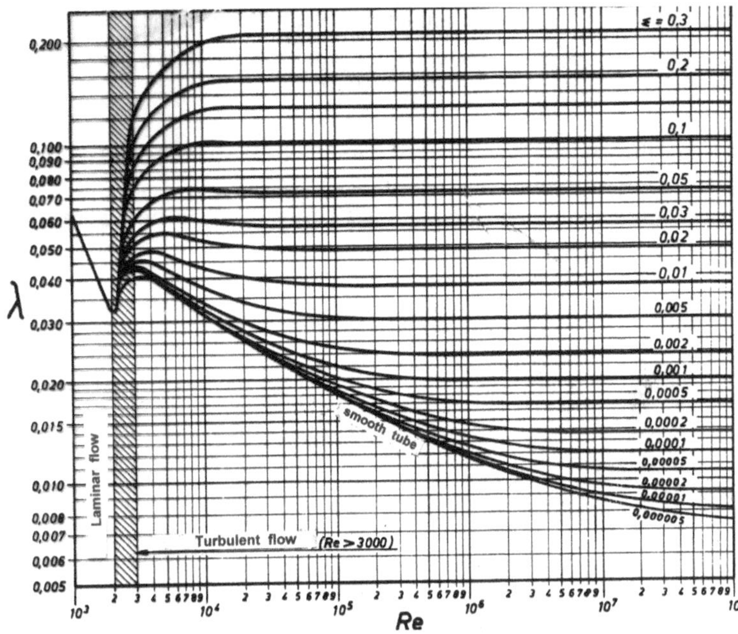

Fig 8.2 Moody's diagram for the friction factor

For Re = ∞ from 8.16, and by indicating the value of corresponding λ with λ_0, we have

$$\lambda_0 = \frac{1}{4\left(\log_e \varepsilon - 0.568\right)^2}. \tag{8.17}$$

We write (8.16) as follows:

$$\lambda_i = \frac{1}{4\log_e^2\left(\dfrac{\varepsilon}{3.7} + \dfrac{2.51}{\mathrm{Re}\sqrt{\lambda_{(i-1)}}}\right)}. \tag{8.18}$$

This is the computation process.

Given that $i = 1,2,\ldots\ldots,$ the values of λ_i are computed as a function of ε, Re and λ_{i-1} until the value of λ_i practically coincides with the value of λ_{i-1} (λ_0 is obtained through (8.17). Convergence is quick.

Further considerations can be made.

The analysis of Fig. 8.2 shows that the value of λ is influenced by the value of the number of Reynolds until the latter is smaller than a certain value that depends on the value of ε.

Basically, Re does not seem to influence the value of λ when

$$\mathrm{Re}\,\varepsilon \geq 1000. \tag{8.19}$$

In that case the value of λ_0 computed through (8.17) diagrammed in Fig. 8.3 can be taken as the value of λ.

(8.17) or the diagram can be used if (8.19) is true, or committing a error by defect of 3−4% at the most, if

$$\mathrm{Re}\,\varepsilon \geq 300. \tag{8.20}$$

Interestingly, for commercial warm machined steel pipes it is possible to write that

$$\varepsilon = \frac{4.4 \times 10^{-5}}{d_i} \tag{8.21}$$

with d_i expressed in m.

Then, from (8.20) and recalling the significance of Re, we obtain

$$G \geq 6.8\mu \times 10^6 \tag{8.22}$$

where G is in kg/m^{2}s, whereas the dynamic viscosity μ is expressed in kg/ms.

Figure 8.4 is obtained based on the dynamic viscosity of the water.

If the mass velocity of water is equal or greater than the values obtainable from this diagram as a function of temperature, it is possible to use (8.17) or the diagram in Fig. 8.3 to compute λ.

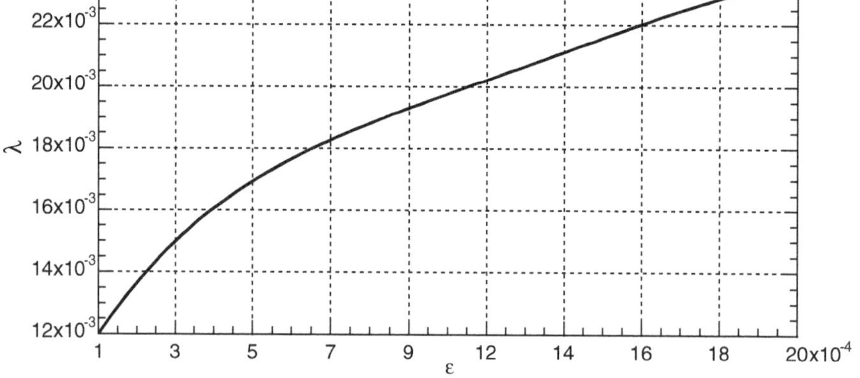

Fig. 8.3 Factor λ for a perfectly rough tube

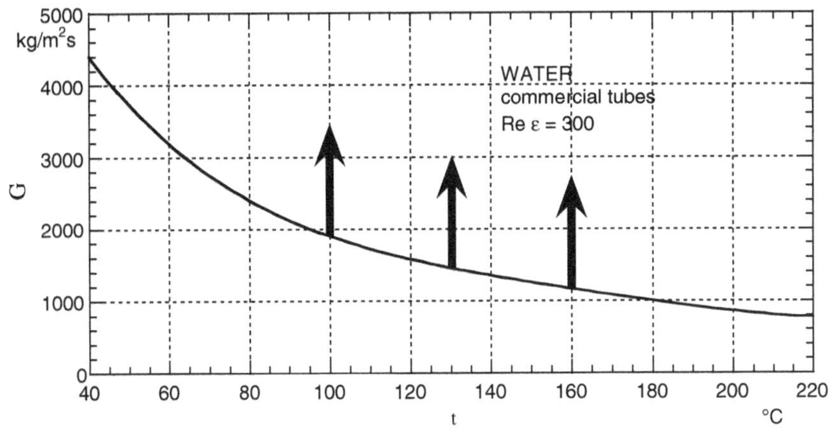

Fig. 8.4

Note, though, that often the values of G and μ are such that the tube can be considered to be basically smooth.

If the tube is smooth, (8.15) is reduced to the following:

$$\frac{1}{\sqrt{\lambda}} = -2\log_e \frac{2.51}{\mathrm{Re}\sqrt{\lambda}}. \tag{8.23}$$

The diagram in Fig. 8.5 was built based on (8.23).

The tube can be assumed to be practically smooth, and factor λ can be obtained through the diagram mentioned earlier with a error by defect equal to 3−4% at the most if

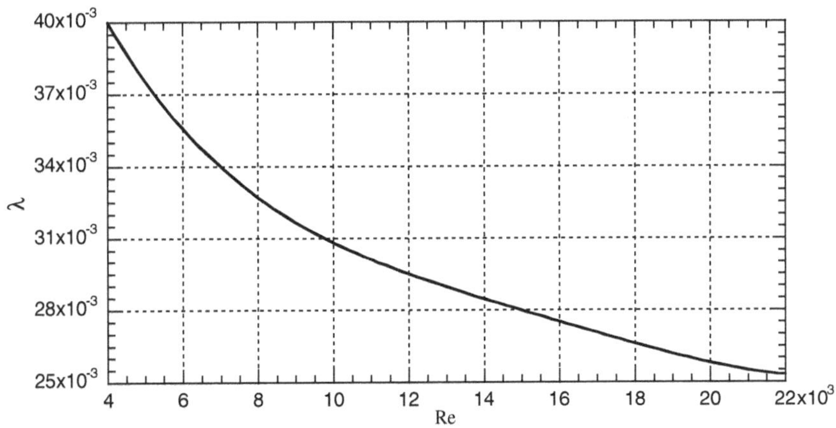

Fig. 8.5 Factor λ for a practically smooth tube

$$\mathrm{Re}\,\varepsilon \leq 10. \tag{8.24}$$

In the case of commercial warm machined tubes, (8.24) satisfies the following condition:

$$G \leq 0.23\mu \times 10^6. \tag{8.25}$$

Equation (8.25) may be satisfied by gases with low velocity.

For instance, based on (8.25) for air we obtain Fig. 8.6; if the value of G is under those of the curve, it is possible to derive λ from Fig. 8.5. Note that the limit values of G are matched by velocities ranging between 4 and 5 m/s depending on variations in temperature.

The value of the pressure drop Δp depends on the reference temperature. In fact, the value of ρ in (8.3) depends on this temperature; in both (8.5) and (8.6) the temperature is included explicitly. Normally, the adopted temperature of reference is the average between the entrance and exit temperatures of the fluid.

It is possible, though, to compute a more realistic temperature of reference if we consider the flue gas inside the tubes of smoke tube boilers transferring heat to the surrounding boiling water.

Based on Sect. 7.5.2 we know that in this case

$$t'_2 = t''_1 + e^{-\gamma}\left(t'_1 - t''_1\right) \tag{8.26}$$

where t'_1 and t'_2 are the inlet and outlet temperatures of the heating fluid (flue gas) and t''_1 is the inlet temperature of the heated fluid. Moreover,

$$\gamma = \frac{US_0}{M'c'_{pm}} \tag{8.27}$$

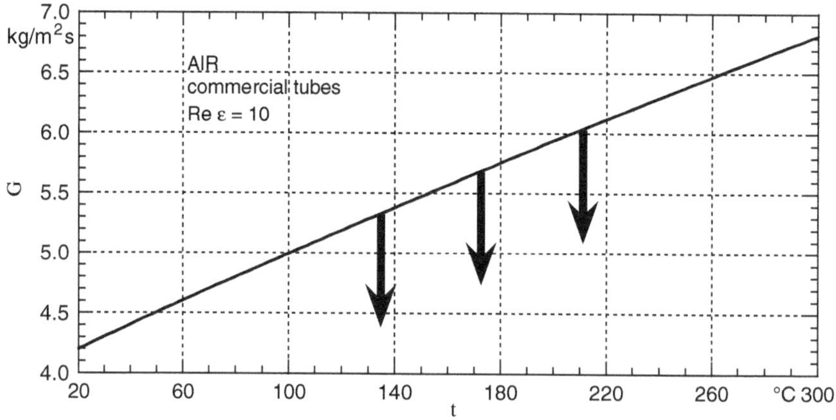

Fig. 8.6

where U stands for the overall heat transfer coefficient, M' for the mass flow rate of flue gas and c'_{pm} for the mean specific isobaric heat, and indicating the surface of the considered passage with S_0. The efficiency of the heat exchange is considered to be close to one.

Note that the presentation in Sect. 7.5.2 is based on U being constant; we stick to this position even as far as the following considerations.

Factor γ is proportional to the surface of the tubes. We may write that

$$\gamma = KS_0 \tag{8.28}$$

where K is a constant equal to $U/M'c'_{pm}$.

(8.26) can therefore be written as follows:

$$t'_2 = t''_1 + e^{-KS_0}\left(t'_1 - t''_1\right). \tag{8.29}$$

Now, if we consider any portion of the total surface and indicate it with S, the generic temperature reached by the gas after coming in contact with surface S is given by:

$$t' = t''_1 + e^{-KS}\left(t'_1 - t''_1\right). \tag{8.30}$$

Referring with more clarity to a numerical example with $t''_1 = 200\,°\mathrm{C}$, $t'_1 = 1000\,°\mathrm{C}$ and $t'_2 = 400\,°\mathrm{C}$, the pattern of the gas temperature as a function of the surface it got in contact with is represented by curve in Fig. 8.7. The temperature clearly decreases with a smaller and smaller gradient along the path of the gas.

Let us examine (8.6). Given that the mass velocity is constant and that the variations in temperature have negligible influence on factor λ, we determine that the pressure drop is proportional only to the absolute temperature of the gas.

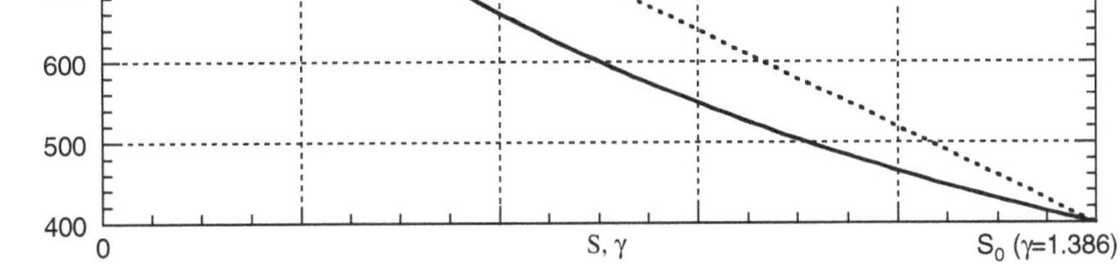

Fig. 8.7 Flue gas temperature inside smoke tubes or through a steam generating tube bank

Considering the pattern of the temperature of the latter, as shown in Fig. 8.7, one makes a rather big error by referring to the average between the inlet and outlet temperatures of the passage. In fact, as far as Δp, this equals substitution of curve with the dashed line.

Thus, it is necessary to refer to the correctly calculated mean value of the absolute temperature. To that extent note that based on (8.30) the mean value of t' is given by:

$$t'_m = \frac{1}{S_0}\int_0^{S_0}\left[t''_1 + e^{-KS}\left(t'_1 - t''_1\right)\right] dS. \tag{8.31}$$

Then, resolving the integral

$$t'_m = t''_1 - \frac{t'_1 - t''_1}{KS_0}\left(e^{-KS_0} - 1\right). \tag{8.32}$$

Based on (8.29)

$$e^{-KS_0} = \frac{t'_2 - t''_1}{t'_1 - t''_1}; \tag{8.33}$$

$$KS_0 = \log_e \frac{t'_1 - t''_1}{t'_2 - t''_1}. \tag{8.34}$$

Based on (8.32) we obtain:

$$t'_m = t''_1 - \frac{t'_1 - t''_1}{\log_e \dfrac{t'_1 - t''_1}{t'_2 - t''_1}}\left(\frac{t'_2 - t''_1}{t'_1 - t''_1} - 1\right); \tag{8.35}$$

and after a series of steps:

$$t'_m = t''_1 + \frac{t'_1 - t'_2}{\log_e \dfrac{t'_1 - t''_1}{t'_2 - t''_1}}. \tag{8.36}$$

(8.36) correctly computes the mean temperature of the gas, thus its mean absolute temperature to introduce in (8.5) or in (8.6) for the computation of Δp.

Note that (8.36) can also be written as follows:

$$t'_m = t'_1 \left(\frac{t''_1}{t'_1} + \frac{1 - \frac{t'_2}{t'_1}}{\log_e \frac{1 - \frac{t''_1}{t'_1}}{\frac{t'_2}{t'_1} - \frac{t''_1}{t'_1}}} \right); \tag{8.37}$$

finally,

$$t'_m = \varphi t'_1 \tag{8.38}$$

where the dimensionless factor φ which corresponds to the expression in parenthesis in (8.37) can be directly taken from the diagram in Fig. 8.8 as a function of t'_2/t'_1 and t''_1/t'_1. The superior line represents the values of φ corresponding to the linear pattern of the temperature (like the straight line in Fig. 8.7). This highlights the influence of the actual pattern of the temperature on the value of t'_m.

Of course, the error made by adopting the average between the extreme temperatures would be increasingly greater as the ratio between them were to farther away from one.

The friction factor λ depends on the type of tube or duct surface (absolute roughness). It is also a function of the diameter and, through Re, a function of the fluid velocity and its kinematic viscosity.

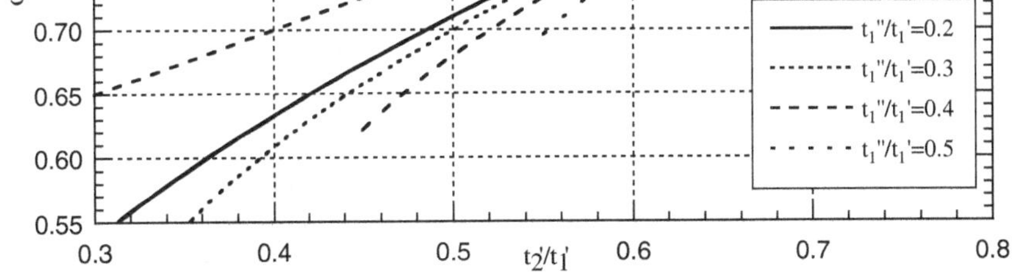

Fig. 8.8 Factor φ for the calculation of the mean flue gas temperature

The pressure drop per length unit of the tube is a function of the absolute roughness of the surface, of the inside diameter, of the velocity, of the dynamic viscosity and of the density.

If R indicates absolute roughness, we have

$$\frac{\Delta p}{L} = f\,(R, V, d_i, \mu, \rho)\,. \tag{8.39}$$

On the other hand, for a given fluid μ and ρ are a function of temperature only. Thus, we can write that

$$\frac{\Delta p}{L} = f\,(R, V, d_i, t)\,. \tag{8.40}$$

We determine that once a type of surface (for instance, commercial tubes) and a reference temperature are set, the pressure drop by length unit of a given fluid is only a function of velocity and diameter.

This way it is possible to create diagrams to obtain $\Delta p/L$ directly.

The diagrams in Figs. 8.9 and 8.10 refer to commercial steel tubes and water flowing through them at 20°C. The pressure drop referred to the length unit of the tube can be obtained based on two of the quantities Q, V and d_i (the volumetric flow rate is expressed in m^3/h).

If the temperature is different from 20°C, the actual pressure drop Δp can be obtained by first approximation from the following equation and by indicating the pressure drop gained from the diagrams with Δp^*, thus

$$\Delta p = \Delta p^* \left(1 - 1.04 \frac{\sqrt{t-20}}{100\, V^{0.4}}\right) \frac{\rho}{1000}; \tag{8.41}$$

ρ stands for the water density in kg/m^3, V for the velocity in m/s and t for the temperature in °C.

Figures 8.11, as well as 8.12, shows the diagrams relative to $\Delta p/L$ for commercial tubes or ducts in metal sheet with air under normal conditions.

If the temperature is different from 0°C, the actual pressure drop Δp can be obtained by first approximation from the following equation and by indicating the pressure drop gained from the diagrams with Δp^*, thus

$$\Delta p = \Delta p^* \left(1 + 2.55 \frac{\sqrt{t}}{100 V^{0.6}}\right) \frac{273}{273 + t} \tag{8.42}$$

where V stands for the velocity in m/s and t for the temperature in °C.

Note that the reference to a commercial tube indicates a new tube. If the tube shows increasing roughness during runtime caused by wear and tear the pressure drop increases. The increase is small for low values of the number of Reynolds, but it can be considerable for high values of Re.

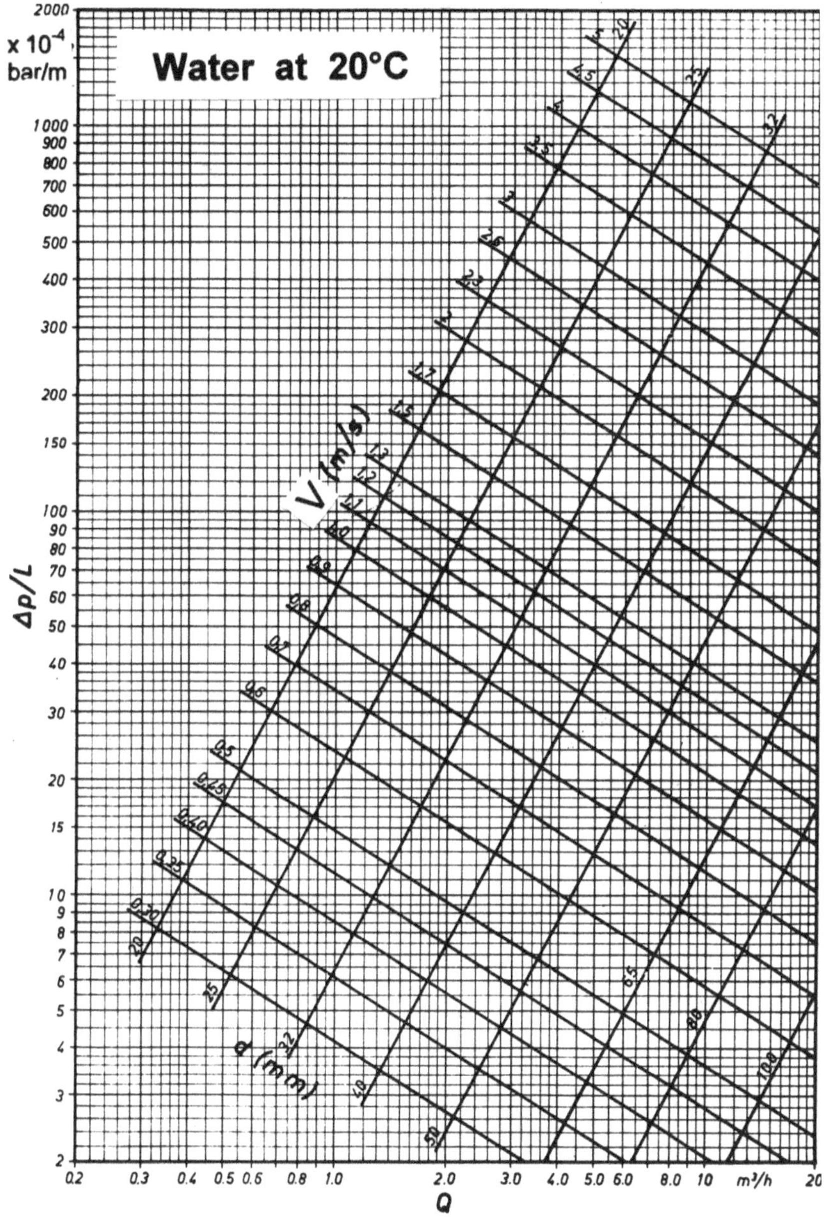

Fig. 8.9 Pressure drop for water at 20°C (Q=0.2–20 mc/h)

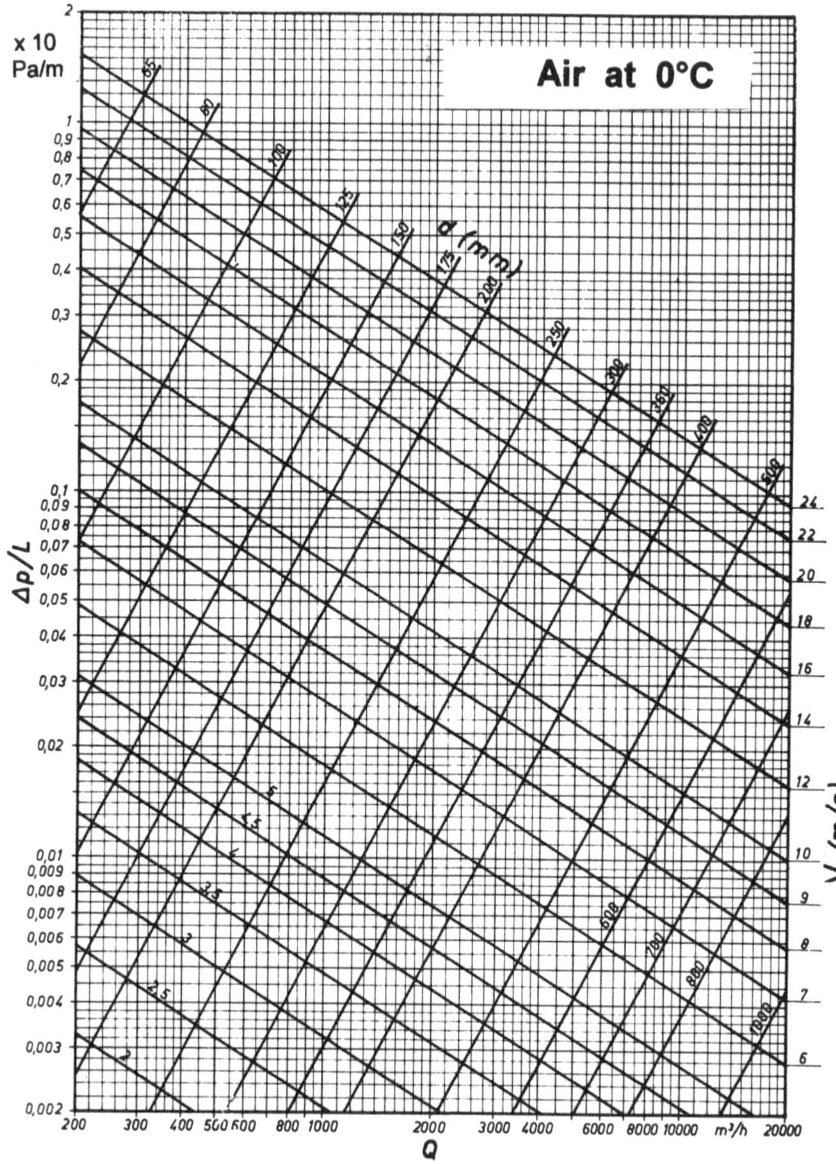

Fig. 8.10 Pressure drop for water at 20°C (Q=20–2000 mc/h)

Finally, note that in the case of ducts without a round cross section, the diagrams above must be used in reference to velocity and diameter, i.e. the quantities that are crucial for the value of Δp (8.40). The value of the volumetric flow rate Q on the abscissa corresponds to the values of V and d_i only if the cross section is round, but not if d_i is the hydraulic diameter.

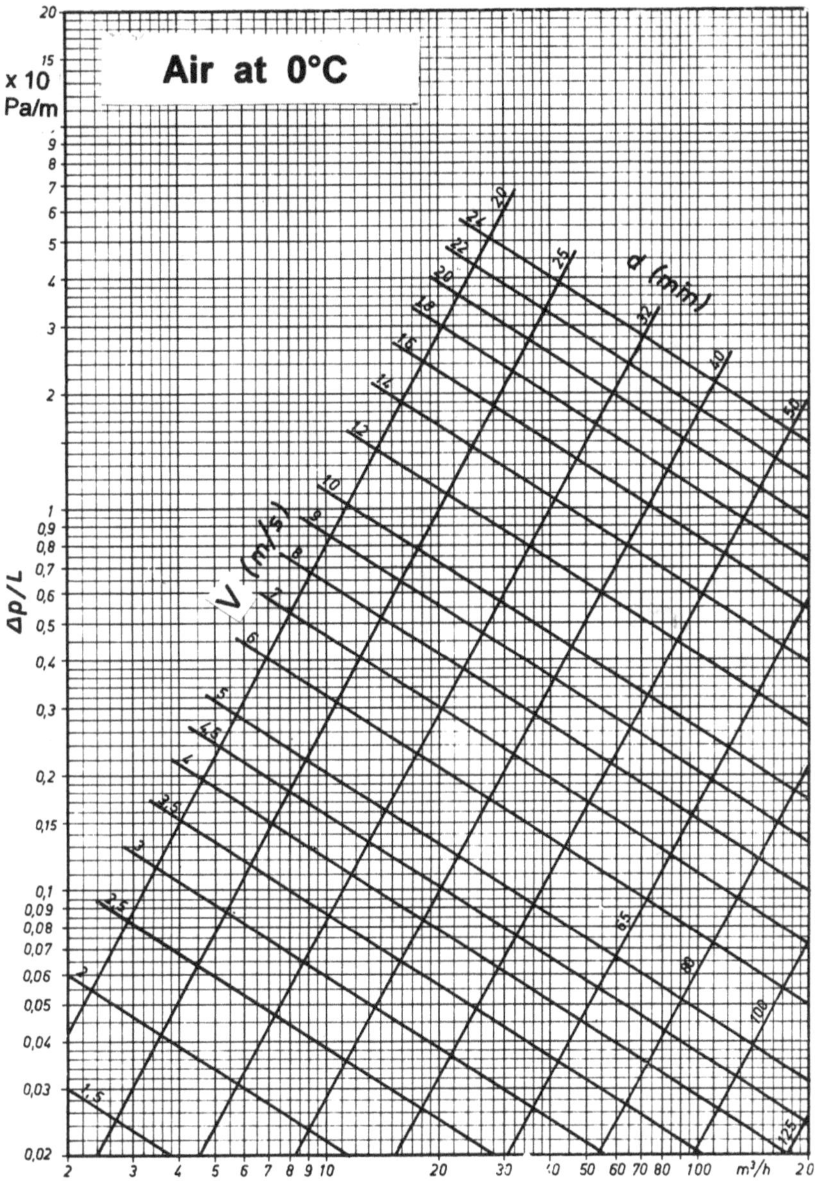

Fig. 8.11 Pressure drop for air at 0°C (Q=2–200 mc/h)

8.2.2 Laminar Motion

If the motion is laminar, in other words if Re $\leq$ 2000, we establish that, contrary to turbulent motion, pressure drops are not influenced by the density of the fluid, but

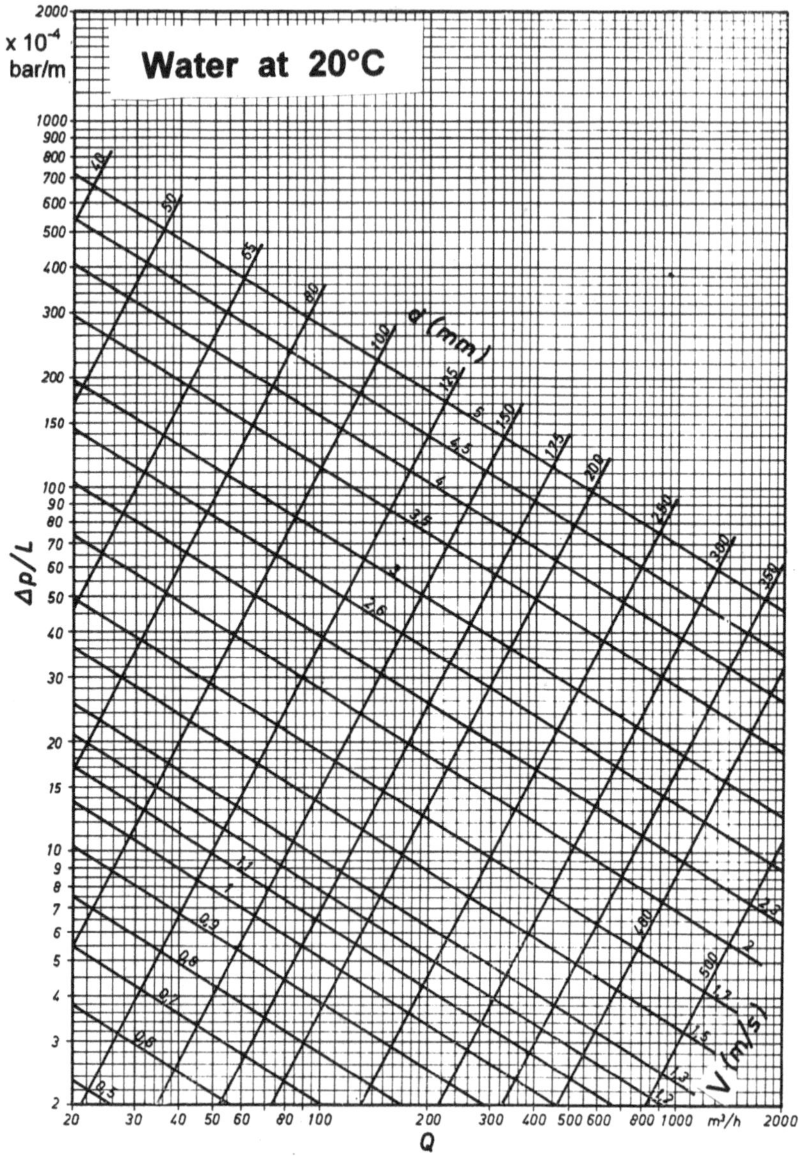

Fig 8.12 Pressure drop for air at 0°C (Q=200–20,000 mc/h)

by its viscosity instead. In addition, they are proportional to the velocity of the fluid and not to its square, as is the case with turbulent motion. In other words, the drops are proportional to the product μV.

This is not surprising if we remember that the force that must be applied to a plane moving in parallel to a fixed plane with the interposition between the two planes of a viscous fluid (and this is what we are referring to) is equal to

$$F = \mu S \frac{V}{h} \tag{8.43}$$

where S is the surface of the moving plane and h is the distance between the two planes. Even this force is proportional to the product μV.

Given that in laminar motion, where the phenomena characterizing turbulent motion are missing, the forces of friction due to the viscosity of the fluid are crucial the entity of pressure drops is directly conditioned by viscosity.

Specifically, the pressure drop in a straight tube is obtained through the following equation by Hagen-Poiseuille

$$\Delta p = \frac{32\mu V}{d_i} \frac{L}{d_i} \tag{8.44}$$

with an obvious significance of the symbols.

Note that (8.44) can also be written like this:

$$\Delta p = \frac{64\mu}{V d_i \rho} \frac{L}{d_i} \frac{\rho V^2}{2}; \tag{8.45}$$

and recalling the significance of the number of Reynolds,

$$\Delta p = \frac{64}{\mathrm{Re}} \frac{L}{d_i} \frac{\rho V^2}{2}. \tag{8.46}$$

Recalling (8.1) we establish that even with laminar motion it is possible to refer to (8.1) relative to turbulent motion provided that

$$\lambda = \frac{64}{\mathrm{Re}} \tag{8.47}$$

This is shown in Moody's diagram in Fig. 8.2; in fact, the only line in the diagram for the field concerning laminar motion corresponds to (8.47). This facilitates a comparison between the two types of motion with respect to pressure drops. If pressure drops with laminar motion are rather low, this depends on the low velocities in question, but certainly not on the ideal factor friction λ which is rather high.

Note that if the velocity is high but the viscosity is so high that $\mathrm{Re} < 2000$ the motion is laminar and the pressure drop is high.

Sect. 8.3 will focus on concentrated pressure drops, including drops in curves, yet still with reference to turbulent motion.

Therefore, a quick reference to pressure drops in curves with laminar motion is due taking coils as an instance.

Figure 8.13 based on Drew's studies makes it possible to compute the ratio between the friction factor in a curve at 180° and the corresponding one of a straight tube. In other words, the pressure drop in the curve is equal to the product of the ratio

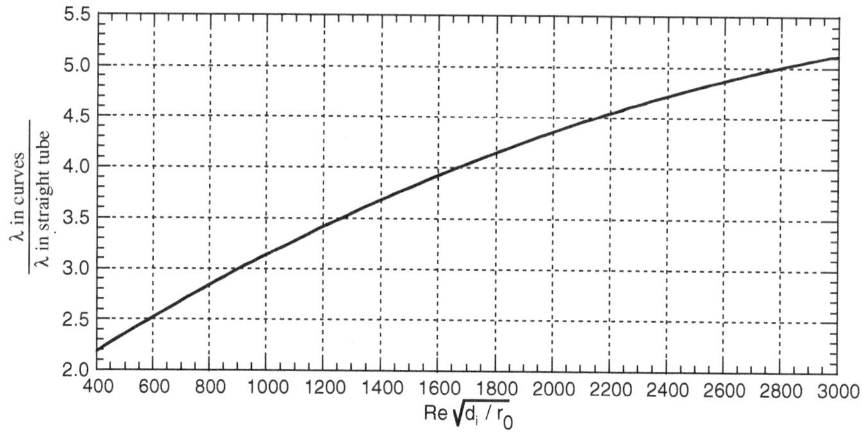

Fig. 8.13 Values of λ in curves with laminar motion

in question by the pressure drop in a straight tube of equal length. In the parameter on the x-axis r_0 it is the bending radius.

Note that in curves the laminar motion may be present with values of the number of Reynolds greater than 3000 which produce turbulent motion in straight tubes.

8.3 Concentrated Pressure Drops

Concentrated pressure drops are caused by inlets and outlets, curves, changes in the cross sections, offtakes, and so on.

They are computed as follows:

$$\Delta p = \zeta \frac{\rho V^2}{2} \tag{8.48}$$

where ζ is a factor that shall be explained later on.

If ρ is in kg/m^3 and V in m/s, the pressure drop Δp is in Pa.

Similarly to distributed pressure drops, if we refer to mass velocity G expressed in kg/m^{2}s, from (8.48) we have:

$$\Delta p = \zeta \frac{G^2}{2\rho}. \tag{8.49}$$

In the case of gas, by introducing density under normal conditions ρ_0 from (8.49), and similarly to (8.5), we obtain

$$\Delta p = 1.855\zeta \frac{G^2}{p\rho_0} \frac{T}{1000}. \tag{8.50}$$

By assimilating pressure p to atmospheric pressure (1.013 bar), from (8.50) we obtain the following equation:

$$\Delta p = 1.83\zeta \frac{G^2}{\rho_0} \frac{T}{1000}. \tag{8.51}$$

Factor ζ in the previous equations can be considered to be practically independent from the number of Reynolds when the latter is greater than 3000−4000.

For instance, Fig. 8.14 shows curves standing for ζ for the inlet and outlet of a tube from a tank for which we adopt $\zeta = 0.5$ and $\zeta = 1$, as we shall see later on.

Inlet and outlet from a tank are, in fact, threshold instances of sudden sectional changes.

Two cases stand out during analysis of this issue in general terms. In the first case, the cross section abruptly decreases along the pass of the fluid. The value of ζ is obtained from curve A of Fig. 8.15 as a function of ratio d_i/d_i' between the diameter of the reduced cross section and the diameter of the original cross section. If the fluid enters the tube exiting from a tank, we may consider that $d_i' = \infty$, thus $d_i/d_i' = 0$ which finally leads to $\zeta = 0.5$.

In the second instance the cross section widens, the value of ζ is obtained from curve B of Fig. 8.15. If the fluid enters in a tank in a similar way, we can consider that $d_i/d_i' = 0$, thus $\zeta = 1$. In that case all the kinetic energy of the fluid is lost.

In both instances the velocity to consider in (8.49) is the one in the reduced cross section.

In general (here the equation can also be used with a duct with a non-circular cross section), factor ζ for abrupt reduction of the cross section is given by:

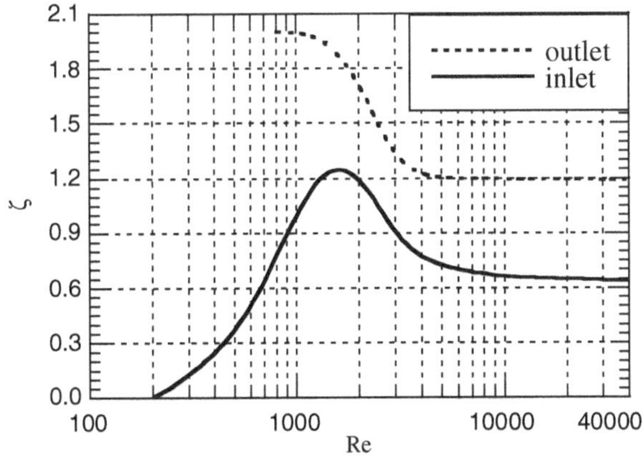

Fig 8.14 Factor ζ for inlet and outlet in a tank

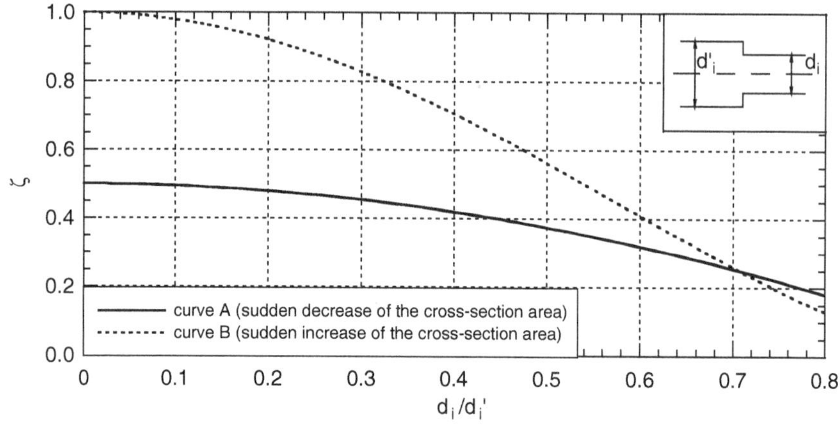

Fig. 8.15 Factor ζ for sudden variations of the cross-sectional area

$$\zeta = 0.5\left(1 - \frac{A}{A'}\right) \tag{8.52}$$

where A stands for the smaller and A' for the greater cross section.

In the case of widening of the cross section we have:

$$\zeta = \left(1 - \frac{A}{A'}\right)^2, \tag{8.53}$$

including the same significance of symbols.

If the cross section is gradually reduced with an angle ω smaller that 30° (Fig. 8.16) instead, it is possible to assume that $\zeta = 0.05$ regardless of the value of d_i/d'_i or A/A'. If the cross section gradually widens, the values obtained from curve B in Fig. 8.15 or from (8.53) are multiplied by the corrective factor β the values of which are shown in Table 8.1.

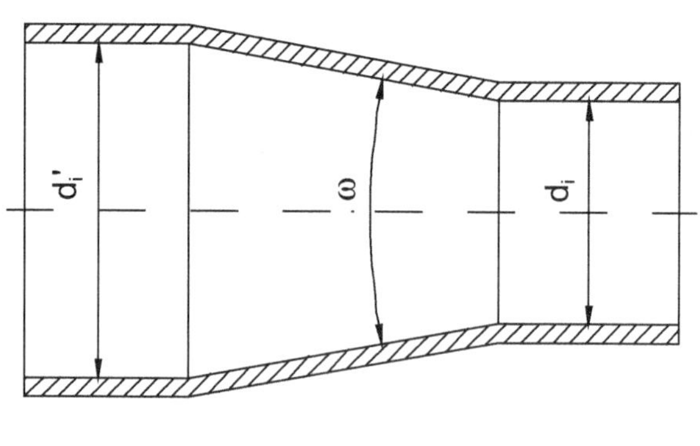

Fig. 8.16

Table 8.1 Factor β for gradual increase of the cross-section area

ω	3°	5° – 8°	10°	14°	20°	30°	45°	60°	≥90°
β	0.18	0.14	0.16	0.25	0.45	0.70	0.95	1.10	1.00

In the case of inlet from a tank, if the tube enters part of the tank itself (Fig. 8.17), we can assume that $\zeta = 0.8$.

For pressure drops through a drilled diaphragm (Fig. 8.18) factor ζ is obtained with reference to curve B in Fig. 8.15 and by multiplying the value obtained this way by the corrective factor β given by:

$$\beta = 2.8\left[1 - \left(\frac{d_i}{d'_i}\right)^2\right]. \tag{8.54}$$

More in general, the value of ζ computed based on (8.53) is multiplied by the corrective factor β given by

$$\beta = 2.8\left(1 - \frac{A}{A'}\right). \tag{8.55}$$

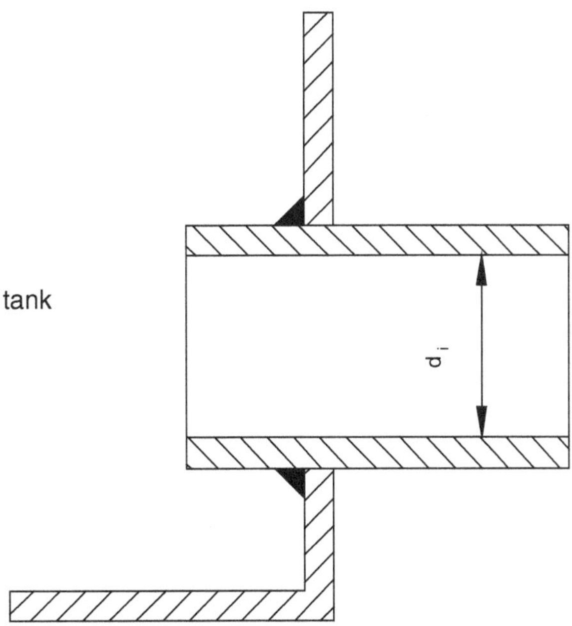

Fig. 8.17

Fig. 8.18

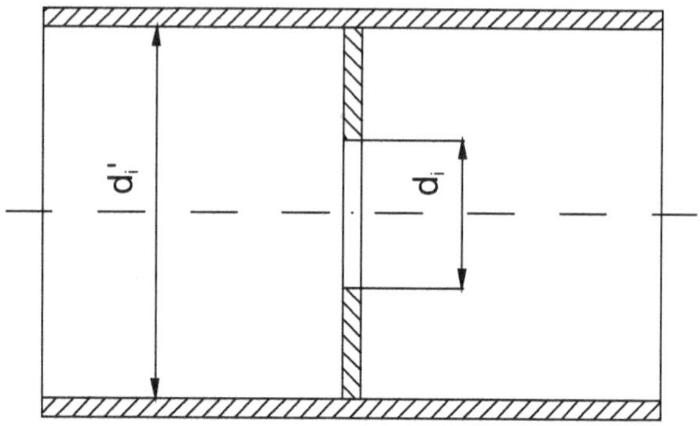

The velocity to consider in (8.48) or in the equations obtained from it is always the one referring to the reduced cross-sectional area.

Besides the bending angle, factor ζ relative to pressure drops in the curves is also a function of the ratio between the bending radius referred to the axis of the tube and its inside diameter.

The diagram in Fig. 8.19 leads to its value for 45°, 90°, 135° and 180° curves. The interpolation is done for intermediate angles.

In the case of ducts it is worthwhile knowing the values relative to elbows, too (Fig. 8.20). In the absence of baffles they are taken from Table 8.2.

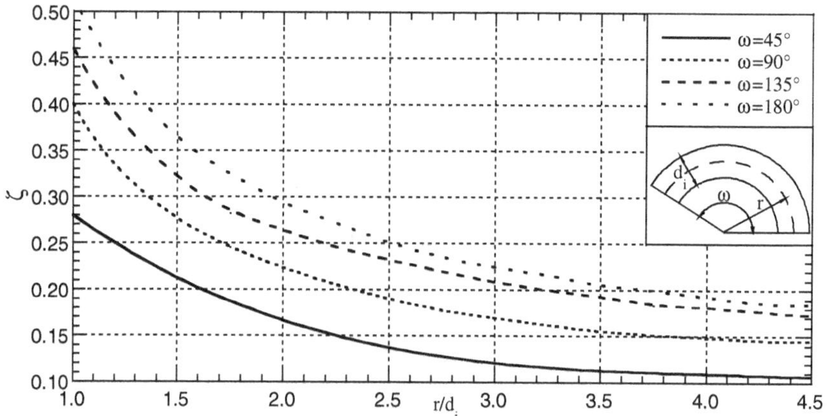

Fig. 8.19 Factor ζ for curves

Table 8.2 Factor ζ for elbows without baffles

ω	15°	30°	45°	60°	90°
ζ	0.02	0.07	0.18	0.36	1.00

Fig. 8.20

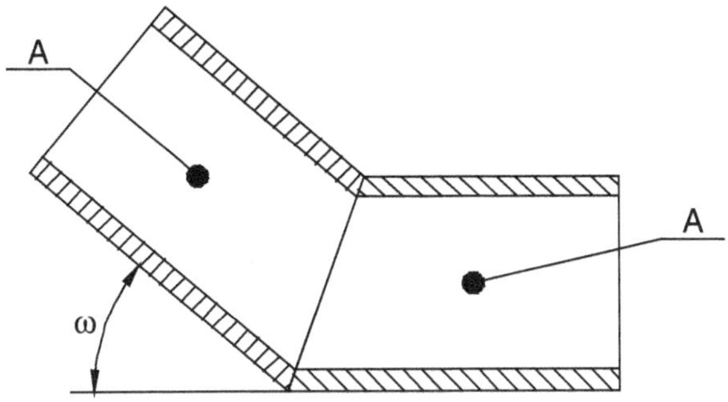

If the elbows are equipped with baffles instead, it is possible to refer to Table 8.2 by multiplying the values by 0.4.

As far as the T and Y-shaped offtakes, the values of ζ are not available for all the numerous potential instances, both in terms of the different values of volumetric flow rates in both branches of the offtake and in terms of the different diameters of the concurrent tubes.

The diagram in Fig. 8.21 makes it possible to obtain ζ for T-shaped offtakes with concurrent tubes of equal diameter as a function of the ratio between the flow rates and two possible flux directions.

The diagram in Fig. 8.22, instead, makes it possible to obtain ζ for Y-shaped offtakes with concurrent tubes of equal diameter as a function of angle ω between the two spread out branches under the assumption that the flow rate in both branches is the same and for the two possible flux directions.

The velocity to consider in (8.48) and the like is the one corresponding to volumetric flow rate Q.

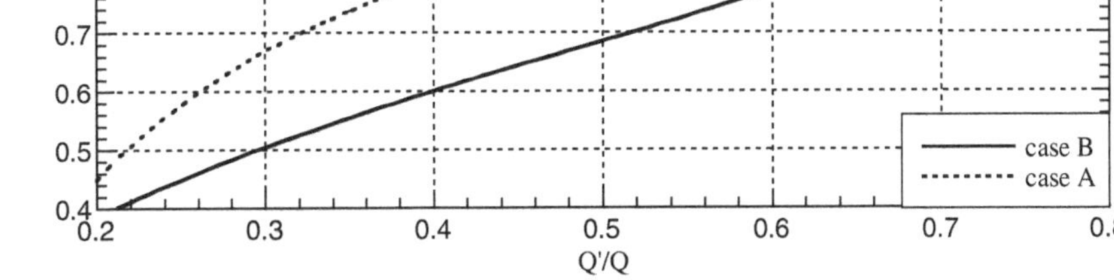

Fig. 8.21 Factor ζ for T pieces

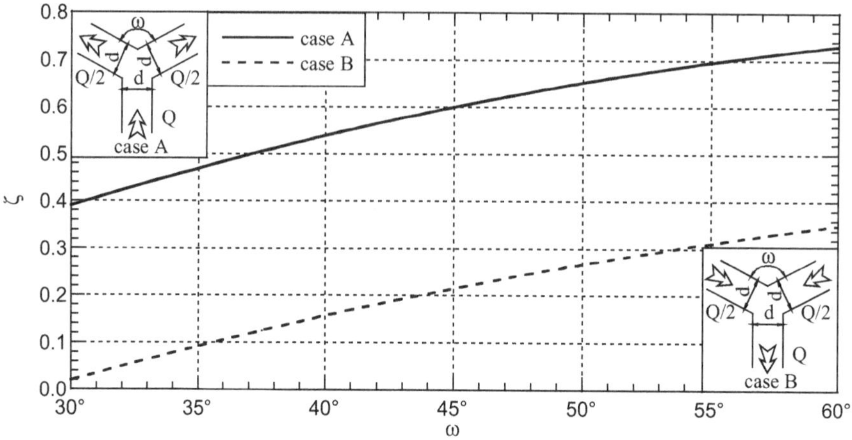

Fig. 8.22 Factor ζ for Y pieces

Other authors suggest the following values for T-shaped offtakes with equal diameters and distinguishing the two values of ζ relative to continuous piping and the one at 90° that we indicate with ζ and ζ', respectively. In case B of Fig. 8.21

for $Q' = 0$	$\zeta = 0.04$	$\zeta' = 0$
for $Q' = 0{,}5Q$	$\zeta = 0.40$	$\zeta' = 0.30$
for $Q' = Q$	$\zeta = 0$	$\zeta' = 0.90$

In case A of Fig. 8.21

for $Q' = 0$	$\zeta = 0.04$	$\zeta' = 0$
for $Q' = 0{,}5Q$	$\zeta = 0.01$	$\zeta' = 0.90$
for $Q' = Q$	$\zeta = 0$	$\zeta' = 1.30$

The values of both ζ and ζ' refer to the velocity corresponding to Q.

As far as the valves, there are many more kinds. Definitely reliable data can only be obtained by consulting with the producing company.

The same is true for gate valves.

Finally, flow rate meters, slide regulation valves and expansion compensators are inserted in the ducts. It is impossible to provide general information on the relative pressure drops.

If the flow rate meter is a Venturi type meter, computation is still possible by examining its shape and by applying the described values of ζ, even though in this case it is still recommended to turn to the specialist company for available experimental data on their instruments.

In conclusion, note the widespread criterion to assimilate the generic concentrated drop to the one in a portion of straight tube of a certain length, called virtual length or equivalent length.

A comparison between (8.48) and (8.1) shows the requirement for equal pressure drops to be

$$\lambda \frac{L_e}{d_i} = \zeta, \tag{8.56}$$

where L_e stands for the virtual length (equivalent length).

Then

$$L_e = \frac{\zeta}{\lambda} d_i. \tag{8.57}$$

The criterion above actually consists of computing a virtual length equal to a certain number of diameters for every concentrated drop. Given that N_e is this number, from (8.57) we obtain

$$L_e = N_e d_i; \tag{8.58}$$

then

$$N_e = \frac{\zeta}{\lambda}. \tag{8.59}$$

The criterion is interesting because it simplifies the computation of the pressure drop of a tube or a duct. In fact, once the values of N_e for the different "flux perturbations" are known, one calculates a fictitious length L' given by:

$$L' = L + d_i \sum N_e \tag{8.60}$$

where L stands for the length of the tube with a constant diameter d_i.

Δp is computed based on L' through one the equations in Sect. 8.2.1 or deriving the value of $\Delta p/L'$ directly from the diagrams.

This takes both distributed and concentrated pressure drops into account.

Note that to safeguard the validity of the method (i.e., to simplify computation) requires a constant value for N_e which is typical of any type of perturbation. This is what is actually done.

But based on (8.59) N_e is clearly a function of λ and ζ; thus, it is influenced by the number of Reynolds, as well as the roughness.

Calculation based on constant values of N_e is therefore acceptable only for an approximate evaluation of Δp, especially if the concentrated pressure drops have a considerable impact on the total.

8.4 Pressure Drops through Tube Banks

If an external fluid hits a tube bank, it is subject to a pressure drop which can be computed as follows:

$$\Delta p = f_d f_a N \rho \frac{V^2}{2} \tag{8.61}$$

where f_d and f_a stand for two factors that will be discussed later on and N for a number of tube rows crossed by fluid.

With the density ρ in kg/m^3 and the velocity V in m/s the pressure drop Δp is in Pa.

For gas it is more convenient to refer to mass velocity G and to density ρ_0 under normal conditions.

Similarly to (8.5) and (8.50), we obtain the following:

$$\Delta p = 1.855\, f_d f_a N \frac{G^2}{p \rho_0} \frac{T}{1000}; \tag{8.62}$$

G in kg/m^{2}s, p in bar and the absolute temperature T in K.

As usual, assimilating pressure p to atmospheric pressure we obtain the following through (8.62):

$$\Delta p = 1.83\, f_d f_a N \frac{G^2}{\rho_0} \frac{T}{1000}. \tag{8.63}$$

It is equally possible to introduce velocity V_0 under normal conditions. In that case (8.61) leads to

$$\Delta p = 1.855\, f_d f_a N \rho_0 \frac{V_0^2}{p} \frac{T}{1000}; \tag{8.64}$$

and, assimilating pressure p to atmospheric pressure

$$\Delta p = 1.83\, f_d f_a N \rho_0 V_0^2 \frac{T}{1000}. \tag{8.65}$$

Factor f_d intervenes only if the number of crossed rows is below 10. In fact, for $N \geq 10$ we have $f_d = 1$; for $N<10$ the values of f_d can be obtained through Fig. 8.23.

The arrangement factor f_a is a function of the number of Reynolds, of the outside diameter, of the transversal and longitudinal pitch of the tubes, and finally of the type of arrangement.

Let us consider Fig. 8.24 showing an arrangement with in-line tubes where the transversal pitch is indicated by s_t and the longitudinal one by s_l. Fig. 8.25 helps to obtain f_a for different values of the ratios s_t/d_o and s_l/d_o as a function of Re.

Fig. 8.26 shows a layout with staggered tubes instead, and the diagram in Fig. 8.27 helps to determine the arrangement factor f_a.

Note that except for special situations, s_l/d_o, s_t/d_o and Re being equal, the values of f_a in staggered arrangements are always higher than those relative to in-line tubes.

On the other hand, as we know, the staggered arrangement implies higher overall heat transfer coefficients.

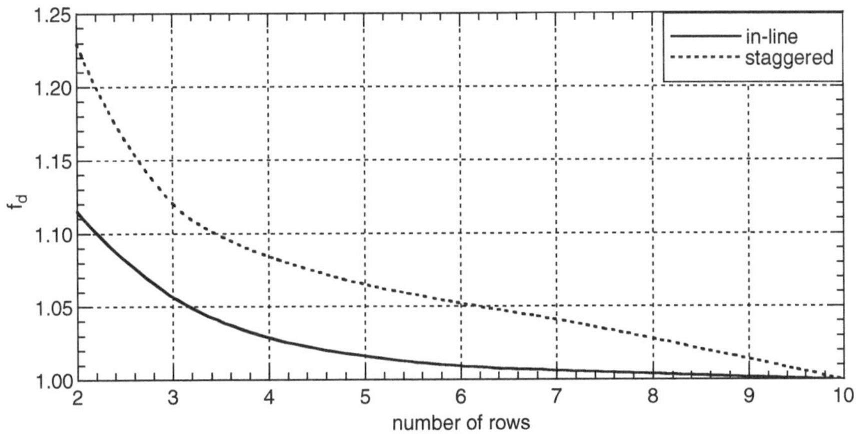

Fig. 8.23 Factor f_d for the pressure drop trough a tube bank

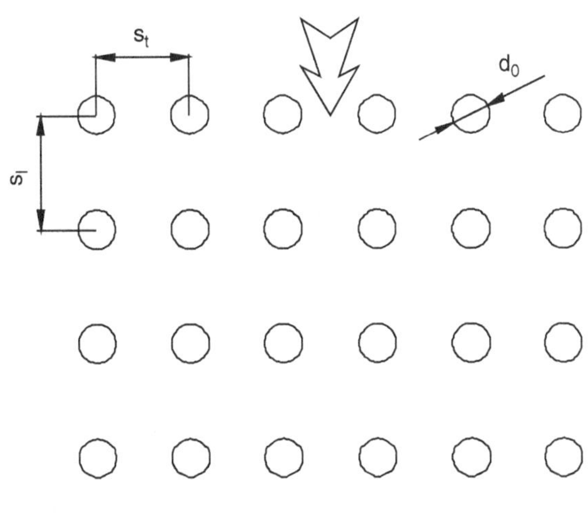

Fig. 8.24 Arrangement with in-line tubes

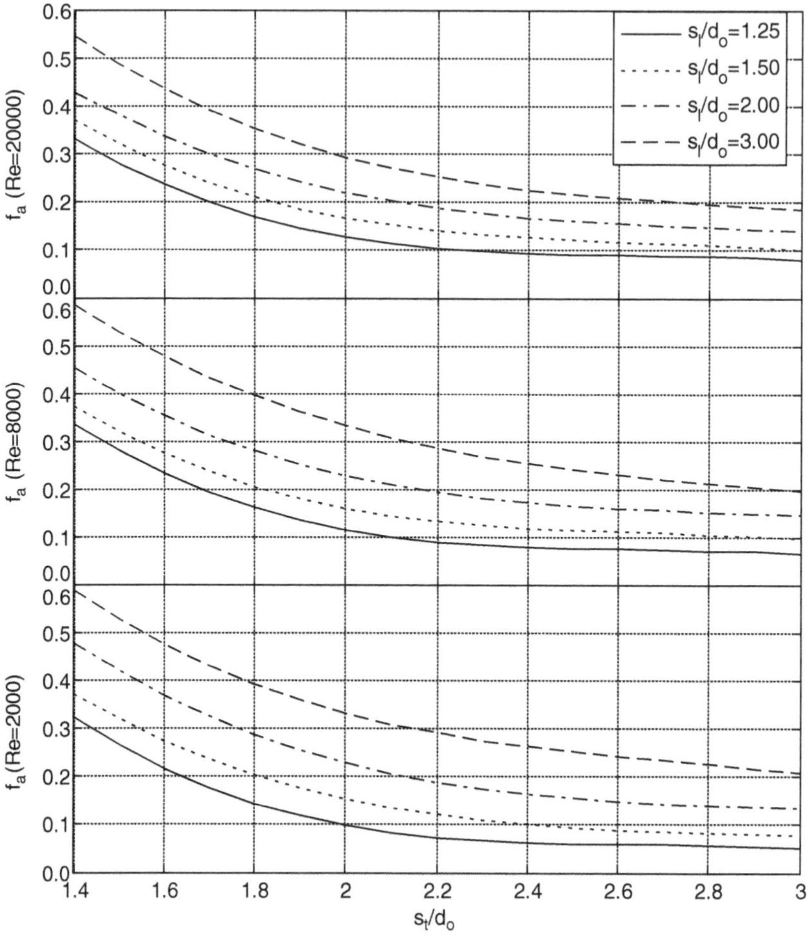

Fig. 8.25 Arrangements factor f_a for in-line tubes

The criteria for adopting one over the other arrangement was already discussed in Sect. 5.9 highlighting the parameters to consider in order to evaluate the problem correctly and to simultaneously take both heat transfer and pressure drops into account. Certain constructive and running requirements notwithstanding, it is a question of adopting the optimal solution for a specific project.

In conclusion, even though it is not recommended as far as heat transfer, sometimes the gas flow is parallel to the axis of the tubes and not transversal, as we assumed so far. In this case we consider the cross-sectional areas shown by the hatching lines in Fig. 8.28 (with staggered arrangement an in-line tubes) and the relative wet perimeter (bold line in the figure) and compute the hydraulic diameter through 8.7.

The pressure drop is computed based on Sect. 8.2.1.

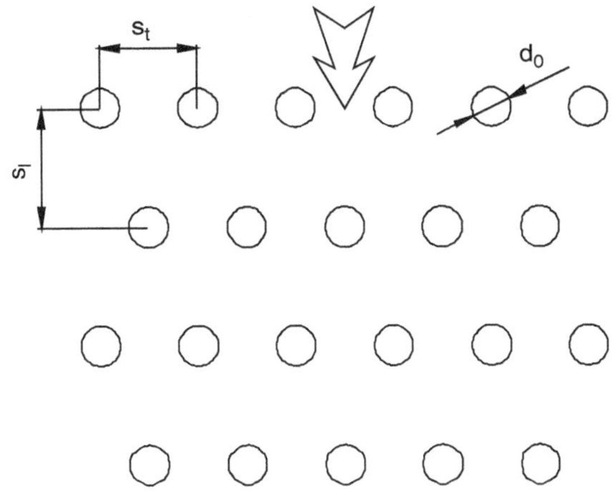

Fig. 8.26 Arrangement with staggered tubes

8.5 Pressure Drop in Finned Tubes

Pressure drops inside finned tubes are computed based on Sects. 8.2.1 and 8.3.

The pressure drops relative to the fluid hitting the tubes from outside require the following rather complex calculation.

Based on the fact that all lengths are in m, $d_o$indicates the outside diameter of the tube without fins and d_f the diameter inclusive of the fins; if H is the height of the fins, we have $d_f = d_o - 2H$.

N_f indicates the number of fins per meter of tube, x_f the thickness of the fins and b the space between fins, given that $b = 1/N_f - x_f$.

N_t indicates the number of finned tubes for every row of tubes and N_r the number of rows; s_t and s_l indicate the transversal and longitudinal pitch of the tubes.

If A_v indicates the cross-section area of the space available to the tube bank and A indicates the passage area of the external fluid:

$$A = A_v - N_t \left(d_o + 2N_f H x_f\right) L \tag{8.66}$$

given that L is the length of the individual tube.

Based on the flow rate of the external fluid and on area A, it is possible to compute mass velocity G of the fluid and, based on $d_o$and on viscosity, the number of Reynolds.

Therefore, we compute factor C_2 given by

$$C_2 = 0.07 + 8\ \text{Re}^{-0.45}. \tag{8.67}$$

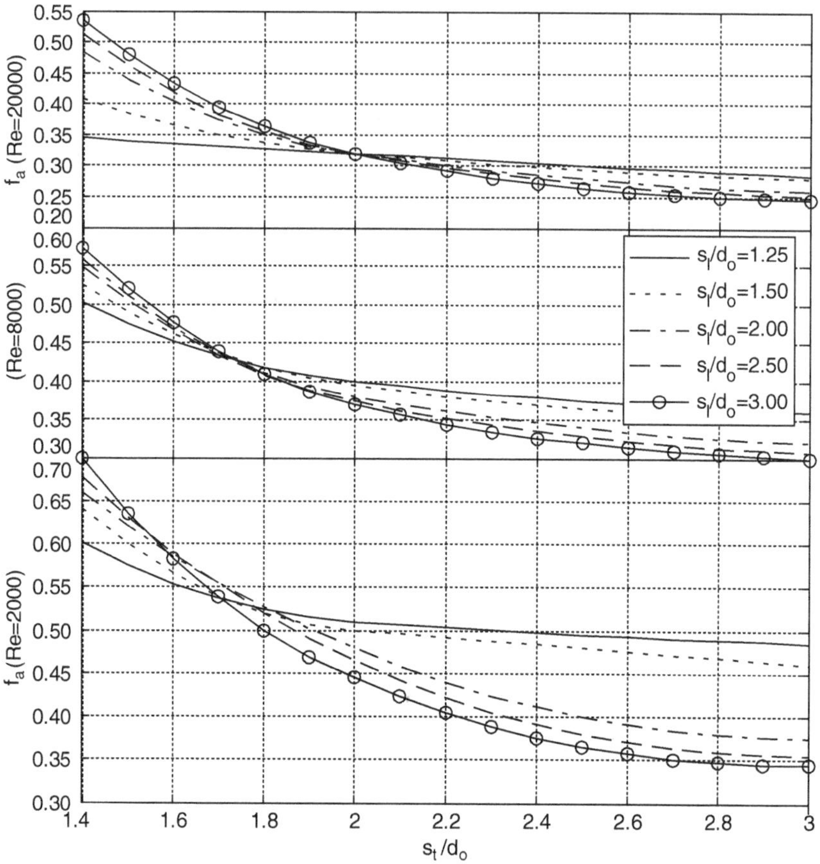

Fig. 8.27 Arrangement factor f_a for staggered tubes

Fig. 8.28

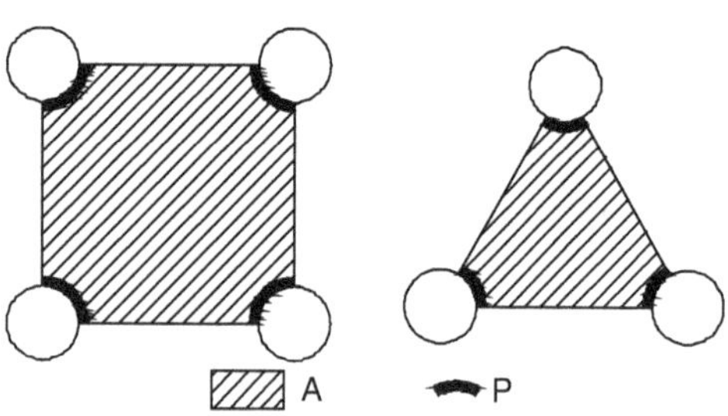

Then, we compute factor C_4 equal to

$$C_4 = 0.08\left(0.15\frac{s_t}{d_o}\right)^{-1.1(H/b)^{0.15}} \tag{8.68}$$

if the tubes are in-line, or

$$C_4 = 0.11\left(0.05\frac{s_t}{d_o}\right)^{-0.7(H/b)^{0.2}} \tag{8.69}$$

if the tubes are staggered.

Finally, we compute factor C_6 given by

$$C_6 = 1.6 - \left(0.75 - 1.5e^{-0.7N_r}\right)e^{-0.2(s_l/s_t)^2} \tag{8.70}$$

if the tubes are in-line or

$$C_6 = 1.1 + \left(1.8 - 2.1e^{-0.15N_r^2}\right)e^{-2s_l/s_t} - (0.7 - 0.8e^{-0.15N_r^2})e^{-0.6s_l/s_t} \tag{8.71}$$

if they are staggered.

If $N_r > 6$ for the computation of C_4 e C_6 (and only to this extent) we must assume that $N_r = 6$.

We compute factor F given by

$$F = C_2C_4C_6\sqrt{d_f/d_o} \tag{8.72}$$

At this point we consider the density of the fluid, at inlet and outlet temperature from the bank, respectively, by indicating them with ρ_1 andρ_2, then we compute the mean density ρ_m. Factor B is given by

$$B = \frac{1+\left(\frac{A}{A_v}\right)^2}{4N_r}\rho_m\left(\frac{1}{\rho_2}-\frac{1}{\rho_1}\right). \tag{8.73}$$

Finally, the pressure drop is given by

$$\Delta p = 2\left(F+B\right)\frac{N_rG^2}{\rho_m}. \tag{8.74}$$

Appendix A
Thermal Characteristics of Materials

Table A.1 Steel thermal conductivity (W/mK)

	t (°C)							
Steel type	0	100	200	300	400	500	600	700
Carbon steel	53.2	50.7	48.2	45.7	43.2	40.7	–	–
P 11/12	43.6	43.6	42.9	41.4	39.1	36.2	32.5	–
P 22	35.8	36.9	37.2	36.7	35.4	33.3	30.4	–
P 304/316/321	14.6	16.0	17.5	18.9	20.4	21.9	23.3	24.8
P 310	12.3	13.9	15.6	17.3	18.9	20.6	22.3	23.9
TP 409/410	24.2	25.1	26.0	26.8	27.5	28.2	28.8	29.3
TP 430	23.0	23.7	24.4	25.0	25.5	25.9	26.3	26.6

Table A.2 Thermal conductivity of metals and alloys (W/mK) (According to R.W. Powel and others)

	t (°C)					
Metal type or alloy	0	100	200	300	400	500
Aluminium	202	206	215	230	249	268
Beryllium	–	167	146	126	105	–
Cast iron	55	52	48	45	43	–
Copper (pure)	388	377	372	367	363	358
Lead	35	33	31	31	–	–
Magnesium	–	146	146	146	138	–
Nickel	62	59	57	55	–	–
Electrolytic nickel	–	–	71	62	58	60
Platinum	–	67	72	73	75	77
Silver	419	412	–	–	–	–
Tin	62	59	57	–	–	–
Titanium	–	15.9	15.5	15.2	15.0	14.8
Zinc	112	111	107	102	93	–
Bronze (8 Sn, 0.3 P, 91.7 Cu)	–	43	63	–	–	–
Monel metal	–	24.4	27.4	30.5	33.5	36.6
80–20 Ni Cr	–	13.4	15.8	17.4	18.8	20.9
TiC+40% Ni	–	20.1	22.0	23.6	25.2	26.5
TiC+40% Co	–	18.8	20.7	22.3	23.8	25.2

D. Annaratone, *Engineering Heat Transfer*, DOI 10.1007/978-3-642-03932-4,

Table A.3 Some values of thermal conductivity of metals (W/mK)

Metal type	t (°C)	k	Metal type	t (°C)	k
Antimony	0	18.3	Constantan	18	22.6
	100	16.8		100	26.7
Bismuth	18	8.1	Gold	18	292
	100	6.8		100	294
Brass	20	92–109	Mercury	0	8.4
	100	105–128		100	14.5
Cadmium	18	93	Wolfram	0	167
	100	90		100	120

Table A.4 Thermal conductivity of metal oxides (W/mK) (According to R.W. Powel)

	t (°C)									
Oxide	100	200	300	400	500	600	700	800	900	1000
BeO	218	172	126	84	58	42	31	24.4	19.8	18.0
Al_2O_3	29.3	20.9	16.3	12.9	10.7	9.21	8.02	7.21	6.63	6.28
ThO_2	8.37	7.21	6.22	5.35	4.53	3.84	3.31	2.97	2.67	2.50
ZrO	1.67	1.70	1.72	1.74	1.76	1.78	–	1.88	–	1.93
ZrO_2	1.88	1.88	1.88	1.88	1.92	1.93	2.00	2.09	2.16	2.17
$2MgOSiO_2$	5.47	4.42	3.84	3.26	2.96	2.67	2.50	2.36	2.27	2.19
$3Al_2O_32SiO_2$	6.40	5.58	5.00	4.54	4.24	3.95	3.84	3.72	3.66	3.60
TiO_2	6.74	5.12	4.30	3.78	3.49	3.28	3.20	3.14	3.14	3.14
UO_2	9.19	7.68	6.40	5.43	4.65	4.19	3.84	3.60	3.40	3.27
NiO	12.6	9.54	8.26	7.33	6.51	5.87	5.35	4.88	4.42	4.07
Al_2O_3MgO	14.6	13.6	12.3	9.77	8.49	7.67	7.09	6.74	6.28	5.93
CaO	15.5	11.2	9.54	8.79	8.37	8.14	7.91	7.79	7.79	7.74
ZnO	–	16.7	15.5	13.2	9.54	7.12	6.05	5.44	–	–
MgO	34.4	26.4	21.7	17.2	14.4	12.1	10.1	8.26	7.21	6.71
ZrO_2SiO_2	–	5.86	5.44	5.02	4.77	4.48	4.28	4.07	3.95	3.77

Table A.5 Thermal conductivity of solid materials (W/mK)

Material	t (°C)	k	Material	t (°C)	k
Soft rubber	30	0.175	Charcoal	0	0.058
Hard rubber	25–50	0.159		100	0.073
Graphite	20	164	Run silica	100	1.46
Synthetic graphite	79	15.6		400	1.72
Fossil coal	20–100	0.18	Leather	30	0.159
	800	2.51	Linoleum	20	0.186
	1000	2.93	Paper	20	0.139
Petroleum coke	100	2.09	Cardboard	–	0.189
	400	2.67	Paraffin	23	0.267
	800	3.28	Celluloid	30	0.209
Coal dust (dry)	30	0.11	Boiler scale (rich in calcium sulphate)	300	0.6–2.3
Coke dust	20	0.15			
			Boiler scale (rich in silicates)	300	0.08–0.17

Table A.6 Thermal conductivity of building, refractory and little insulating materials

Material	ρ_0 (*) (kg/m^3)	t (°C)	k (W/mK)
Asphalt	2115	20	0.744
Basalt	3200	100	1.77
Bricks (hand shaped, dry)	1568	25	0.39
Bricks (mach. shaped, dry)	1620	50	0.49
Bricks (refractory)			
Alumina (92–99% Al_2O_3 by weight)	–	427	3.12
Alumina (64–65% Al_2O_3 by weight)	–	1315	4.67
Chrome (32% Cr_2O_3)	3200	200	1.16
		650	1.47
		1315	1.73
Kaolin	432	500	0.26
		1150	0.45
Silicon carbide	2065	600	18.5
		800	15.9
		1000	13.8
		1200	12.1
		1400	10.9
Clay	1985	400	1.26
		600	1.29
		800	1.34
Magnesite (86.8 % MgO, 6.3% Fe_2O_3, 3% CaO, 2.6% SiO_2 by weight)	2530	204	3.81
		650	2.77
		1200	1.90
Silica (95% by weight)	1650	200	0.65
	1700	600	1.02
	1900	1000	1.38
Zirconium	–	400	1.62
		1400	2.02
Calcium carbonate	2595	30	2.25
Calcium sulfate (artificial)	1355	40	0.38
Chalk	1540	–	0.69
Clinker	–	0–700	0.47
Coke, petroleum	–	100	5.88
		500	5.00
Concrete (stone)	–	–	0.76–0.93
Concrete (cinder)	–	–	0.35
Dolomite	2675	50	1.73
Ebonite	–	–	0.17
Glass	–	–	0.35–1.25
borosilicate type	2230	30–75	1.10
soda glass	–	–	0.5–0.75
window glass	–	–	0.5–1.05
Granite	–	–	1.73–3.98

Table A.6 (continued)

Material	ρ_0 (*) (kg/m^3)	t (°C)	k (W/mK)
Ice	920	0	2.25
Lava	–	–	0.85
Limestone (15.3 H_2O in vol.)	1650	24	0.93
Marble, white	–	–	2.94
Mica (Indian)	–	100	0.79
		200	0.71
		400	0.71
		600	0.81
Mica (Canadian)	–	100	0.63
		200	0.42
		400	0.19
		600	0.20
Porcelain	–	200	1.52
Pumice stone	–	21–66	0.24
Sand, dry	1515	20	0.33
Sandstone	2240	40	1.83
Snow	556	0	0.47
Sulphur, monoclinic	–	100	0.16
Wood, across grain			
oak	825	15	0.21
maple	716	50	0.19
pine, white	545	15	0.15
Pine wood, parallel to grain	545	21	0.35

(*) apparent density at room temperature

Table A.7 Thermal conductivity of insulating materials

Material	ρ_0 (*) (kg/m^3)	t (°C)	k (W/mK)	Material	ρ_0 (*) (kg/m^3)	t (°C)	k (W/mK)
Aerogel (silica)..................	136	120	0.022	Gypsum in powdered form	416–544	–	0.074–0.086
		290	0.045	Kapok	14	20	0.034
Ashes (wood)....	–	0–100	0.071	Linen	–	30	0.086
Balsa wood	120	30	0.043–0.052	Mineral wool	150	30	0.039
Cork (plate).......................	160	30	0.043		315	30	0.042
Cotton wool	80	30	0.041	loose, fibrous	96	–	0.0375
Diatomaceous earth, natural, across strata	444	204	0.088		160	–	0.0389
		871	0.133		224	–	0.0403
Diatomaceous earth, natural, parallel to strata	444	204	0.140	block with binder	267	–	0.0537
		871	0.183	Pumice............................	300	20	0.093–0.23
Eiderdown	79	80	0.0244	Sawdust	192	21	0.052
Felt, wool	330	30	0.052	Silk	100	–	0.045
Flax fibers between paper ..	78	–	0.040	Sugar-cane fiber	210–240	–	0.048
Gypsum (cellular)	128	–	0.050	Wadding	80	30	0.042
	192	–	0.064	Wallboard (insulating type)	237	21	0.048
	288	–	0.085	Wool (animal)............	110	30	0.036
	384	–	0.111				

(*) apparent density at room temperature

Table A.8 Thermal conductivity of liquids

Material	t (°C)	k (W/mK)	Material	t (°C)	k (W/mK)
Acetic acid 100%	20	0.171	Hexane	30	0.138
50%	20	0.346		60	0.135
Acetone	30	0.176	Kerosene	20	0.149
	75	0.164		75	0.140
Allyl alcohol	25–30	0.180	Mercury	28	8.36
Ammonia	15–20	0.50	Methyl alcohol 100%	20	0.215
Amyl acetate	10	0.144	60%	20	0.329
Aniline	0–20	0.173	20%	20	0.492
Benzene	30	0.159	Nitrobenzene	30	0.164
	60	0.151		100	0.152
Benzol	20	0.153	Octane	30	0.144
Butyl acetate	25–30	0.147		60	0.140
Chlorobenzene	10	0.144	Oils, olive	20	0.168
Chloroform	30	0.138		100	0.164
Diesel oil	20	0.126	Pentane	30	0.135
	50	0.121		75	0.128
	80	0.116	Petroleum	15	0.151
Ethyl acetate	20	0.175	Sodium	100	84.8
Ether	30	0.138		210	79.6
	75	0.135	Sulfuric acid 90%	30	0.363
Ethylene glycol	0	0.265	60%	30	0.433
Gasoline	30	0.135	30%	30	0.519
Glycerol 100%	20	0.284	Sulfur dioxide	-15	0.222
60%	20	0.381		30	0.192
20%	20	0.481	Toluene	30	0.148
Heptane	30	0.140		75	0.145
	60	0.137	Vaseline	15	0.183

Table A.9 Thermal conductivity of gases and vapors

Substance	t (°C)	k (W/mK)	Substance	t (°C)	k (W/mK)
Acetone	0	0.0100	Ether	0	0.0133
	46	0.0128		46	0.0171
	100	0.0171		100	0.0227
	184	0.0254		184	0.0327
Acetylene	0	0.0187		212	0.0362
	50	0.0242	Ethylene	0	0.0175
	100	0.0297		50	0.0227
Air	0	0.0233		100	0.0277
	100	0.0302	Helium	0	0.141
	200	0.0360		100	0.171
	300	0.0430	Heptane	100	0.0178
	400	0.0488		200	0.0194
	500	0.0547	Hexane	0	0.0125

Table A.9 (continued)

Substance	t (°C)	k (W/mK)	Substance	t (°C)	k (W/mK)
Ammonia	0	0.0218		20	0.0138
	100	0.0332	Hydrogen	0	0.169
	200	0.0485		100	0.207
	300	0.0666		200	0.244
	400	0.0902		300	0.280
Argon	0	0.0164		400	0.316
	100	0.0213		500	0.350
	200	0.0256		600	0.385
	300	0.0296		700	0.421
Carbon dioxide	0	0.0139	Mercury	200	0.0341
	100	0.0221	Methane	0	0.0302
	200	0.0314		100	0.0453
	300	0.0384		200	0.0628
	400	0.0454		300	0.0802
	500	0.0523		400	0.0988
	600	0.0581		500	0.117
	700	0.0640		600	0.138
	800	0.0700		700	0.163
	900	0.0744	Methyl acetate	0	0.0102
	1000	0.0802		20	0.0118
Carbon monoxide	0	0.0221	Nitric oxide	0	0.0239
	100	0.0291		50	0.0279
	200	0.0361	Nitrogen	0	0.0233
	300	0.0419		100	0.0302
	400	0.0477		200	0.0360
	500	0.0535		300	0.0419
	600	0.0581		400	0.0465
	700	0.0628		500	0.0523
	800	0.0675		600	0.0558
	900	0.0721		700	0.0605
Chlorine	0	0.0074		800	0.0640
Chloroform	0	0.0066	Oxygen	0	0.0233
	46	0.0080		100	0.0314
	100	0.0100		200	0.0384
	184	0.0133		300	0.0454
Dichlorodifluoromethane	0	0.0083		400	0.0523
	50	0.0111		500	0.0581
	100	0.0138		600	0.0640
	150	0.0168		700	0.0700
Ethane	0	0.0183	Propane	0	0.0150
	100	0.0303		100	0.0261
Ethyl alcohol	20	0.0154	Sulfur dioxide	0	0.0086
	100	0.0215		100	0.0119

Table A.10 Density and specific heat of metals

Metal	ρ (kg/m^3)	t (°C)	c (J/kgK)	Metal	ρ (kg/m^3)	t (°C)	c (J/kgK)
Aluminium	2700	0	882	Mercury	13300–13600	0–20	139 (*)
		100	932	Molybdenum	9000	100	260
		200	981	Nickel	8800	0	429
		300	1031			100	474
		400	1081			200	520
		500	1131			300	566
Bras	8400–8700	20-100	384 (*)			400	542
Copper	8300–8900	0	385			500	548
		100	394	Platinum	21400	100	136
		200	404	Silver	10500	0	233
		300	414			100	239
		400	423			200	245
		500	433			300	251
Chromium	7100	0	427			400	256
		100	473			500	262
		300	523	Steel	7850	100	440
Gold	19300	0	128			300	582
		100	131			600	754
		200	134			0–400	523 (*)
		300	137			17–680	553 (*)
		400	140			0–1530	687 (*)
		500	143	Tin	7300	0	224
Lead	11300	0	128	Wolfram	18800	0–1000	144 (*)
		100	132	Zinc	7100	0	384
		200	136			100	401
		300	140			200	418
		400	137			300	436
		500	137			400	453
Magnesium	1740	100	1076			500	513

(*) mean values

Table A.11 Density and specific heat of building, refractory and insulating materials

Material	ρ (kg/m^3)	t (°C)	c (J/kgK)	Material	ρ (kg/m^3)	t (°C)	c (J/kgK)
Asphalt	2120	20	920	Chalk	1250	0–100	837(*)
Basalt	2700	0–100	858(*)	Clay	1800–2600	0–100	880(*)
Bricks				Concrete	2270	20	883
hand made	1570	27–49	740(*)	Cork	119	0	1716
mach. made	1620	0–100	920(*)	Corundum	3780	0–300	900(*)
Bricks, refractory				Glass	2400–2600	17	795
magnesite	2600	0–400	1140(*)	Granite	2500–3000	20–100	850(*)
		0–800	1295(*)	Kapok	150	20	1357
		0–1200	1465(*)	Limestone	2555	15–100	908(*)
silica	1700–1900	0–400	1060(*)	Marble	2700	25–100	880(*)
		0–800	1245(*)	Mineral wool	200	0	657
		0–1200	1355(*)	Porcelain	2300–2500	15–912	1080(*)
silicon carbide	3200	18–900	1060(*)	Silk	100	0	1385
zirconium		0–600	575(*)	Wedding	80	30	1335
		0–1400	730(*)	Wool	136	0–100	1360(*)
Cement	820–1950		1130	Wood, oak	820	0–100	2385(*)
				Wood, pine	550	0–100	2720(*)

(*) mean values

Table A.12 Density and mean specific heat of liquids

Liquid	ρ (kg/m^3)	t (°C)	c (J/kgK)	Liquid	ρ(kg/m^3)	t (°C)	c (J/kgK)
Acetal		19–99	2177	Isobutane		0	2300
Alcohol	794	0	2290	Isoheptane		0–50	2100
		40	2713	Isopentane		0	2143
Allyl alcohol		21–96	2784	Kerosene		0–100	2093
Benzaldehyde		22–172	1792	Olive oil		0–100	1675
Benzol	879	0–100	1674	Paraffin oil		0–100	2177
Bromobenzene		20–100	967	Petroleum	800	0–100	2093
Coal tar oil		15–90	1423	Phenol	1060	14–26	2349
Cyclohexane		10–18	1805	Propane		0	2412
Dichloroacetic acid		21–196	1457	Stearic acid		75–137	2328
Dichlorodifluoromethane		–43	879	Sulfuric acid	1838	0–100	1382
Diesel oil		20–100	2052	Tetrachloroethane		20	1122
Ethylene chloride		20	1252	Tetrachloroethylene		20	904
Formic acid	1220	20–100	2194	Trichloroethane		20	1114
Gasoline	680	0–100	2093	Trichloroethylene		20	976
Glycerol	1269	0–100	2428	Trinitrotoluene			1403

Table A.13 Isobaric specific heat of gases and steam (J/kgK)

Gas	ρ_0 (*)	t (°C) 0	100	200	300	400	500	600	700	800	900	1000
Air	1.293	1004	1010	1026	1046	1068	1091	1114	1137	1156	1169	1185
Carbon monoxide	1.250	1038	1048	1062	1082	1105	1132	1159	1179	1199	1216	1233
Carbon dioxide	1.977	813	915	989	1061	1114	1159	1194	1224	1248	1269	1286
Ethylene	1.261	1484	1839	2184	2477	2729	2948	3137	3303	3449	3579	–
Methane	0.717	2149	2447	2786	3165	3480	3802	4076	4327	4532	4748	4935
Nitrogen	1.2505	1038	1045	1055	1071	1091	1115	1138	1162	1182	1199	1215
Oxygen	1.4290	914	935	964	996	1025	1049	1069	1087	1099	1113	1122
Steam (**)	0.804	–	1880	1932	1989	2051	2119	2187	2260	2328	2396	2458

Hydrogen	ρ_0 (*)	t (°C) 0	100	200	300	400	500	600
	0.0899	14252	14485	14531	14578	14624	14671	14811

(*) density under normal conditions (kg/Nm^3) (**) at atmospheric pressure

Table A.14 Dynamic viscosity of liquids

Liquid	t (°C)	μ (kg/ms) ×10^{-3}	Liquid	t (°C)	μ (kg/ms) ×10^{-3}	Liquid	t (°C)	μ (kg/ms) ×10^{-3}
Acetic acid	20	1.25	Freon-22	–10	0.28	Nitric acid	20	1.15
	50	0.82		20	0.24		70	0.65
	100	0.45	Glycerol	80	35	Nitrobenzene	50	1.20
Acetone	0	0.41		100	14		100	0.54
	50	0.25		130	3.5		150	0.28
Ammonia	−20	0.20	Heptane	20	0.42	Pentane	0	0.29
Benzene	20	0.67		50	0.31	Phenol	50	3.70
	50	0.45		80	0.24		100	1.03
Chloroform	20	0.57	Hexane	20	0.33		150	0.34
	50	0.44		50	0.25	Sodium	100	0.72
Ethyl alcohol	0	1.80	Kerosene	20	2.45		200	0.47
	50	0.70		50	1.40	Sulphuric acid	50	11
Formic acid	20	1.90		100	0.62		100	3.1
	50	1.12	Methanol	0	0.80		150	1.0
	80	0.70		50	0.41	Toluene	0	0.75
Freon-12	−10	0.33	Methyl chloride	–30	0.57		100	0.28
	20	0.27						

Table A.15 Dynamic viscosity of gases and steam at atmospheric pressure (kg/ms)

t(°C)	Air $\times 10^{-6}$	CO_2	H_2	CO	CH_4	C_2H_4	O_2	N_2	Steam
0	17.1	13.9	8.43	16.6	10.2	9.41	19.0	16.5	–
20	18.1	14.8	8.82	17.6	10.8	10.0	–	–	–
40	19.1	15.7	9.22	18.5	11.4	10.7	21.0	18.3	–
60	20.0	16.7	9.61	19.4	12.1	11.4	–	–	–
80	20.9	17.6	9.90	20.2	12.6	12.0	23.0	20.0	–
100	21.8	18.5	10.3	21.0	13.2	12.6	–	–	12.8
120	22.6	19.4	10.6	21.8	13.8	13.1	24.5	21.8	13.5
140	23.3	20.3	11.0	22.6	14.4	13.7	–	–	14.3
160	24.2	21.2	11.4	23.2	14.9	14.3	26.0	23.5	15.1
180	25.0	22.1	11.7	23.9	15.5	14.9	–	–	15.9
200	25.9	22.9	12.1	24.7	16.1	15.4	28.0	25.0	16.6
250	27.8	24.9	12.9	26.4	17.3	16.6	29.5	27.0	18.3
300	29.6	26.8	13.9	27.8	18.5	17.8	32.0	28.7	20.1
350	31.2	28.6	14.6	29.4	19.8	19.1	–	30.5	21.9
400	32.8	30.5	15.4	31.0	21.0	20.3	35.5	32.5	23.5
450	34.4	32.2	16.2	32.4	22.1	21.5	–	–	25.2
500	35.9	33.8	16.9	33.9	23.1	22.6	39.0	36.0	26.8
600	38.8	37.0	18.2	36.6	25.1	24.5	42.5	39.0	30.2
700	41.7	40.0	19.6	38.8	27.1	26.5	–	–	33.4
800	44.3	42.9	21.0	41.2	28.9	28.3	48.5	45.0	36.4
900	46.8	45.5	22.1	43.3	30.7	30.1	–	–	39.4
1000	49.3	48.0	23.0	45.4	32.3	–	55.0	54.0	42.3

Table A.16 Number of Prandtl of gases and steam (According to values of c_p,μ,K in previous tables)

Gas	t (°C) 0	100	200	300	400	500	600	700	800	900
Air	0.737	0.729	0.738	0.720	0.718	0.716	–	–	–	–
Carbon dioxide	0.813	0.766	0.721	0.740	0.748	0.749	0.760	0.765	0.765	0.776
Carbon monoxide	0.780	0.756	0.727	0.718	0.718	0.717	0.730	0.728	0.732	0.730
Hydrogen	0.711	0.721	0.721	0.724	0.713	0.708	0.700	–	–	–
Methane	0.726	0.713	0.714	0.730	0.740	0.751	0.741	0.719	–	–
Nitrogen	0.735	0.723	0.733	0.734	0.762	0.767	0.795	–	–	–
Oxygen	0.745	0.707	0.709	0.702	0.702	0.704	0.709	0.706	–	–
Steam (*)	–	0.984	0.959	0.939	0.921	0.908	0.900	0.892	–	–

(*) at atmospheric pressure

Table A.17 Enthalpy of gases (kJ/kg) (Computed values)

t (°C)	Air	CO_2	N_2	O_2	SO_2	CO	NO	OH
50	50.1	42.2	51.9	45.6	31.7	51.8	49.0	87.3
100	100.7	86.6	104.1	92.0	64.5	104.2	98.7	174.7
150	151.8	132.9	156.8	139.3	98.3	157.0	149.0	262.1
200	203.3	181.1	209.8	187.4	133.1	210.3	199.9	349.7
250	255.3	231.1	263.2	236.2	168.7	264.2	251.4	437.5
300	307.7	282.8	317.0	285.8	205.3	318.5	303.5	525.6
350	360.6	336.0	371.2	336.1	242.7	373.4	356.1	614.2
400	414.0	390.7	425.8	387.0	280.9	428.7	409.3	703.2
450	467.9	446.8	480.9	438.6	319.8	484.6	463.0	792.7
500	522.2	504.1	536.4	490.8	359.4	540.9	517.2	882.9
550	577.0	562.6	592.4	543.5	399.7	597.8	571.9	973.9
600	632.2	622.2	648.8	596.7	440.6	655.1	627.0	1065.6
650	687.9	682.6	705.7	650.4	482.1	713.0	682.6	1158.2
700	744.1	744.0	763.1	704.6	524.1	771.4	738.6	1251.7
750	800.8	806.0	821.0	759.1	566.7	830.2	795.1	1346.3
800	857.9	868.7	879.4	814.0	609.7	889.6	851.9	1442.0
850	915.5	932.0	938.3	869.3	653.0	949.5	909.1	1538.9
900	973.5	995.6	997.7	924.8	696.8	1009.9	966.6	1637.1

Table A.18 Enthalpy of flue gas (kJ/kg) (Computed values)

m (*) (%)	t (°C) 100	200	300	400	500	600	700	800	900	1000	1100	1200
0	98.8	200.6	305.5	413.1	523.3	636.1	751.2	868.5	987.8	1108.9	1232	1356
1	99.8	202.7	308.4	417.0	528.2	641.9	758.0	876.3	996.7	1119.0	1243	1369
2	100.8	204.7	311.4	420.9	533.1	647.8	764.8	884.1	1005.6	1129.0	1254	1381
3	101.9	206.7	314.4	424.9	538.0	653.6	771.6	892.0	1014.5	1139.0	1265	1394
4	102.9	208.7	317.4	428.8	542.9	659.4	778.5	899.8	1023.3	1149.0	1277	1406
5	103.9	210.7	320.4	432.7	547.7	665.3	785.3	907.6	1032.2	1159.0	1288	1419
6	104.9	212.7	323.3	436.7	552.6	671.1	792.1	915.5	1041.1	1169.1	1299	1431
7	106.0	214.7	326.3	440.6	557.5	676.9	798.9	923.3	1050.0	1179.1	1310	1444
8	107.0	216.8	329.3	444.5	562.4	682.8	805.7	931.1	1058.9	1189.1	1322	1456
9	108.0	218.8	332.3	448.4	567.2	688.6	812.5	939.0	1067.8	1199.1	1333	1469
10	109.0	220.8	335.2	452.4	572.1	694.4	819.4	946.8	1076.7	1209.2	1344	1481
11	110.1	222.8	338.2	456.3	577.0	700.3	826.2	954.6	1085.6	1219.2	1355	1494
12	111.1	224.8	341.2	460.2	581.9	706.1	833.0	962.5	1094.5	1229.2	1366	1506

(*) mass moisture percentage

Table A.19 Black level of metals

Metal	t (°C)	B	Metal	t (°C)	B
Aluminium			cast plate, rough	23	0.82
polished	100	0.095	steel plate, rough	40–370	0.94–0.97
rough polish	100	0.18	Lead		
commercial sheet	100	0.09	pure, unoxidized	127–227	0.057–0.075
heavily oxidized	90–500	0.20–0.31	oxidized at 150 °C	200	0.63
aluminium oxide	280–830	0.63–0.26	Magnesium		
Brass			magnesium oxide	280–1700	0.55–0.20
highly polished			Mercury	0–100	0.09–0.12
73.2 Cu, 26.7 Zn	245–355	0.028–0.031	Molybdenum, polished	100	0.071
82.9 Cu, 17.0 Zn	277	0.030	Monel metal		
polished	38–315	0.10	oxidized at 600 °C	200–600	0.41–0.46
dull plate	50–350	0.22	Nickel		
oxidized by heating			pure, polished	227–377	0.07–0.087
at 600 °C	200–600	0.61–0.59	plate, oxidized by		
Chromium, polished	40–1100	0.08–0.36	heating at 600 °C	200–600	0.37–0.48
Copper			nickel oxide	650–1250	0.59–0.86
polished	100	0.052	Platinum		
plate heated			pure, polished plate	227–627	0.054–0.104
at 600 °C	200–600	0.57	wire	227–1380	0.073–0.182
cuprous oxide	800–1100	0.66–0.54	Silver		
Gold, pure,			pure, polished	227–627	0.020–0.032
highly polished	230–630	0.018–0.035	polished	100	0.052

Table A.19 (continued)

Metal	t (°C)	B	Metal	t (°C)	B
Iron and steel			Stainless steel		
(not including stainless)			type 301	230–940	0.51–0.70
metallic surfaces			type 316	230–1150	0.26–0.66
steel, polished	100	0.066	type 347	230–900	0.49–0.65
iron, polished	430–1030	0.14–0.38	Tantalum filament	1320–3000	0.19–0.31
iron, roughly polished	100	0.17	Thorium oxide	280–830	0.58–0.21
cast iron, polished	200	0.21	Tin, bright	50	0.06
oxidized surfaces			Tungsten		
iron plate,			filament	3300	0.39
completely rusted	20	0.69	polished coat	100	0.066
rolled sheet steel	21	0.66	Zinc		
oxidized iron	100	0.74	commercial pure,		
steel, oxidized at			polished	227–327	0.045–0.053
600 °C	200–600	0.79	oxidized by heating		
iron oxide	500–1200	0.85–0.89	at 400 °C	400	0.11
			zinc, galvanized sheet	100	0.21

Table A.20 Black level of building, refractory and miscellaneous materials

Material	t (°C)	B	Material	t (°C)	B
Bricks	22	0.93	Carborundum (87% SiC)	1010–1400	0.92–0.82
Bricks, refractory			Cloth, wool, silk, cotton		0.78
96.7% Al_2O_3, 2.7% SiO_2	1000	0.30	Concrete tiles	1000	0.63
porosity 37.9%	1100	0.28	Glass, smooth	22	0.94
	1300	0.23	Gypsum (0.5 mm thick.)	21	0.903
	1500	0.19	Marble, polished	22	0.93
90% Al_2O_3	1000	0.63	Oak, planed	21	0.90
porosity 16%	1200	0.53	Paints		
	1300	0.53	black shiny lacquer,		
	1500	0.53	spayed on iron	24	0.875
98% SiO_2	1000	0.62	flat black lacquer	38–93	0.96–0.98
without iron oxide	1200	0.535	oil paints, all colors	100	0.92–0.96
	1400	0.49	aluminium lacquer	21	0.39
	1500	0.46	radiator paint, white	100	0.79
38% SiO_2, 58% Al_2O_3	1000	0.61	radiator paint, bronze	100	0.51
0.9% Fe_2O_3	1200	0.52	Paper	19	0.92–0.94
	1400	0.47	Plaster, rough lime	10–90	0.91
	1500	0.45	Porcelain	22	0.92
50 SiO_2, 36% Al_2O_3	1000	0.73	Quartz		
1.65% Fe_2O_3	1200	0.68	rough, fused	21	0.93
	1400	0.62	opaque	300–840	0.92–0.68
	1500	0.62	Roofing paper	21	0.91
magnesite	1000	0.38	Rubber		
Carbon			hard, glossy plate	23	0.94
rough plate	100–320	0.77–0.72	soft, grey	24	0.86
graphitized	100–500	0.76–0.71	Serpentine, polished	23	0.90
lampblack, rough deposit	100–500	0.84–0.78	Water	0–100	0.95–0.96
graphite	250–510	0.98	Zirconium silicate	240–830	0.92–0.52

Appendix B
Corrective Factors for the Design Computation in Real Cases

The following Tables list the values of χ_p and χ_c (see Sect. 7.5) that allow computation of the real value of the mean temperature difference Δt_m

(A) Fluids with Cross Flow (Fig. 7.4)

Table B.1 Fluids with cross-flow – Corrective factor χ_c ($\psi = 0.03 - 0.42$)

	ψ													
β	0.03	0.06	0.09	0.12	0.15	0.18	0.21	0.24	0.27	0.30	0.33	0.36	0.39	0.42
0.1	0.898	0.923	0.938	0.949	0.957	0.965	0.969	0.972	0.976	0.980	0.982	0.984	0.986	0.988
0.2	0.801	0.847	0.876	0.896	0.912	0.925	0.936	0.946	0.953	0.959	0.965	0.969	0.972	0.976
0.3	.....	0.770	0.811	0.840	0.864	0.883	0.900	0.914	0.925	0.936	0.944	0.951	0.959	0.965
0.4	.....	0.694	0.743	0.781	0.812	0.839	0.861	0.880	0.896	0.910	0.921	0.932	0.942	0.949
0.5	.....	.....	0.673	0.719	0.758	0.790	0.817	0.842	0.863	0.882	0.898	0.912	0.925	0.934
0.6	.....	.....	.....	0.653	0.698	0.737	0.771	0.801	0.827	0.851	0.871	0.889	0.905	0.918
0.7	.....	.....	.....	.....	0.635	0.680	0.719	0.756	0.787	0.816	0.840	0.863	0.882	0.900
0.8	.....	.....	.....	.....	.....	0.618	0.664	0.705	0.743	0.777	0.808	0.834	0.857	0.880
0.9	.....	.....	.....	.....	.....	.....	0.601	0.649	0.694	0.734	0.770	0.803	0.830	0.856
1.0	.....	.....	.....	.....	.....	.....	.....	0.587	0.638	0.684	0.728	0.765	0.799	0.830
1.1	.....	.....	.....	.....	.....	.....	.....	.....	0.573	0.628	0.679	0.725	0.765	0.801
1.2	.....	.....	.....	.....	.....	.....	.....	.....	.....	0.560	0.622	0.676	0.725	0.768
1.3	.....	.....	.....	.....	.....	.....	.....	.....	.....	.....	0.551	0.618	0.677	0.729
1.4	.....	.....	.....	.....	.....	.....	.....	.....	.....	.....	.....	0.545	0.619	0.683
1.5	.....	.....	.....	.....	.....	.....	.....	.....	.....	.....	.....	.....	0.544	0.625
1.6	.....	.....	.....	.....	.....	.....	.....	.....	.....	.....	.....	.....	.....	0.548

Table B.2 Fluids with cross-flow – Corrective factor χ_c ($\psi = 0.45 - 0.87$)

β	ψ 0.45	0.48	0.51	0.54	0.57	0.60	0.63	0.66	0.69	0.72	0.75	0.78	0.81	0.84	0.87
0.2	0.980	0.982	0.986	0.988	–	–	–	–	–	–	–	–	–	–	–
0.3	0.969	0.974	0.978	0.980	0.984	0.986	0.988	–	–	–	–	–	–	–	–
0.4	0.957	0.963	0.969	0.974	0.978	0.982	0.984	0.988	–	–	–	–	–	–	–
0.5	0.944	0.953	0.961	0.967	0.972	0.976	0.980	0.984	0.988	–	–	–	–	–	–
0.6	0.931	0.942	0.949	0.959	0.965	0.970	0.976	0.980	0.984	0.988	–	–	–	–	–
0.7	0.914	0.929	0.940	0.949	0.957	0.965	0.972	0.976	0.982	0.986	0.988	–	–	–	–
0.8	0.898	0.914	0.927	0.940	0.949	0.959	0.967	0.972	0.978	0.982	0.986	–	–	–	–
0.9	0.878	0.898	0.914	0.929	0.942	0.951	0.961	0.969	0.974	0.980	0.984	0.988	–	–	–
1.0	0.857	0.880	0.900	0.918	0.932	0.944	0.955	0.965	0.972	0.978	0.984	0.988	–	–	–
1.1	0.832	0.861	0.883	0.905	0.921	0.936	0.949	0.959	0.969	0.974	0.982	0.986	–	–	–
1.2	0.806	0.837	0.866	0.891	0.910	0.927	0.942	0.955	0.965	0.972	0.978	0.984	0.988	–	–
1.3	0.774	0.812	0.845	0.875	0.898	0.918	0.934	0.949	0.961	0.969	0.976	0.982	0.988	–	–
1.4	0.737	0.784	0.822	0.856	0.883	0.907	0.927	0.942	0.955	0.967	0.974	0.982	0.986	–	–
1.5	0.692	0.748	0.796	0.835	0.868	0.896	0.918	0.936	0.951	0.963	0.972	0.980	0.986	–	–
1.6	0.637	0.708	0.765	0.812	0.851	0.882	0.909	0.929	0.946	0.959	0.970	0.978	0.984	–	–
1.7	0.561	0.656	0.728	0.785	0.830	0.868	0.898	0.921	0.940	0.955	0.967	0.976	0.984	0.988	–
1.8	.....	0.584	0.680	0.751	0.808	0.851	0.885	0.912	0.934	0.951	0.965	0.974	0.982	0.988	–
1.9	.....	.....	0.617	0.711	0.779	0.832	0.871	0.903	0.927	0.946	0.961	0.972	0.980	0.986	–
2.0	.....	.....	.....	0.656	0.745	0.809	0.856	0.892	0.920	0.942	0.957	0.970	0.980	0.986	–
2.2	.....	.....	.....	.....	0.640	0.747	0.817	0.868	0.903	0.931	0.951	0.965	0.976	0.984	–
2.4	.....	.....	.....	.....	.....	0.638	0.761	0.834	0.883	0.918	0.942	0.961	0.972	0.982	–
2.6	.....	.....	.....	.....	.....	.....	0.659	0.785	0.856	0.901	0.932	0.955	0.970	0.980	0.988
2.8	.....	.....	.....	.....	.....	.....	.....	0.704	0.819	0.882	0.921	0.947	0.967	0.978	0.986
3.0	.....	.....	.....	.....	.....	.....	.....	.....	0.761	0.856	0.907	0.940	0.961	0.976	0.986

(B) Heat Exchangers (Figs. 7.5 and 7.6)

Table B.3 Heat exchanger – A and B type – 2 passages of the external fluid – Factor χ_c

β	ψ 0.15	0.20	0.25	0.30	0.35	0.40	0.45	0.50	0.55	0.60	0.65	0.70	0.75	0.80	0.85
0.2	0.654	0.821	0.882	0.920	0.942	0.956	0.968	0.976	0.982	0.988	–	–	–	–	–
0.4	.....	.....	0.543	0.771	0.848	0.895	0.925	0.948	0.962	0.973	0.982	0.988	–	–	–
0.6	.....	.....	.....	.....	0.642	0.792	0.861	0.906	0.936	0.956	0.970	0.979	0.988	–	–
0.8	.....	.....	.....	.....	.....	.....	0.750	0.843	0.898	0.933	0.956	0.970	0.982	0.988	–
1.0	.....	.....	.....	.....	.....	.....	.....	0.732	0.843	0.900	0.939	0.962	0.976	0.985	–
1.2	.....	.....	.....	.....	.....	.....	.....	.....	0.748	0.858	0.914	0.948	0.970	0.982	–
1.4	.....	.....	.....	.....	.....	.....	.....	.....	.....	0.789	0.884	0.933	0.962	0.979	0.988
1.6	.....	.....	.....	.....	.....	.....	.....	.....	.....	0.638	0.838	0.914	0.953	0.973	0.988
1.8	.....	.....	.....	.....	.....	.....	.....	.....	.....	.....	0.764	0.887	0.942	0.970	0.985
2.0	.....	.....	.....	.....	.....	.....	.....	.....	.....	.....	0.569	0.851	0.928	0.965	0.982

Table B.4 Heat exchanger – A and B type – 3 passages of the external fluid – Factor χ_c

	ψ														
β	0.15	0.20	0.25	0.30	0.35	0.40	0.45	0.50	0.55	0.60	0.65	0.70	0.75	0.80	0.85
0.2	0.730	0.843	0.898	0.928	0.948	0.962	0.970	0.979	0.985	0.988	–	–	–	–	–
0.4	......	......	0.673	0.801	0.866	0.906	0.933	0.950	0.965	0.976	0.982	0.988	–	–	–
0.6	......	......	......	......	0.711	0.818	0.877	0.914	0.942	0.959	0.973	0.982	0.988	–	–
0.8	.....	.....	.....	.....	.....	0.621	0.785	0.861	0.909	0.939	0.959	0.973	0.982	–	–
1.0	.....	.....	.....	.....	.....	.....	0.538	0.773	0.861	0.911	0.945	0.965	0.979	0.988	–
1.2	.....	.....	.....	.....	.....	.....	.....	0.530	0.785	0.874	0.922	0.953	0.973	0.985	–
1.4	.....	.....	.....	.....	.....	.....	.....	.....	0.614	0.816	0.895	0.939	0.965	0.979	–
1.6	.....	.....	.....	.....	.....	.....	.....	.....	.....	0.711	0.858	0.922	0.956	0.976	0.988
1.8	.....	.....	.....	.....	.....	.....	.....	.....	.....	.....	0.799	0.900	0.948	0.973	0.985
2.0	.....	.....	.....	.....	.....	.....	.....	.....	.....	.....	0.682	0.869	0.936	0.968	0.985

Table B.5 Heat exchanger – A and B type – 4 passages of the external fluid – Factor χ_c

β	ψ 0.15	0.20	0.25	0.30	0.35	0.40	0.45	0.50	0.55	0.60	0.65	0.70	0.75	0.80	0.85
0.2	0.752	0.853	0.900	0.931	0.948	0.962	0.970	0.979	0.985	0.988	–	–	–	–	–
0.4	....	....	0.702	0.813	0.871	0.911	0.936	0.953	0.968	0.976	0.982	0.988	–	–	–
0.6	....	....	....	....	0.730	0.826	0.882	0.920	0.942	0.962	0.973	0.982	0.988	–	–
0.8	....	....	....	....	....	0.657	0.796	0.869	0.911	0.942	0.962	0.973	0.985	–	–
1.0	....	....	....	....	....	....	0.605	0.785	0.866	0.914	0.945	0.965	0.979	0.988	–
1.2	....	....	....	....	....	....	....	0.601	0.796	0.879	0.925	0.956	0.973	0.985	–
1.4	....	....	....	....	....	....	....	....	0.654	0.826	0.900	0.942	0.968	0.982	–
1.6	....	....	....	....	....	....	....	....	....	0.732	0.863	0.925	0.959	0.976	0.988
1.8	....	....	....	....	....	....	....	....	....	....	0.808	0.903	0.948	0.973	0.988
2.0	....	....	....	....	....	....	....	....	....	....	0.709	0.874	0.936	0.968	0.985

Table B.6 Heat exchanger – A and B type – 5 passages of the external fluid – Factor χ_c

	ψ														
β	0.15	0.20	0.25	0.30	0.35	0.40	0.45	0.50	0.55	0.60	0.65	0.70	0.75	0.80	0.85
0.2	0.761	0.856	0.903	0.931	0.950	0.962	0.973	0.979	0.985	0.988	–	–	–	–	–
0.4	.....	.....	0.713	0.818	0.877	0.911	0.936	0.953	0.968	0.976	0.982	0.988	–	–	–
0.6	.....	.....	.....	0.505	0.741	0.831	0.884	0.920	0.945	0.962	0.973	0.982	0.988	–	–
0.8	.....	.....	.....	.....	.....	0.671	0.801	0.871	0.914	0.942	0.962	0.973	0.985	–	–
1.0	.....	.....	.....	.....	.....	.....	0.625	0.792	0.869	0.917	0.948	0.965	0.979	0.988	–
1.2	.....	.....	.....	.....	.....	.....	.....	0.623	0.801	0.882	0.928	0.956	0.973	0.985	–
1.4	.....	.....	.....	.....	.....	.....	.....	.....	0.669	0.828	0.903	0.942	0.968	0.982	–
1.6	.....	.....	.....	.....	.....	.....	.....	.....	.....	0.741	0.866	0.925	0.959	0.979	0.988
1.8	.....	.....	.....	.....	.....	.....	.....	.....	.....	0.469	0.813	0.906	0.950	0.973	0.988
2.0	.....	.....	.....	.....	.....	.....	.....	.....	.....	.....	0.719	0.877	0.939	0.970	0.985

Table B.7 Heat exchanger – C and D type – 2 passages of the external fluid – Factor χ_c

β	ψ 0.10	0.15	0.20	0.25	0.30	0.35	0.40	0.45	0.50	0.55	0.60	0.65	0.70	0.75	0.80	0.85
0.2	0.761	0.853	0.900	0.931	0.948	0.962	0.970	0.979	0.982	0.988	–	–	–	–	–	–
0.4	......	0.594	0.735	0.821	0.874	0.909	0.933	0.950	0.965	0.973	0.982	0.988	–	–	–	–
0.6	.....	.....	0.470	0.639	0.748	0.828	0.879	0.914	0.939	0.956	0.970	0.979	0.985	–	–	–
0.8	.....	.....	.....	.....	0.541	0.690	0.794	0.861	0.903	0.933	0.956	0.970	0.979	0.988	–	–
1.0	.....	.....	.....	.....	.....	0.479	0.650	0.775	0.853	0.903	0.936	0.959	0.973	0.982	–	–
1.2	.....	.....	.....	.....	.....	.....	0.440	0.630	0.775	0.861	0.911	0.945	0.965	0.979	0.988	–
1.4	.....	.....	.....	.....	.....	.....	.....	0.430	0.642	0.794	0.877	0.925	0.956	0.973	0.985	–
1.6	.....	.....	.....	.....	.....	.....	.....	.....	0.447	0.682	0.828	0.903	0.945	0.968	0.982	–
1.8	.....	.....	.....	.....	.....	.....	.....	.....	.....	0.504	0.748	0.869	0.928	0.962	0.979	–
2.0	.....	.....	.....	.....	.....	.....	.....	.....	.....	0.316	0.610	0.821	0.909	0.953	0.976	0.988

Table B.8 Heat exchanger – C and D type – 3 passages of the external fluid – Factor χ_c

	ψ															
β	0.10	0.15	0.20	0.25	0.30	0.35	0.40	0.45	0.50	0.55	0.60	0.65	0.70	0.75	0.80	0.85
0.2	.675	.818	.882	.917	.942	.956	.968	.976	.982	.985	–	–	–	–	–	–
0.4	.....	.434	.652	.780	.851	.895	.925	.945	.959	.970	.979	.985	.988	–	–	–
0.6	.....	.....	.....	.492	.682	.792	.858	.900	.931	.950	.965	.976	.985	.988	–	–
0.8	.....	.....	.....	.....	.381	.594	.748	.835	.890	.925	.950	.965	.976	.985	–	–
1.0	.....	.....	.....	.....	.....	.322	.528	.721	.826	.887	.928	.953	.970	.982	.988	–
1.2	.....	.....	.....	.....	.....	.....	.....	.499	.721	.835	.898	.936	.962	.976	.985	–
1.4	.....	.....	.....	.....	.....	.....	.....	.286	.513	.746	.856	.914	.950	.970	.982	–
1.6	.....	.....	.....	.....	.....	.....	.....	.....	.299	.578	.792	.887	.936	.965	.979	–
1.8	.....	.....	.....	.....	.....	.....	.....	.....	.....	.342	.677	.846	.917	.956	.976	.988
2.0	.....	.....	.....	.....	.....	.....	.....	.....	.....	.....	.459	.780	.895	.984	.973	.988

Table B.9 Heat exchanger – C and D type – 4 passages of the external fluid – Factor χ_c

	ψ															
β	0.10	0.15	0.20	0.25	0.30	0.35	0.40	0.45	0.50	0.55	0.60	0.65	0.70	0.75	0.80	0.85
0.2	0.623	0.804	0.874	0.914	0.939	0.953	0.965	0.973	0.982	0.985	0.988	–	–	–	–	–
0.4	.....	.....	0.606	0.761	0.841	0.887	0.920	0.942	0.959	0.970	0.979	0.985	0.988	–	–	–
0.6	.....	.....	.....	0.397	0.648	0.778	0.851	0.895	0.928	0.948	0.965	0.976	0.982	0.988	–	–
0.8	.....	.....	.....	.....	.....	0.538	0.728	0.826	0.884	0.922	0.948	0.965	0.976	0.985	–	–
1.0	.....	.....	.....	.....	.....	.....	0.448	0.698	0.816	0.882	0.925	0.950	0.968	0.979	0.988	–
1.2	.....	.....	.....	.....	.....	.....	.....	0.408	0.696	0.826	0.892	0.933	0.959	0.976	0.985	–
1.4	.....	.....	.....	.....	.....	.....	.....	.....	0.426	0.726	0.848	0.911	0.948	0.970	0.982	–
1.6	.....	.....	.....	.....	.....	.....	.....	.....	.....	0.511	0.778	0.879	0.933	0.962	0.979	–
1.8	.....	.....	.....	.....	.....	.....	.....	.....	.....	0.257	0.644	0.835	0.914	0.953	0.976	0.988
2.0	.....	.....	.....	.....	.....	.....	.....	.....	.....	.....	0.357	0.764	0.890	0.945	0.970	0.985

Table B.10 Heat exchanger – C and D type – 5 passages of the external fluid –Factor χ_C

β \ ψ	0.10	0.15	0.20	0.25	0.30	0.35	0.40	0.45	0.50	0.55	0.60	0.65	0.70	0.75	0.80	0.85
0.2	0.587	0.796	0.871	0.911	0.936	0.953	0.965	0.973	0.979	0.985	0.988	–	–	–	–	–
0.4	....	....	0.576	0.752	0.835	0.884	0.920	0.942	0.956	0.968	0.976	0.985	0.988	–	–	–
0.6	....	....	....	....	0.629	0.771	0.846	0.892	0.925	0.948	0.965	0.976	0.982	0.988	–	–
0.8	....	....	....	....	....	0.498	0.717	0.821	0.882	0.920	0.945	0.965	0.976	0.985	–	–
1.0	....	....	....	....	....	....	0.386	0.686	0.811	0.879	0.922	0.950	0.968	0.979	0.988	–
1.2	....	....	....	....	....	....	....	0.340	0.684	0.821	0.890	0.933	0.959	0.976	0.985	–
1.4	....	....	....	....	....	....	....	....	0.358	0.715	0.843	0.909	0.948	0.968	0.982	–
1.6	....	....	....	....	....	....	....	....	....	0.463	0.771	0.877	0.931	0.962	0.979	0.988
1.8	....	....	....	....	....	....	....	....	....	....	0.623	0.831	0.911	0.953	0.976	0.988
2.0	....	....	....	....	....	....	....	....	....	....	0.288	0.755	0.887	0.942	0.970	0.985

Table B.11 Heat exchanger – E end F type – 2 passages of the external fluid – Factor χ_c

	ψ														
β	0.15	0.20	0.25	0.30	0.35	0.40	0.45	0.50	0.55	0.60	0.65	0.70	0.75	0.80	0.85
0.2	0.753	0.854	0.901	0.931	0.949	0.963	0.972	0.980	0.984	0.988	–	–	–	–	–
0.4	.....	.....	0.702	0.814	0.873	0.910	0.936	0.953	0.967	0.976	0.984	0.988	–	–	–
0.6	.....	.....	.....	.....	0.732	0.827	0.883	0.920	0.944	0.961	0.974	0.982	0.988	–	–
0.8	.....	.....	.....	.....	.....	0.657	0.796	0.868	0.912	0.942	0.961	0.974	0.984	–	–
1.0	.....	.....	.....	.....	.....	.....	0.606	0.785	0.868	0.916	0.946	0.967	0.980	0.988	–
1.2	.....	.....	.....	.....	.....	.....	.....	0.602	0.796	0.880	0.927	0.955	0.974	0.986	–
1.4	.....	.....	.....	.....	.....	.....	.....	.....	0.655	0.825	0.901	0.942	0.967	0.982	–
1.6	.....	.....	.....	.....	.....	.....	.....	.....	.....	0.732	0.864	0.925	0.959	0.978	–
1.8	.....	.....	.....	.....	.....	.....	.....	.....	.....	.....	0.809	0.903	0.949	0.974	0.988
2.0	.....	.....	.....	.....	.....	.....	.....	.....	.....	.....	0.709	0.875	0.938	0.969	0.986

Table B.12 Heat exchanger – G and H type – 2 passages of the external fluid – Factor χ_c

	ψ															
β	0.10	0.15	0.20	0.25	0.30	0.35	0.40	0.45	0.50	0.55	0.60	0.65	0.70	0.75	0.80	0.85
0.2	0.624	0.804	0.875	0.914	0.938	0.955	0.967	0.974	0.982	0.986	–	–	–	–	–	–
0.4	.....	.....	0.607	0.762	0.840	0.889	0.921	0.942	0.959	0.970	0.978	0.984	–	–	–	–
0.6	.....	.....	.....	0.399	0.649	0.777	0.851	0.896	0.927	0.949	0.965	0.976	0.984	–	–	–
0.8	.....	.....	.....	......	......	0.538	0.728	0.825	0.883	0.921	0.947	0.965	0.976	0.986	–	–
1.0	.....	.....	.....	.....	.....	......	0.450	0.698	0.816	0.883	0.925	0.951	0.969	0.980	0.988	–
1.2	.....	.....	.....	.....	.....	.....	.....	0.411	0.697	0.825	0.892	0.934	0.959	0.976	0.986	–
1.4	.....	.....	.....	.....	.....	.....	.....	.....	0.429	0.726	0.849	0.912	0.947	0.970	0.984	–
1.6	.....	.....	.....	.....	.....	.....	.....	.....	.....	0.513	0.777	0.880	0.932	0.963	0.980	–
1.8	.....	.....	.....	.....	.....	.....	.....	.....	.....	0.264	0.644	0.835	0.914	0.955	0.976	0.988
2.0	.....	.....	.....	.....	.....	.....	.....	.....	.....	.....	0.362	0.764	0.889	0.944	0.972	0.986

(C) Coils (Figs. 7.7 and 7.8)

Table B.13 Coil with parallel flow – 2 sections – Corrective factor χ_p

	ψ										
β	0.20	0.25	0.30	0.35	0.40	0.45	0.50	0.55	0.60	0.65	0.70
0.2	1.061	1.032	1.020	–	–	–	–	–	–	–	–
0.4	.....	.....	1.109	1.048	1.028	–	–	–	–	–	–
0.6	.....	.....	.....	.....	1.084	1.042	1.025	–	–	–	–
0.8	.....	.....	.....	.....	.....	.....	1.049	1.027	–	–	–
1.0	.....	.....	.....	.....	.....	.....	.....	1.048	1.026	–	–
1.2	.....	.....	.....	.....	.....	.....	.....	.....	1.041	1.021	–
1.4	.....	.....	.....	.....	.....	.....	.....	.....	1.054	1.032	–
1.6	.....	.....	.....	.....	.....	.....	.....	.....	.....	1.045	1.021
1.8	.....	.....	.....	.....	.....	.....	.....	.....	.....	.....	1.028
2.0	.....	.....	.....	.....	.....	.....	.....	.....	.....	.....	1.038

Table B.14 Coil with parallel flow – 3 sections – Corrective factor χ_p

	ψ										
β	0.20	0.25	0.30	0.35	0.40	0.45	0.50	0.55	0.60	0.65	0.70
0.2	1.028	–	–	–	–	–	–	–	–	–	–
0.4	.....	.....	1.062	1.023	–	–	–	–	–	–	–
0.6	.....	.....	.....	.....	1.048	1.020	–	–	–	–	–
0.8	.....	.....	.....	.....	.....	1.115	1.025	–	–	–	–
1.0	.....	.....	.....	.....	.....	.....	.....	1.026	–	–	–
1.2	.....	.....	.....	.....	.....	.....	.....	1.128	1.022	–	–
1.4	.....	.....	.....	.....	.....	.....	.....	.....	1.053	–	–
1.6	.....	.....	.....	.....	.....	.....	.....	.....	.....	–	–
1.8	.....	.....	.....	.....	.....	.....	.....	.....	.....	1.085	–
2.0	.....	.....	.....	.....	.....	.....	.....	.....	.....	.....	1.024

Table B.15 Coil with counterflow – 2 sections – Corrective factor χ_c

β	ψ 0.05	0.10	0.15	0.20	0.25	0.30	0.35	0.40	0.45	0.50	0.55	0.60	0.65	0.70
0.2	0.870	0.934	0.958	0.971	0.979	–	–	–	–	–	–	–	–	–
0.4	....	0.798	0.883	0.924	0.947	0.963	0.972	–	–	–	–	–	–	–
0.6	....	....	0.723	0.837	0.894	0.928	0.949	0.964	0.974	–	–	–	–	–
0.8	....	....	....	0.630	0.792	0.869	0.913	0.941	0.959	0.971	–	–	–	–
1.0	....	....	....	....	....	0.747	0.848	0.903	0.936	0.958	0.971	–	–	–
1.2	....	....	....	....	....	....	0.704	0.835	0.900	0.937	0.960	0.974	–	–
1.4	....	....	....	....	....	....	....	0.672	0.835	0.905	0.943	0.965	0.978	–
1.6	....	....	....	....	....	....	....	....	0.666	0.847	0.917	0.952	0.971	–
1.8	....	....	....	....	....	....	....	....	....	0.701	0.872	0.933	0.963	0.979
2.0	....	....	....	....	....	....	....	....	....	....	0.771	0.904	0.951	0.974

Table B.16 Coil with counterflow – 3 sections – Corrective factor χ_c

	ψ												
β	0.05	0.10	0.15	0.20	0.25	0.30	0.35	0.40	0.45	0.50	0.55	0.60	0.65
0.2	0.950	0.972	–	–	–	–	–	–	–	–	–	–	–
0.4	0.843	0.922	0.952	0.968	0.977	–	–	–	–	–	–	–	–
0.6	.....	0.809	0.894	0.933	0.955	0.969	0.978	–	–	–	–	–	–
0.8	.....	.....	0.747	0.861	0.914	0.945	0.963	0.974	–	–	–	–	–
1.0	.....	.....	.....	.....	0.819	0.895	0.935	0.958	0.972	–	–	–	–
1.2	.....	.....	.....	.....	.....	0.760	0.876	0.929	0.956	0.972	–	–	–
1.4	.....	.....	.....	.....	.....	.....	0.668	0.862	0.927	0.958	0.974	–	–
1.6	.....	.....	.....	.....	.....	.....	.....	.....	0.856	0.931	0.963	0.978	–
1.8	.....	.....	.....	.....	.....	.....	.....	.....	.....	0.867	0.942	0.969	–
2.0	.....	.....	.....	.....	.....	.....	.....	.....	.....	.....	0.896	0.956	0.978

Table B.17 Coil with counterflow – 4 sections – Corrective factor χ_c

	ψ											
β	0.05	0.10	0.15	0.20	0.25	0.30	0.35	0.40	0.45	0.50	0.55	0.60
0.2	0.972	–	–	–	–	–	–	–	–	–	–	–
0.4	0.919	0.958	0.973	–	–	–	–	–	–	–	–	–
0.6	.....	0.900	0.942	0.963	0.974	–	–	–	–	–	–	–
0.8	.....	......	0.869	0.924	0.952	0.969	0.978	–	–	–	–	–
1.0	.....	.....	.....	0.816	0.901	0.942	0.963	0.976	–	–	–	–
1.2	.....	.....	.....	.....	.....	0.870	0.931	0.960	0.975	–	–	–
1.4	.....	.....	.....	.....	.....	.....	0.824	0.922	0.959	0.975	–	–
1.6	.....	.....	.....	.....	.....	.....	.....	0.740	0.919	0.961	0.978	–
1.8	.....	.....	.....	.....	.....	.....	.....	.....	.....	0.924	0.967	–
2.0	.....	.....	.....	.....	.....	.....	.....	.....	.....	.....	0.940	0.974

Table B.18 Coil with counterflow – 6 sections – Corrective factor χ_c

β	ψ 0.05	0.10	0.15	0.20	0.25	0.30	0.35	0.40	0.45	0.50	0.55
0.4	0.966	–	–	–	–	–	–	–	–	–	–
0.6	0.912	0.958	0.974	–	–	–	–	–	–	–	–
0.8	.....	0.894	0.945	0.967	0.979	–	–	–	–	–	–
1.0	.....	.....	.....	0.922	0.957	0.974	–	–	–	–	–
1.2	.....	.....	.....	.....	0.876	0.944	0.969	–	–	–	–
1.4	.....	.....	.....	.....	.....	.....	0.923	0.966	–	–	–
1.6	.....	.....	.....	.....	.....	.....	.....	0.889	0.964	–	–
1.8	.....	.....	.....	.....	.....	.....	.....	.....	0.806	0.966	–
2.0	.....	.....	.....	.....	.....	.....	.....	.....	.....	.....	0.972

Table B.19 Coil with counterflow – 8 sections – Corrective factor χ_c

	ψ								
β	0.05	0.10	0.15	0.20	0.25	0.30	0.35	0.40	0.45
0.6	0.952	0.976	–	–	–	–	–	–	–
0.8	.....	0.942	0.969	–	–	–	–	–	–
1.0	.....	.....	.....	0.957	0.976	–	–	–	–
1.2	.....	.....	.....	.....	0.932	0.969	–	–	–
1.4	.....	.....	.....	.....	.....	.....	0.957	–	–
1.6	.....	.....	.....	.....	.....	.....	.....	0.938	0.979
1.8	.....	.....	.....	.....	.....	.....	.....	.....	0.892

(D) Tube Bank with Several Passages of the External Fluid (Figs. 7.9 and 7.10)

Table B.20 Tube bank with 2 passages of the external fluid – Parallel flow – Factor χ_p

	ψ											
β	0.25	0.30	0.35	0.40	0.45	0.50	0.55	0.60	0.65	0.70	0.75	0.80
0.2	1.020	1.016	1.012	1.010	–	–	–	–	–	–	–	–
0.4	....	....	1.039	1.026	1.018	1.014	1.010	–	–	–	–	–
0.6	....	....	....	1.060	1.041	1.026	1.016	1.012	–	–	–	–
0.8	....	....	....	....	....	1.053	1.030	1.018	1.012	–	–	–
1.0	....	....	....	....	....	....	1.056	1.028	1.016	1.010	–	–
1.2	....	....	....	....	....	....	....	1.049	1.024	1.014	–	–
1.4	....	....	....	....	....	....	....	1.107	1.037	1.018	1.010	–
1.6	....	....	....	....	....	....	....	....	1.064	1.026	1.012	–
1.8	....	....	....	....	....	....	....	....	1.150	1.037	1.016	–
2.0	....	....	....	....	....	....	....	....	....	1.056	1.020	1.010

Table B.21 Tube bank with 3 passages of the external fluid – Parallel flow – Factor χ_p

	ψ											
β	0.20	0.25	0.30	0.35	0.40	0.45	0.50	0.55	0.60	0.65	0.70	0.75
0.2	1.039	1.016	1.010	–	–	–	–	–	–	–	–	–
0.4	....	....	1.088	1.026	1.014	1.010	–	–	–	–	–	–
0.6	....	....	....	....	1.058	1.022	1.014	–	–	–	–	–
0.8	....	....	....	....	....	1.141	1.028	1.014	–	–	–	–
1.0	....	....	....	....	....	....	....	1.028	1.014	–	–	–
1.2	....	....	....	....	....	....	....	1.132	1.024	1.012	–	–
1.4	....	....	....	....	....	....	....	....	1.058	1.018	–	–
1.6	....	....	....	....	....	....	....	....	....	1.030	1.012	–
1.8	....	....	....	....	....	....	....	....	....	1.083	1.016	–
2.0	....	....	....	....	....	....	....	....	....	....	1.026	1.010

Table B.22 Tube bank with 2 passages of the external fluid – Counterflow – Factor χ_c

	ψ												
β	0.15	0.20	0.25	0.30	0.35	0.40	0.45	0.50	0.55	0.60	0.65	0.70	0.75
0.2	0.980	0.980	0.984	0.986	–	–	–	–	–	–	–	–	–
0.4	0.972	0.963	0.967	0.972	0.978	0.982	0.986	–	–	–	–	–	–
0.6	......	0.957	0.949	0.953	0.963	0.970	0.978	0.982	0.988	–	–	–	–
0.8	......	......	0.946	0.936	0.944	0.955	0.967	0.974	0.982	0.986	–	–	–
1.0	......	......	......	0.938	0.927	0.936	0.951	0.965	0.974	0.982	0.988	–	–
1.2	......	......	......	......	0.932	0.918	0.932	0.949	0.965	0.976	0.984	–	–
1.4	......	......	......	......	......	0.927	0.912	0.932	0.953	0.969	0.980	0.988	–
1.6	......	......	......	......	......	......	0.920	0.910	0.936	0.959	0.974	0.984	–
1.8	......	......	......	......	......	......	......	0.909	0.914	0.946	0.967	0.980	0.988
2.0	......	......	......	......	......	......	......	......	0.898	0.927	0.959	0.976	0.986

Table B.23 Tube bank with 3 passages of the external fluid – Counterflow – Factor χ_c

β	ψ 0.05	0.10	0.15	0.20	0.25	0.30	0.35	0.40	0.45	0.50	0.55	0.60	0.65	0.70
0.2	0.887	0.963	0.978	0.986	–	–	–	–	–	–	–	–	–	–
0.4	.....	0.856	0.940	0.963	0.976	0.982	0.988	–	–	–	–	–	–	–
0.6	.....	.....	0.781	0.918	0.951	0.967	0.976	0.984	0.988	–	–	–	–	–
0.8	.....	.....	.....	.....	0.894	0.940	0.961	0.972	0.982	0.986	–	–	–	–
1.0	.....	.....	.....	....	0.509	0.866	0.931	0.957	0.972	0.980	0.986	–	–	–
1.2	.....	.....	.....	.....	.....	0.622	0.834	0.925	0.955	0.972	0.982	0.988	–	–
1.4	.....	.....	.....	.....	.....	.....	0.753	0.796	0.925	0.959	0.974	0.984	–	–
1.6	.....	.....	.....	.....	.....	.....	.....	.....	0.788	0.932	0.963	0.978	0.988	–
1.8	.....	.....	.....	.....	.....	.....	.....	.....	......	0.837	0.944	0.970	0.984	–
2.0	.....	.....	.....		.....	.....	.....	.....	......	......	0.891	0.957	0.978	0.988

Table B.24 Tube bank with 4 passages of the external fluid – Counterflow – Factor χ_c

	ψ											
β	0.10	0.15	0.20	0.25	0.30	0.35	0.40	0.45	0.50	0.55	0.60	0.65
0.2	0.988	–	–	–	–	–	–	–	–	–	–	–
0.4	0.980	0.980	0.984	0.988	–	–	–	–	–	–	–	–
0.6	.....	0.972	0.972	0.978	0.984	0.988	–	–	–	–	–	–
0.8	.....	.....	0.965	0.967	0.974	0.980	0.986	–	–	–	–	–
1.0	.....	.....	.....	0.959	0.959	0.970	0.978	0.984	–	–	–	–
1.2	.....	.....	.....	.....	0.959	0.953	0.967	0.978	0.984	–	–	–
1.4	.....	.....	.....	.....	.....	0.970	0.949	0.965	0.978	0.986	–	–
1.6	.....	.....	.....	.....	.....	.....	.....	0.946	0.967	0.980	0.988	–
1.8	.....	.....	.....	.....	.....	.....	.....	.....	0.946	0.970	0.984	–
2.0	.....	.....	.....	.....	.....	.....	.....	.....	.....	0.953	0.976	0.988

Appendix C
Corrective Factors for the Verification Computation in Real Cases

The following Tables list the values of ψ or corrective factors ϕ_p and ϕ_c (see Sect. 7.6.3) that allow calculation of the actual value of ψ in real instances.

(A) Fluids with Cross Flow (Fig. 7.4)

Table C.1 Factor ψ for cross flow ($\gamma = 0.3 - 1.5$)

	γ												
β	.30	.40	.50	.60	.70	.80	.90	1.0	1.1	1.2	1.3	1.4	1.5
0.20	0.747	0.681	0.621	0.568	0.520	0.477	0.438	0.403	0.372	0.343	0.317	0.294	0.272
0.30	0.751	0.686	0.628	0.577	0.531	0.490	0.454	0.420	0.390	0.363	0.338	0.316	0.295
0.40	0.754	0.691	0.635	0.586	0.542	0.503	0.468	0.437	0.408	0.382	0.359	0.338	0.318
0.50	0.757	0.696	0.642	0.595	0.553	0.516	0.482	0.453	0.426	0.401	0.379	0.359	0.340
0.60	0.760	0.701	0.649	0.603	0.563	0.528	0.496	0.468	0.442	0.419	0.398	0.379	0.362
0.70	0.763	0.705	0.655	0.612	0.573	0.539	0.509	0.483	0.458	0.437	0.417	0.399	0.382
0.80	0.766	0.710	0.662	0.620	0.583	0.551	0.522	0.497	0.474	0.453	0.435	0.418	0.402
0.90	0.769	0.714	0.668	0.627	0.592	0.562	0.535	0.511	0.489	0.470	0.452	0.436	0.421
1.0	0.771	0.719	0.674	0.635	0.602	0.572	0.547	0.524	0.503	0.485	0.469	0.454	0.440
1.1	0.774	0.723	0.680	0.642	0.610	0.583	0.558	0.537	0.517	0.500	0.485	0.471	0.458
1.2	0.777	0.727	0.685	0.650	0.619	0.593	0.569	0.549	0.531	0.515	0.500	0.487	0.475
1.3	0.780	0.731	0.691	0.657	0.627	0.602	0.580	0.561	0.544	0.529	0.515	0.503	0.491
1.4	0.782	0.735	0.696	0.663	0.635	0.612	0.591	0.572	0.556	0.542	0.529	0.518	0.507
1.5	0.785	0.739	0.702	0.670	0.643	0.621	0.601	0.584	0.568	0.555	0.543	0.532	0.522
1.6	0.788	0.743	0.707	0.676	0.651	0.629	0.611	0.594	0.580	0.568	0.556	0.546	0.537
1.7	0.790	0.747	0.712	0.683	0.658	0.638	0.620	0.605	0.591	0.580	0.569	0.560	0.551
1.8	0.793	0.751	0.717	0.689	0.666	0.646	0.629	0.615	0.602	0.591	0.581	0.573	0.565

Table C.2 Factor ψ for cross flow ($\gamma = 1.6 - 2.8$)

	γ												
β	1.6	1.7	1.8	1.9	2.0	2.1	2.2	2.3	2.4	2.5	2.6	2.7	2.8
0.20	0.252	0.234	0.218	0.203	0.189	0.177	0.165	0.154	0.144	0.135	0.127	0.119	0.111
0.30	0.277	0.260	0.244	0.229	0.216	0.203	0.192	0.181	0.172	0.162	0.154	0.146	0.138
0.40	0.300	0.284	0.269	0.255	0.242	0.230	0.219	0.208	0.199	0.190	0.181	0.173	0.166
0.50	0.323	0.308	0.293	0.280	0.268	0.256	0.245	0.235	0.226	0.217	0.209	0.201	0.194
0.60	0.346	0.331	0.317	0.304	0.293	0.282	0.271	0.262	0.253	0.244	0.236	0.229	0.222
0.70	0.367	0.353	0.340	0.328	0.317	0.307	0.297	0.288	0.279	0.271	0.263	0.256	0.249
0.80	0.388	0.375	0.363	0.351	0.341	0.331	0.322	0.313	0.305	0.297	0.290	0.283	0.277
0.90	0.408	0.396	0.384	0.373	0.364	0.354	0.346	0.338	0.330	0.323	0.316	0.309	0.303
1.0	0.427	0.416	0.405	0.395	0.386	0.377	0.369	0.361	0.354	0.348	0.341	0.335	0.329
1.1	0.446	0.435	0.425	0.416	0.407	0.399	0.392	0.384	0.378	0.372	0.366	0.360	0.355
1.2	0.464	0.454	0.444	0.436	0.428	0.420	0.413	0.407	0.401	0.395	0.389	0.384	0.379
1.3	0.481	0.472	0.463	0.455	0.448	0.441	0.434	0.428	0.422	0.417	0.412	0.407	0.403
1.4	0.498	0.489	0.481	0.473	0.467	0.460	0.454	0.449	0.443	0.439	0.434	0.430	0.425
1.5	0.514	0.505	0.498	0.491	0.485	0.479	0.473	0.468	0.464	0.459	0.455	0.451	0.447
1.6	0.529	0.521	0.514	0.508	0.502	0.497	0.492	0.487	0.483	0.479	0.475	0.471	0.468
1.7	0.544	0.537	0.530	0.524	0.519	0.514	0.510	0.505	0.501	0.498	0.494	0.491	0.488
1.8	0.558	0.551	0.545	0.540	0.535	0.531	0.526	0.523	0.519	0.516	0.513	0.510	0.507

(B) Heat Exchangers (Figs. 7.5 and 7.6)

Table C.3 Factor ϕ_p for A and B types of Fig. 7.5 – 2 passages of the external fluid

	γ													
β	0.4	0.6	0.8	1.0	1.2	1.4	1.6	1.8	2.0	2.2	2.4	2.6	2.8	3.0
0.2	–	–	0.989	0.979	0.967	0.951	0.934	0.915	0.896	0.879	0.862	0.848	0.836	0.827
0.4	–	0.991	0.981	0.968	0.952	0.934	0.916	0.899	0.884	0.872	0.862	0.855	0.851	0.849
0.6	–	0.988	0.977	0.962	0.946	0.930	0.915	0.903	0.893	0.885	0.881	0.879	0.880	0.882
0.8	0.995	0.986	0.974	0.960	0.945	0.932	0.921	0.912	0.907	0.904	0.903	0.905	0.908	0.913
1.0	0.994	0.985	0.973	0.960	0.947	0.937	0.929	0.924	0.921	0.922	0.924	0.928	0.933	0.939
1.2	0.994	0.984	0.973	0.961	0.950	0.943	0.938	0.935	0.936	0.938	0.942	0.947	0.954	0.960
1.4	0.993	0.984	0.973	0.963	0.954	0.949	0.947	0.946	0.948	0.952	0.957	0.963	0.970	0.977
1.6	0.993	0.984	0.974	0.965	0.959	0.955	0.955	0.956	0.960	0.964	0.970	0.976	0.983	0.990
1.8	0.993	0.984	0.975	0.968	0.963	0.961	0.962	0.965	0.969	0.974	0.980	0.987	0.993	–
2.0	0.993	0.984	0.976	0.970	0.967	0.967	0.969	0.972	0.977	0.983	0.989	0.995	–	1.01

Table C.4 Factor ϕ_p for A and B types of Fig. 7.5 – 3 passages of the external fluid

	γ													
β	0.4	0.6	0.8	1.0	1.2	1.4	1.6	1.8	2.0	2.2	2.4	2.6	2.8	3.0
0.2	–	–	0.987	0.976	0.961	0.943	0.922	0.900	0.876	0.853	0.831	0.811	0.793	0.778
0.4	–	0.990	0.979	0.963	0.944	0.923	0.901	0.879	0.859	0.842	0.827	0.814	0.804	0.797
0.6	0.996	0.987	0.973	0.956	0.937	0.917	0.898	0.881	0.866	0.854	0.845	0.838	0.833	0.831
0.8	0.994	0.985	0.970	0.953	0.935	0.918	0.903	0.890	0.880	0.873	0.867	0.864	0.863	0.864
1.0	0.994	0.983	0.969	0.953	0.937	0.923	0.911	0.902	0.896	0.891	0.889	0.889	0.890	0.893
1.2	0.993	0.982	0.968	0.953	0.940	0.929	0.920	0.914	0.910	0.909	0.909	0.910	0.913	0.917
1.4	0.992	0.981	0.968	0.955	0.944	0.935	0.929	0.926	0.924	0.924	0.926	0.928	0.932	0.936
1.6	0.992	0.981	0.969	0.958	0.948	0.942	0.938	0.936	0.936	0.937	0.940	0.943	0.947	0.952
1.8	0.992	0.981	0.970	0.960	0.953	0.948	0.946	0.945	0.946	0.948	0.952	0.955	0.960	0.964
2.0	0.992	0.981	0.971	0.963	0.957	0.954	0.953	0.954	0.955	0.958	0.962	0.965	0.969	0.974

Table C.5 Factor ϕ_p for A and B types of Fig. 7.5 – 4 passages of the external fluid

	γ													
β	0.4	0.6	0.8	1.0	1.2	1.4	1.6	1.8	2.0	2.2	2.4	2.6	2.8	3.0
0.2	–	–	0.986	0.975	0.960	0.940	0.918	0.894	0.869	0.843	0.819	0.796	0.776	0.758
0.4	–	0.990	0.978	0.961	0.941	0.918	0.895	0.872	0.850	0.830	0.812	0.798	0.785	0.775
0.6	0.995	0.986	0.972	0.954	0.933	0.912	0.892	0.873	0.856	0.842	0.830	0.821	0.814	0.809
0.8	0.994	0.984	0.969	0.951	0.932	0.913	0.896	0.882	0.870	0.860	0.853	0.848	0.845	0.843
1.0	0.993	0.982	0.967	0.950	0.933	0.917	0.904	0.893	0.885	0.879	0.875	0.873	0.871	0.872
1.2	0.993	0.981	0.966	0.951	0.936	0.923	0.913	0.906	0.900	0.897	0.895	0.894	0.895	0.896
1.4	0.992	0.981	0.967	0.953	0.940	0.930	0.922	0.917	0.914	0.912	0.912	0.913	0.914	0.917
1.6	0.992	0.980	0.967	0.955	0.944	0.937	0.931	0.928	0.926	0.926	0.926	0.928	0.930	0.933
1.8	0.991	0.980	0.968	0.957	0.949	0.943	0.939	0.937	0.937	0.937	0.939	0.941	0.943	0.946
2.0	0.991	0.980	0.969	0.960	0.953	0.949	0.946	0.946	0.946	0.947	0.949	0.951	0.954	0.957

Table C.6 Factor ϕ_c for C and D types of Fig. 7.5 – 2 passages of the external fluid

	γ													
β	0.4	0.6	0.8	1.0	1.2	1.4	1.6	1.8	2.0	2.2	2.4	2.6	2.8	3.0
0.2	–	–	1.01	1.02	1.04	1.06	1.09	1.13	1.18	1.24	1.31	1.39	1.49	1.61
0.4	–	1.01	1.02	1.04	1.07	1.10	1.14	1.20	1.26	1.34	1.43	1.53	1.64	1.78
0.6	–	1.01	1.03	1.05	1.08	1.12	1.17	1.22	1.29	1.36	1.44	1.53	1.63	1.74
0.8	1.01	1.02	1.03	1.06	1.09	1.13	1.17	1.22	1.28	1.34	1.40	1.47	1.54	1.62
1.0	1.01	1.02	1.04	1.06	1.09	1.12	1.16	1.21	1.25	1.30	1.35	1.39	1.44	1.49
1.2	1.01	1.02	1.04	1.06	1.09	1.12	1.15	1.19	1.22	1.25	1.29	1.32	1.35	1.38
1.4	1.01	1.02	1.04	1.06	1.08	1.11	1.14	1.16	1.19	1.21	1.23	1.25	1.27	1.29
1.6	1.01	1.02	1.04	1.06	1.08	1.10	1.12	1.14	1.16	1.17	1.19	1.20	1.21	1.21
1.8	1.01	1.02	1.04	1.05	1.07	1.09	1.11	1.12	1.13	1.14	1.15	1.15	1.16	1.16
2.0	1.01	1.02	1.04	1.05	1.07	1.08	1.09	1.10	1.11	1.11	1.12	1.12	1.12	1.12

Table C.7 Factor ϕ_c for C and D types of Fig. 7.5 – 3 passages of the external fluid

β	γ 0.4	0.6	0.8	1.0	1.2	1.4	1.6	1.8	2.0	2.2	2.4	2.6	2.8	3.0
0.2	–	–	1.01	1.03	1.05	1.07	1.11	1.15	1.21	1.28	1.36	1.47	1.59	1.74
0.4	–	1.01	1.02	1.05	1.07	1.11	1.16	1.23	1.30	1.39	1.50	1.62	1.76	1.93
0.6	–	1.01	1.03	1.06	1.09	1.14	1.19	1.25	1.33	1.42	1.51	1.62	1.74	1.87
0.8	1.01	1.02	1.04	1.06	1.10	1.14	1.20	1.25	1.32	1.39	1.47	1.55	1.64	1.73
1.0	1.01	1.02	1.04	1.07	1.10	1.14	1.19	1.24	1.29	1.35	1.40	1.46	1.52	1.58
1.2	1.01	1.02	1.04	1.07	1.10	1.14	1.18	1.22	1.26	1.30	1.34	1.38	1.42	1.45
1.4	1.01	1.02	1.04	1.07	1.10	1.13	1.16	1.19	1.22	1.25	1.28	1.30	1.32	1.34
1.6	1.01	1.02	1.04	1.07	1.09	1.12	1.14	1.17	1.19	1.21	1.22	1.24	1.25	1.26
1.8	1.01	1.02	1.04	1.06	1.08	1.11	1.13	1.14	1.16	1.17	1.18	1.19	1.20	1.20
2.0	1.01	1.02	1.04	1.06	1.08	1.09	1.11	1.12	1.13	1.14	1.15	1.15	1.15	1.15

Table C.8 Factor ϕ_c for C and D types of Fig. 7.5 – 4 passages of the external fluid

	γ													
β	0.4	0.6	0.8	1.0	1.2	1.4	1.6	1.8	2.0	2.2	2.4	2.6	2.8	3.0
0.2	–	–	1.01	1.03	1.05	1.08	1.11	1.16	1.22	1.29	1.39	1.50	1.63	1.79
0.4	–	1.01	1.03	1.05	1.08	1.12	1.17	1.24	1.32	1.41	1.53	1.66	1.81	1.99
0.6	–	1.02	1.03	1.06	1.10	1.14	1.20	1.27	1.35	1.44	1.54	1.66	1.79	1.93
0.8	1.01	1.02	1.04	1.07	1.10	1.15	1.20	1.27	1.34	1.41	1.50	1.59	1.68	1.78
1.0	1.01	1.02	1.04	1.07	1.11	1.15	1.20	1.25	1.31	1.37	1.43	1.49	1.56	1.62
1.2	1.01	1.02	1.04	1.07	1.11	1.14	1.18	1.23	1.27	1.32	1.36	1.40	1.44	1.49
1.4	1.01	1.02	1.04	1.07	1.10	1.13	1.17	1.20	1.23	1.27	1.30	1.32	1.35	1.37
1.6	1.01	1.02	1.04	1.07	1.10	1.12	1.15	1.18	1.20	1.22	1.24	1.26	1.27	1.29
1.8	1.01	1.02	1.04	1.07	1.09	1.11	1.13	1.15	1.17	1.18	1.20	1.21	1.21	1.22
2.0	1.01	1.02	1.04	1.06	1.08	1.10	1.12	1.13	1.14	1.15	1.16	1.16	1.17	1.17

Table C.9 Corrective factor ϕ_p for E and F types of Fig. 7.6

	γ													
β	0.4	0.6	0.8	1.0	1.2	1.4	1.6	1.8	2.0	2.2	2.4	2.6	2.8	3.0
0.2	–	–	0.986	0.975	0.959	0.940	0.918	0.894	0.868	0.843	0.818	0.796	0.775	0.757
0.4	–	0.989	0.978	0.961	0.941	0.918	0.895	0.871	0.849	0.830	0.812	0.797	0.785	0.775
0.6	0.995	0.986	0.972	0.954	0.933	0.912	0.891	0.872	0.856	0.841	0.830	0.821	0.814	0.809
0.8	0.994	0.984	0.969	0.951	0.931	0.913	0.896	0.882	0.869	0.860	0.853	0.847	0.844	0.842
1.0	0.993	0.982	0.967	0.950	0.933	0.917	0.904	0.893	0.885	0.879	0.875	0.872	0.871	0.871
1.2	0.993	0.981	0.966	0.951	0.936	0.923	0.913	0.905	0.900	0.896	0.894	0.894	0.894	0.895
1.4	0.992	0.981	0.967	0.952	0.940	0.930	0.922	0.917	0.914	0.912	0.911	0.912	0.913	0.915
1.6	0.992	0.980	0.967	0.955	0.944	0.936	0.931	0.927	0.926	0.925	0.926	0.927	0.929	0.931
1.8	0.991	0.980	0.968	0.957	0.949	0.943	0.939	0.937	0.936	0.937	0.938	0.940	0.942	0.944
2.0	0.991	0.980	0.969	0.960	0.953	0.949	0.946	0.945	0.945	0.946	0.948	0.950	0.952	0.955

Table C.10 Corrective factor ϕ_c for G and H types of Fig. 7.6

	γ													
β	0.4	0.6	0.8	1.0	1.2	1.4	1.6	1.8	2.0	2.2	2.4	2.6	2.8	3.0
0.2	–	–	1.01	1.03	1.05	1.08	1.11	1.16	1.22	1.29	1.39	1.50	1.63	1.79
0.4	–	1.01	1.03	1.05	1.08	1.12	1.17	1.24	1.32	1.41	1.53	1.66	1.81	1.99
0.6	–	1.02	1.03	1.06	1.10	1.14	1.20	1.27	1.35	1.44	1.54	1.66	1.79	1.93
0.8	1.01	1.02	1.04	1.07	1.10	1.15	1.20	1.27	1.34	1.41	1.50	1.59	1.68	1.78
1.0	1.01	1.02	1.04	1.07	1.11	1.15	1.20	1.25	1.31	1.37	1.43	1.49	1.56	1.62
1.2	1.01	1.02	1.04	1.07	1.11	1.14	1.18	1.23	1.27	1.32	1.36	1.40	1.44	1.48
1.4	1.01	1.02	1.04	1.07	1.10	1.13	1.17	1.20	1.23	1.27	1.30	1.32	1.35	1.37
1.6	1.01	1.02	1.04	1.07	1.10	1.12	1.15	1.18	1.20	1.22	1.24	1.26	1.27	1.29
1.8	1.01	1.02	1.04	1.07	1.09	1.11	1.13	1.15	1.17	1.18	1.20	1.21	1.21	1.22
2.0	1.01	1.02	1.04	1.06	1.08	1.10	1.12	1.13	1.14	1.15	1.16	1.16	1.17	1.17

(C) Coils (Figs. 7.7 and 7.8)

Table C.11 Coil with 2 sections – Fluids in parallel flow – Corrective factor ϕ_p

β	γ 0.8	1.0	1.2	1.4	1.6	1.8	2.0	2.2	2.4	2.6	2.8	3.0
0.4	–	–	–	.983	.980	.978	.976	.976	.977	–	–	–
0.6	–	.990	.987	.984	.983	.983	.984	–	–	–	–	–
0.8	–	.990	.988	.987	.987	–	–	–	–	–	1.014	1.020
1.0	.993	.991	.990	.990	–	–	–	–	1.012	1.019	1.026	1.033
1.2	.993	.992	.992	–	–	–	–	1.014	1.021	1.028	1.035	1.043
1.4	.994	.993	–	–	–	1.007	1.014	1.021	1.028	1.035	1.043	1.050
1.6	.994	–	–	–	–	1.012	1.019	1.026	1.034	1.041	1.048	1.055
1.8	.995	–	–	–	1.009	1.016	1.023	1.031	1.038	1.045	1.052	1.059
2.0	–	–	–	1.006	1.012	1.019	1.027	1.034	1.042	1.049	1.055	1.062

Table C.12 Coil with 2 sections – Fluids in counterflow – Corrective factor ϕ_c

	γ												
β	0.6	0.8	1.0	1.2	1.4	1.6	1.8	2.0	2.2	2.4	2.6	2.8	3.0
0.2	–	–	–	–	–	–	1.04	1.06	1.08	1.10	1.13	1.16	1.20
0.4	–	–	–	1.02	1.03	1.05	1.07	1.09	1.12	1.15	1.19	1.24	1.29
0.6	–	–	1.02	1.03	1.04	1.06	1.08	1.10	1.13	1.17	1.21	1.25	1.30
0.8	–	1.01	1.02	1.03	1.04	1.06	1.08	1.10	1.13	1.16	1.19	1.23	1.27
1.0	–	1.01	1.02	1.03	1.04	1.06	1.08	1.10	1.12	1.14	1.17	1.19	1.22
1.2	1.01	1.01	1.02	1.03	1.04	1.05	1.07	1.08	1.10	1.12	1.13	1.15	1.17
1.4	1.01	1.01	1.02	1.03	1.04	1.05	1.05	1.06	1.08	1.09	1.10	1.11	1.12
1.6	1.01	1.01	1.02	1.03	1.03	1.04	1.05	1.06	1.07	1.07	1.08	1.08	1.09
1.8	1.01	1.01	1.02	1.02	1.03	1.04	1.04	1.05	1.05	1.06	1.06	1.06	1.06
2.0	1.01	1.01	1.02	1.02	1.03	1.03	1.03	1.04	1.04	1.04	1.04	1.04	1.04

Table C.13 Coil with 3 sections – Fluids in counterflow – Corrective factor ϕ_c

	γ										
β	1.0	1.2	1.4	1.6	1.8	2.0	2.2	2.4	2.6	2.8	3.0
0.4	–	–	–	–	–	1.040	1.052	1.067	1.085	1.105	1.128
0.6	–	–	–	1.026	1.035	1.047	1.061	1.077	1.095	1.116	1.140
0.8	–	–	1.020	1.028	1.037	1.048	1.061	1.076	1.092	1.110	1.131
1.0	–	1.014	1.020	1.027	1.036	1.046	1.056	1.068	1.081	1.095	1.110
1.2	1.009	1.014	1.019	1.026	1.033	1.041	1.049	1.058	1.067	1.077	1.086
1.4	1.009	1.013	1.018	1.024	1.029	1.035	1.041	1.047	1.053	1.059	1.064
1.6	1.009	1.013	1.017	1.021	1.025	1.029	1.033	1.037	1.040	1.043	1.046
1.8	1.008	1.011	1.015	1.018	1.021	1.024	1.026	1.028	1.030	1.031	1.032
2.0	1.008	1.010	1.013	1.015	1.017	1.019	1.020	1.021	1.022	1.022	1.022

Table C.14 Coil with 4 sections – Fluids in counterflow Corrective factor ϕ_c

	γ								
β	1.4	1.6	1.8	2.0	2.2	2.4	2.6	2.8	3.0
0.6	–	–	–	–	–	1.044	1.055	1.067	1.081
0.8	–	–	–	1.028	1.035	1.044	1.054	1.065	1.077
1.0	–	–	1.021	1.027	1.033	1.040	1.048	1.056	1.065
1.2	–	1.015	1.019	1.024	1.029	1.034	1.040	1.046	1.052
1.4	–	1.014	1.017	1.021	1.024	1.028	1.032	1.035	1.039
1.6	1.010	1.012	1.015	1.017	1.020	1.022	1.024	1.026	1.028
1.8	1.009	1.011	1.012	1.014	1.015	1.017	1.018	1.019	1.019
2.0	1.008	1.009	1.010	1.011	1.012	1.012	1.013	1.013	1.013

Table C.15 Coil with 5 sections – Fluids in counterflow Corrective factor ϕ_c

	γ				
β	2.2	2.4	2.6	2.8	3.0
0.8	–	–	1.035	1.042	1.050
1.0	–	1.026	1.031	1.037	1.043
1.2	1.019	1.022	1.026	1.030	1.034
1.4	1.016	1.018	1.021	1.023	1.026
1.6	1.013	1.014	1.016	1.017	1.019
1.8	–	1.011	1.012	1.012	1.013

(D) Tube Bank with Several Passages of the External Fluid (Figs. 7.9, 7.10, 7.13 and 7.14)

Table C.16 Tube bank with 2 passages of the external fluid – Parallel flow – Factor ϕ_p

	γ											
β	0.6	0.8	1.0	1.2	1.4	1.6	1.8	2.0	2.2	2,4	2.6	2.8	3.0
0.4	–	–	–	–	–	–	–	–	–	–	–	–	1.03
0.6	–	–	.990	.987	.985	.984	–	–	–	–	–	1.02	1.04
0.8	–	.993	.989	.987	.986	.987	–	–	–	–	1.02	1.03	1.04
1.0	–	.992	.989	.988	.988	.989	–	–	–	1.01	1.02	1.03	1.04
1.2	–	.992	.990	.989	.989	.992	–	–	–	1.01	1.02	1.03	1.04
1.4	.995	.992	.990	.990	.991	–	–	–	1.01	1.02	1.02	1.03	1.04
1.6	.995	.993	.991	.991	.993	–	–	–	1.01	1.02	1.02	1.03	1.03
1.8	.995	.993	.992	.993	–	–	–	1.01	1.01	1.02	1.02	1.03	1.03
2.0	.995	.993	.993	.994	–	–	–	1.01	1.01	1.02	1.02	1.03	1.03

Table C.17 Tube bank with 2 passages of the external fluid – Counterflow – Factor ϕ_c

	γ										
β	0.6	0.8	1.0	1.2	1.4	1.6	1.8	2.0	2.2	2.4	2.6
0.4	–	–	–	1.02	1.03	1.03	1.04	1.05	1.05	–	–
0.6	–	–	1.01	1.02	1.03	1.04	1.05	1.06	1.06	1.06	1.05
0.8	–	1.01	1.02	1.02	1.03	1.04	1.05	1.06	1.06	1.06	1.05
1.0	–	1.01	1.02	1.02	1.03	1.04	1.05	1.05	1.05	1.05	1.04
1.2	–	1.01	1.02	1.02	1.03	1.04	1.04	1.04	1.04	1.04	1.03
1.4	1.01	1.01	1.02	1.02	1.03	1.03	1.04	1.04	1.03	1.03	1.02
1.6	1.01	1.01	1.02	1.02	1.02	1.03	1.03	1.03	1.03	1.02	1.02
1.8	1.01	1.01	1.01	1.02	1.02	1.02	1.02	1.02	1.02	1.02	–
2.0	1.01	1.01	1.01	1.02	1.02	1.02	1.02	1.02	1.01	1.01	–

Table C.18 Tube bank with 3 passages of the external fluid – Counterflow – Factor ϕ_c

β	γ 1.0	1.2	1.4	1.6	1.8	2.0	2.2	2.4	2.6	2.8	3.0
0.4	–	–	–	–	–	1.044	1.059	1.079	1.104	1.137	1.178
0.6	–	–	–	1.027	1.037	1.051	1.068	1.089	1.115	1.148	1.189
0.8	–	–	1.020	1.028	1.039	1.052	1.067	1.086	1.110	1.139	1.175
1.0	–	1.014	1.020	1.028	1.037	1.048	1.062	1.078	1.097	1.120	1.148
1.2	–	1.014	1.019	1.026	1.034	1.043	1.054	1.066	1.081	1.098	1.120
1.4	1.009	1.013	1.018	1.023	1.030	1.037	1.045	1.054	1.065	1.078	1.094
1.6	1.008	1.012	1.016	1.021	1.025	1.030	1.036	1.043	1.051	1.060	1.072
1.8	1.008	1.011	1.014	1.018	1.021	1.025	1.029	1.033	1.039	1.046	1.056
2.0	1.007	1.010	1.012	1.015	1.017	1.020	1.023	1.026	1.030	1.035	1.045

Table C.19 Tube bank with 4 passages of the external fluid – Counterflow – Factor ϕ_c

	γ						
β	1.8	2.0	2.2	2.4	2.6	2.8	3.0
0.8	–	–	–	1.032	1.035	1.038	–
1.0	–	1.021	1.025	1.028	1.031	1.032	1.032
1.2	1.016	1.019	1.022	1.024	1.025	1.025	1.024
1.4	1.014	1.016	1.018	1.019	1.019	1.019	1.018
1.6	1.012	1.013	1.014	1.015	1.014	–	–
1.8	1.010	1.011	1.011	1.011	–	–	–

Table C.20 Air heater (Fig. 7.13) – Cross flow with 4 sections Corrective factor ϕ_p for parallel flow

	γ										
β	0.6	0.8	1.0	1.2	1.4	1.6	1.8	2.0	2.2	2.4	2.6
0.9	0.986	0.973	0.959	0.945	0.932	0.922	0.915	0.911	0.910	0.911	0.913
1.0	0.985	0.972	0.959	0.946	0.935	0.927	0.921	0.919	0.919	0.921	0.924
1.1	0.984	0.972	0.959	0.947	0.938	0.931	0.927	0.926	0.927	0.930	0.934
1.2	0.984	0.972	0.960	0.949	0.941	0.936	0.933	0.933	0.935	0.938	0.944
1.3	0.984	0.972	0.961	0.951	0.944	0.940	0.938	0.939	0.942	0.946	0.952
1.4	0.984	0.972	0.962	0.953	0.947	0.944	0.944	0.945	0.949	0.953	0.959

Table C.21 Air heater (Fig. 7.14) – Cross flow with 6 sections Corrective factor ϕ_p for parallel flow

	γ										
β	0.6	0.8	1.0	1.2	1.4	1.6	1.8	2.0	2.2	2.4	2.6
0.9	0.984	0.969	0.952	0.935	0.919	0.906	0.895	0.887	0.881	0.877	0.876
1.0	0.983	0.968	0.952	0.936	0.922	0.910	0.901	0.894	0.890	0.888	0.888
1.1	0.982	0.968	0.952	0.938	0.925	0.915	0.907	0.902	0.899	0.898	0.899
1.2	0.982	0.968	0.953	0.939	0.928	0.919	0.913	0.909	0.907	0.907	0.909
1.3	0.982	0.968	0.954	0.941	0.931	0.924	0.919	0.916	0.915	0.916	0.918
1.4	0.981	0.968	0.955	0.943	0.935	0.928	0.925	0.923	0.923	0.924	0.927

Bibliography

Annaratone D., *Steam Generators*, Springer Verlag. Heidelberg

Binkley E.R., *Heat Transfer*, ASME

Boussicaud A., *Le Calcul de Pertes de Charge*, Chaud Froid, Plomberie

Boussinesq J., *Théorie Analytique de la Chaleur*, Gauthier-Villars, Paris

Dusimberre G.M., *Numerical Analysis of Heat Flow*, McGraw Hill, New York

Fishenden M., Saunders O.A., *An Introduction to Heat Transfer*, Oxford, New York

Fourier J.B.J., *Théorie analytique de la Chaleur* , Gauthier-Villars

Gumz W., *Kurzes Handbuch der Brennstoff- und Feuerungstechnik*, Springer Verlag, Berlin

Hansen W., *Heizöl Handbuch für Industriefeuerungen*, Springer Verlag, Berlin

Heiligenstädt W., *Wärmetechische Rechnungen für Industrieöfen*, Verlag Stahleisen

Heiligenstädt W., *Regeneratoren, Rekuperatoren, Winderhitzer*, Spamer, Leipzig

Jakob M., *Heat Transfer*, Wiley, New York

Kern O., *Process Heat Transfer*, Mc Graw Hill, New York

Kreith F., *Principi di trasmissione del calore*, Liguori Editore

Ledinegg M., *Dampferzeugung*, Springer Verlag, Berlin

Mac Adams W., *Heat Transmission*, Mc Graw Hill, New York

Necati M., Ozisik D., *Heat Transfer, A Basic Approach*, Mc Graw Hill, New York

Perry J.H., *Chemical Engineer's Handbook*, McGraw-Hill, New York

Properties of Water and Steam in SI-Units, Springer Verlag, Berlin

Schack A., *Der Industrielle Wärmeübergang,* Verlag Stahleisen, Düsseldorf

Trinks W., *Industrial Furnaces*, Wiley, New York

Wilkes G.B., *Heat Insulation*, Wiley, New York

Zinzen A., *Dampfkessel und Feuerungen*, Springer Verlag, Berlin

Annaratone D., *Confronto fra le disposizioni a quinconce e in fila dei fasci tubieri dei generatori di vapore*, Il Calore, 1966

Annaratone D., *Trasmissione del calore e perdita di carico nei fasci tubieri vaporizzatori*, La Termotecnica, 1975

Annaratone D., *Trasmissione del calore per convezione nei focolari delle caldaie a tubi da fumo*, La Termotecnica, 1984

Annaratone D., *Proposta di una formula di calcolo della temperatura di uscita dei gas di combustione dalla camera*. La Termotecnica, 1984

Annaratone D., *Sull'irraggiamento dei gas di combustione*, La Termotecnica, 1986

Annaratone D., *Entalpia dei gas*, La Termotecnica, 1999

Annaratone D., *Adduttanza dell'acqua, del vapor d'acqua, dell'aria e dei gas di combustione*, La Termotecnica, 2001

Boelter L.M.K., Young. H., Iversen H.W., *Nat. Advis. Comm. Aeron. Techn.*, Note 1451, 1948

Bühler H., *Wärmeübergangszahl beim Aufheizen von Stahlteilen im Bleibad*, Stahl u. Eisen, 1955

Burow J.B., *Wärmeaustausch bei der Kondensation des Wasserdampfes an senkrechten Rohren*, J. techn. Physics, 1957

Cary J.M., *The Determination of Local Forced-Convection Coefficients for Spheres*, Trans. ASME, 1953
Churchill S.M., Brier J.C., *Convective Heat Transfer from a Gas Stream at High to a Circular Cylinder Normal to the Flow*, Chem. Engng. Progr. Sympos. Ser., 1955
Corso De S.M., Coit B.L., *Measurement of Total Emissivities of Gas Turbine Combustor Materials*, Trans. ASME, 1955
Doraiswamy L.K., Patel N.B., *Performance of a Pulsed Heat Exchanger*, J. s. i. Instruments, 1959
Dropkin D., Carmi A., *Natural Convection Heat Transfer from a Horizontal Cylinder Rotating in air*, Trams. ASME. 1957
Eckert E,. *Messung der Gesamtstrahlung von Wasserdampf und Kohlensäure in Mischung mit nichtstrahlenden Gasen bei Temperaturen bis zu 1300 °C*, VDI Forschungheft, 1937
Evans S.J., Sarjant B.J., *Heat Transfer and Turbulence in Gases Flowing Inside Tubes*, J. Inst. Fuel, 1951
Grass G., *Wärmeübergang und turbulente strömende Gase in Rohreinlauf*, Allgemeine Wärmetechnik, 1956
Grimison E.D., *Correlation and utilization of new data on flow resistance and heat transfer for cross of gases over tube banks*, Trans. Amer. Soc. Mech. Engrs, 1937
Guerrieri S.A., *Radiation due to Water Vapor and Carbon Dioxide*, Mass. Inst. of Techn., 1932
Haveman H.A., Rarayan Rao N.N., *Heat Transfer in Pulsating Flow*, Nature, 1954
Heilman R.H., *Mellon Inst. Heat Transmission from Bare and Insulating Pipes*, Industr. Engng. Chem., 1924
Held W., *Die Wärmeübertragung zwischen bearbeiteten Oberflächen*, All. Wärmetechnik, 1957
Henning F., *Wärmetechnische Richtwerte*, VDI-Verlag, 1938
Hottel H.C., Mangeldorf H.G., *Heat Transmission by Radiation from Non-luminous Gases*, Trans. Amer. Soc. Mech. Engrs., 1941
Hottel H.C., Egbert R.B., *The Radiation of Furnace Gases*, Trans. Amer. Soc. Mech. Engrs., 1941
Hottel H.C., *Notes on Radiant Heat Transmission among Surfaces Separated by Non-absorbing Media*, Technology Store, Cambridge, 1951
Hughes J.A., *On the Cooling of Cylinders in a Stream of Air*, London 1916
Keyes F.G., *Additional Measurements of Heat Conductivity of Nitrogen, Carbon Dioxide, and Mixtures*, Trans. ASME, 1952
King W.J., *Free Convection*, Mech Eng., 1932
Krischer O., *Wärmeaustausch in Ringspalten bei laminarer und turbulenter Strömung*, Chemie-Ing. Techn., 1961
Konakow P.K., *Wärmeübertragung in Kesselfeuerungen*, Acad. Wiss. URSS, 1952
Linke W., Hufschmidt W., *Wärmeübergang bei pulsierender Strömung*, Chemie-Ing. Techn, 1958
Lorenz H., *Wärmeübergang und Turbulenz*, Phys. Z., 1927
Lyon R.E., *Liquid Metal Heat Transfer Coefficients*, E. Chem. Eng. Prog., 1951
McAdams W.H., Kennel W.E., Addoms J.N., *Heat transfer to Superheated Steam at High Pressures*, Tran. Amer. Soc. Mech. Engrs., 1950
McAdams W.H., *Some Recent Development in Heat Transfer*, Purdue Univ., Eng.Expt. Sta., 1948
Michaud M., *Le Rayonnement thermique des matériaux refractaires à haute température*, Silicate industriels, 1953
Mirkovic Z., *Heat transfer and flow resistance correlation of helical finned tubes in cross flow of staggered tube banks*, Int. Seminar on Heat Exchangers, 1972
Naeser G., Pepperhoff W., *Optische Temperaturmessungen an leuchtenden Flammen*, Arch. Eisehüttenw., 1951
Nusselt W., *Der Wärmeübergang in Rohrleitungen*, Forsch.-Arb. Ing.-Wes., 1910
Nusselt W., *Der Wärmeübergang im Kreuzstrom*, Z VDI, 1911
Nusselt W., *Die Oberflächenkondensation des Wasserdampfes*, Z VDI, 1916
Nusselt W., *Die Gasstrahlung bei der Strömung im Rohr*, Z. VDI, 1926
Pierson O.L., *Experimental Investigation of the Influence of Tube Arrangement on Convection Heat Transfer and Flow Resistance in Cross Flow of Gases over Tube Banks*, Trans. Amer. Soc. Mech. Engrs., 1937

Powel R.W., *Proceeding of the General Discussion on Heat Transfer*, Inst. of Mech. Eng. London 1951

Powel R.W., *Thermal Conductivity of Solid Materials at High Temperatures*, Research, 1954

Prandtl L., *Eine Beziehung zwischen Wärmeaustausch u. Strömungswiderstand der Flüssigkeiten*, Phys., 1910

Profos P., Sharan H.N., *L'influence du pas des tubes et leur arrangement sur l'encrèssement des faisceaux tubulaires*, Rev. Tech. Sulzer, 1960

Reiher H., *Wärmeübergang von strömender Luft an Rohre und Rohrenbündel im Kreuzstrom*, Forsch.-Arb. Ing.-Wes., 1925

Reynolds O., *On the Motion of Water*, Philos. Trans., 1884

Rothman A.J., Bromley L.A., *High Temperature Thermal Conductivity of Gases*, Industr. Engng. Chem., 1955

Schack A., *Über die Strahlung der Feuergase u. ihre praktische Berechnung*, Z. tech. Phys., 1924

Schack A., *Die Strahlung der Feuergase*, Arch. Eisenhüttenw., 1939-40

Schack A., *Strahlung von leuchtenden Flammen*, Z. techn. Phys., 1925

Schmidt E., *Wärmestrahlung technischer Oberflächen bei gewöhnlicher Temperatur*, Gesund. Ing - Reihe, 1927

Schmidt E., *Die Berechnung der Strahlung von Gasräumen*, Z. VDI, 1933

Schmidt E., Furthmann E., *Über die Eigenstrahlung fester Körper*, Mitt. K.-W.-Inst. Eisenforschg. Abh, 1928

Schuster F., *Über Möglichkeiten zur Steigerung der Wärmeübertragung in rekuperativarbeitenden Wärmeaustauschern*, Gaswärme, 1960

Senftleben H., *Messung von Gasstoffwerten*, Arch. Eisenhüttenwes., 1960

Tribus M., *Elementary Heat Transfer*, University of California Press, 1950

Upmalis A., *Die Bemessung der Rippenhöhe bei Rippenrohren für Lufterhitzer*, BWK, 1962

Veron M., *Champs Thermiques et Flux Calorifiques*, Bull. tech. Soc. Franc. Contruct., 1950

Wamsler F., *Die Wärmeabgabe geheizter Körper an Luft*, Forsch.-Arb. Ing.-Wes., 1911

Wenzel H., *Versuche über Tropenkondensation*, Allg. Wärmetechn., 1957

Yudin V.F., Tokhtarova I.S., *Heat transfer and resistance of staggered bundles of tubes with transverse fin in a transverse flow*, Thermal Engineering, 1973

Index

FSC
www.fsc.org
MIX
Papier | Fördert gute Waldnutzung
FSC® C083411

Kolibri 360 GmbH
Ferdinand-Jühlke-Straße 7
99095 Erfurt, Deutschland
produktsicherheit@kolibri360.de